# Nanoparticles in Diagnosis, Drug Delivery and Nanotherapeutics

The integration of nanotechnology with biomaterials, diagnostic tools, analytical equipment, physiotherapy kits, and drug delivery agents has resulted in nanotherapeutics illustrated as a class of medicine with potential of research and development. This book illustrates synthesis, properties, and applications of nanotherapeutics in various healthcare-related issues including treatment of cancer, Alzheimer's disease, targeted drug delivery, anti-HIV-1 nanotherapeutics, antibacterial/antiviral agents, skin therapy, and hyperthermia.

Features:

- Consolidates different aspects of nanoparticles such as synthesis and types of nanotherapeutics in a detailed manner.
- Presents categorical classification of nanoparticles as therapeutics.
- Covers the sustainability of nanotherapeutics.
- Reviews fabrication and advancement of all categories of nanotherapeutics.
- Discusses specific applications such as in cancer therapy, skin treatments, and targeted drug delivery.

This book is aimed at researchers, professionals, and senior undergraduate students in materials and medical science, biomedical engineering, and nanotechnology.

# Emerging Materials and Technologies

Series Editor: Boris I. Kharissov

The *Emerging Materials and Technologies* series is devoted to highlighting publications centered on emerging advanced materials and novel technologies. Attention is paid to those newly discovered or applied materials with potential to solve pressing societal problems and improve quality of life, corresponding to environmental protection, medicine, communications, energy, transportation, advanced manufacturing, and related areas.

The series takes into account that, under present strong demands for energy, material, and cost savings, as well as heavy contamination problems and worldwide pandemic conditions, the area of emerging materials and related scalable technologies is a highly interdisciplinary field, with the need for researchers, professionals, and academics across the spectrum of engineering and technological disciplines. The main objective of this book series is to attract more attention to these materials and technologies and invite conversation among the international R&D community.

**Emerging Applications of Carbon Nanotubes and Graphene**
*Edited by Bhanu Pratap Singh and Kiran M. Subhedar*

**Micro to Quantum Supercapacitor Devices**
Fundamentals and Applications
*Abha Misra*

**Application of Numerical Methods in Civil Engineering Problems**
*M.S.H. Al-Furjan, M. Rabani Bidgoli, Reza Kolahchi, A. Farrokhian, and M.R. Bayati*

**Advanced Functional Metal-Organic Frameworks**
Fundamentals and Applications
*Edited by Jay Singh, Nidhi Goel, Ranjana Verma and Ravindra Pratap Singh*

**Nanoparticles in Diagnosis, Drug Delivery and Nanotherapeutics**
*Edited by Divya Bajpai Tripathy, Anjali Gupta, Arvind Kumar Jain, Anuradha Mishra and Kuldeep Singh*

For more information about this series, please visit: www.routledge.com/Emerging-Materials-and-Technologies/book-series/CRCEMT

# Nanoparticles in Diagnosis, Drug Delivery and Nanotherapeutics

Edited by Divya Bajpai Tripathy, Anjali Gupta,
Arvind Kumar Jain, Anuradha Mishra and
Kuldeep Singh

**CRC Press**
Taylor & Francis Group
Boca Raton London New York

CRC Press is an imprint of the
Taylor & Francis Group, an **informa** business

Designed cover image: © Shutterstock

First edition published 2024
by CRC Press
2385 NW Executive Center Drive, Suite 320, Boca Raton FL 33431

and by CRC Press
4 Park Square, Milton Park, Abingdon, Oxon, OX14 4RN

*CRC Press is an imprint of Taylor & Francis Group, LLC*

ISBN: 9781032327228 (hbk)
ISBN: 9781032327235 (pbk)
ISBN: 9781003316398 (ebk)

DOI: 10.1201/9781003316398

Typeset in Times
by Apex CoVantage, LLC

# Contents

       2.2.1    Polymer-Coated Nanocarriers..................................... 26
       2.2.2    Functionalizing with Ligands....................................... 26
       2.2.3    Nanocarrier Biofunctionalization................................ 27
2.3    Nanocarrier Drug Loading and Properties............................. 27
       2.3.1    Encapsulation .............................................................. 27
       2.3.2    Characterization of Nanomaterials ............................ 27
       2.3.3    Dispersion of Shape and Size...................................... 28
       2.3.4    Surfaces ...................................................................... 29
       2.3.5    Shape/Morphology ..................................................... 29
       2.3.6    Purity, Composition.................................................... 29
       2.3.7    Aggregation/Agglomeration ....................................... 29
2.4    Manufacturing Opportunities for Nanodrugs,
       Nanoadditives, and Nanocarriers ......................................... 29
       2.4.1    Drug Loading Issues in Nanotechnology................... 30
       2.4.2    Stability and Storage .................................................. 30
       2.4.3    The Nanocarriers Complexity.....................................33
       2.4.4    Toxicity.......................................................................33
References ...............................................................................33


Chapter 3    Magnetization Dynamics of Ferromagnetic Nanostructures for
             Spintronics and Biomedical Applications ......................................... 44

             Monika Sharma and Bijoy K. Kuanr

3.1    Introduction .......................................................................... 44
3.2    Magnetization Dynamics in Ferromagnetic Nanostructures .. 46
       3.2.1    Magnetic Damping...................................................... 48
       3.2.2    Uniform Ferromagnetic Resonance Mode................. 49
3.3    Experimental Techniques to Probe Magnetization
       Dynamics................................................................................51
       3.3.1    Brillouin Light Scattering ...........................................51
       3.3.2    Conventional Ferromagnetic Resonance....................52
3.4    Dynamic Measurements of Magnetic Nanostructures............ 54
       3.4.1    Fe/Al/Fe Trilayer Ultra Thin Films........................... 54
       3.4.2    Permalloy Nanostrips..................................................58
       3.4.3    One-Dimensional Magnetic Nanowires..................... 63
3.5    Biomedical Applications ....................................................... 69
3.6    Future Applications ............................................................... 72
3.7    Conclusion(s) .........................................................................73
References ...............................................................................74


Chapter 4    Nanoparticles in CT-Guided Interventional Devices and
             Processes ......................................................................................... 82

             Deepak Gupta, Priya Jadoun, Milan Singh, Aakash Mathur,
             and Asheesh Kumar Gupta

# Preface

Nanomaterials are receiving huge interest due to their nanoscale size and high surface area to volume ratio. Moreover, their unusual properties such as lowered phase transition temps, specific optical properties, increased mechanical strength, altered electrical conductivity, magnetic properties, and self-purification and self-perfection behavior enable them to be commercially exploited in varied industrial sectors like chemical industry, solar hydrogen, building/construction industry, power generation, fuel cell, batteries, sensors, optics, aeronautic industry, automotive engineering, thermoelectric devices, consumer electronics, and pharmaceuticals.

Recently, nanomaterials have emerged as "nanotherapeutics" with a diverse range of applications in the field of medicine and therapy. Size of nanomaterials are very much like various bio-macromolecules that expedite their applicability in *in vivo* as well as *in vitro*. Therefore, by the integration of nanotechnology with biomaterials, numerous diagnostic tools, analytical equipment, physiotherapy kits, and drug delivery agents have been established to date. Hence, nanotherapeutics can be illustrated as a class of medicine with a huge possibility of research and development.

In the current scenario, the pharmaceuticals and medical industry is striving hard to attain better throughput, amended access, and improved quality of cost-effective treatment. Furthermore, chronic neuro-disorders like diabetes, cancer, HIV/AIDS, and cardiac issues have become a challenge to medical researchers. In addition, the use of conventional antibiotics for treating infectious diseases resulted in adverse side effects and multiple drug inhibitions. Target delivery is another major call associated to get therapeutic efficacy. Nanoparticles have become the best solution to all the above problems as they have the capability of controlled and targeted drug delivery with enhanced pharmacokinetic characteristics and bioavailability. Along with the deterrence and cure of the diseases, nanomedicines have latent claims in drug discovery, diagnosis, monitoring therapy, surgery, and gene delivery. Over the last two decades, various nanotherapeutics have been accepted by the FDA to treat hepatitis, cardiovascular diseases, cancer, diabetes, neurological diseases, high cholesterol, autoimmune diseases, Parkinson's disease, and some infectious diseases.

This book illustrates the synthesis, properties, and applications of nanotherapeutics in the field of almost every health care related issue with more emphasis on the treatment of cancer, targeted drug delivery, anti-HIV-1 nanotherapeutics, antibacterial and antiviral agents, treatment of Alzheimer's disease, skin therapy, and hyperthermia.

Chapter 1 and Chapter 2 are based on the introduction of nanomaterials used as drug delivery systems, their manufacturing approaches, and their applications.

Chapter 3 and Chapter 4 mainly emphasize the use of nanoparticles in medical diagnostics and intervention devices. Chapter 5 and Chapter 6 illustrate the use of lipids-based nanoparticles in medical imaging and drug delivery.

Chapter 7 specifically discusses the amino acids functionalized inorganic nanoparticles in diagnostics. Chapter 8 focuses on the special class of nanoparticles called "hybrid nanocomposites". Chapter 9 and Chapter 10 cover the applications of silica and fullerene nanomaterials in drug delivery of anticancer drugs.

**Divya Bajpai Tripathy**
**Anjali Gupta**
**Anuradha Mishra**
**Arvind Kumar Jain**
**Kuldeep Singh**

# About the Editors

**Divya Bajpai Tripathy** is currently a full-time Professor in the Department of Chemistry, School of Basic and Applied Sciences, Galgotias University, Greater Noida, India. She has research and teaching experience of more than 12 years. She has more than 45 research publications in reputed journals, books, book chapters, and conference proceedings and a patent filed to her credit. She has guided 12 master's research students. Currently five doctorate students have been registered under her supervision. She has been principal investigator in a DST-funded research project. She has a research collaboration with Prof. M. A. Quraishi, King Fahd University, Saudi Arabia, Dr. Anjali Gupta, Galgotias University, Greater Noida, and Prof. Anuradha Mishra, Gautam Buddha University, Greater Noida.

**Anjali Gupta** is currently working as Associate Professor in the Department of Chemistry, School of Basic and Applied Sciences, Galgotias University, Greater Noida, India. She has research and teaching experience of around 12 years. She is the recipient of the Young Scientist Award by the Department of Science and Technology and Dr. D.S. Kothari postdoctoral fellowship by University Grants Commission, and Senior Research Fellowship from CSIR, India. Her research area is *in silico* screening and synthesis of naturally occurring bioactive analogs. She has research collaborations with Dr. Divya Bajpai, Galgotias University, Dr. Fahmina Zafar, Jamia Milia Islamia, Delhi, and Dr. Anujit Ghosal, Jawaharlal Nehru University, Delhi. She has eight published patents and around 31 research publications in reputed journals, book chapters, and conference proceedings and one published book to her credit. She has guided one Ph.D. student and is currently guiding three Ph.D. and ten master's research students. She has been the principal investigator in DST-sponsored research projects. She completed her graduation, postgraduation, and doctorate from the University of Delhi, Delhi.

**Arvind Kumar Jain** is currently working as Dean, SBAS, and Dean, Student Welfare, in the School of Basic and Applied Science. He completed his Ph.D. in Chemistry from IIT Roorkee; received an M.Phil. in Industrial Method of Chemistry from University of Roorkee (Now IIT Roorkee) and an M.Sc. in Organic Chemistry from CCS University. His research areas include nanotechnology, photochemistry of II-VI semiconductors, carbon nanotube functionalization and device fabrication, and ZnO synthesis and application in sensing devices. He has more than 15 years of teaching and research experiences. He has many publications in highly indexed SCI journals and has been awarded a travel grant of $1345 from the US NSF to participate in and present a paper to the Third Annual Conference of the Society for the Study of Nanoscience and Emerging Technologies at Tempe Arizona, USA, Nov. 7–10, 2010. He was also a recipient of full travel grant of $2000 from the US NSF to participate in and deliver a lecture to the Second Annual Conference of the Society for the Study of Nanoscience and Emerging Technologies at Technische Universität Darmstadt, Germany, Sept. 29 to Oct. 2, 2010. CNRS Postdoctoral Fellowship at

Institut de chimie de la matiere condensee de Bordeaux, Bordeaux cedex, France, for a period of 16 months (July 2002 to Nov. 2003), CSIR Senior Research Fellowship at Department of Chemistry Indian Institute of Technology Roorkee, Roorkee (May 2001 to June 2002).

**Anuradha Mishra** is currently a Full Professor in the department of Applied Chemistry, School of Vocational Studies and Applied Sciences, Gautam Buddha University, Greater Noida, India. She has been Dean of the School of Vocational Studies and Applied Sciences for almost five years. She has also been the Dean of Academics for three years and the Dean of Planning and Research for more than two years. She has been Nodal Officer of her university for skill development courses in collaboration with the NSDC, Govt. of India and for National Assessment and Accreditation Council (NAAC), University Grants Commission, Government of India, a body responsible for the grading of Indian universities. She has research and teaching experience of more than 25 years and academic administrative experience of more than 15 years in various senior positions. She is a recipient of the coveted commonwealth fellowship award, UK and Research Award for teachers by University Grants Commission, India. Her research area is synthesis of polysaccharides-based biomaterials (green materials) for water remediation. She has worked as Commonwealth Fellow, Green Chemistry Centre of Excellence, Department of Chemistry, University of York, York, UK, and Guest Faculty, Dept. of Agriculture and Environmental Sciences, University of Newcastle, Newcastle upon Tyne, UK. She has research and academic collaboration with Professor James H. Clark, Director, Green Chemistry Centre of Excellence, University of York, UK, and with Dr. Rajani Srinivasan, Department of Chemistry, Geosciences and Environmental Science, College of Science and Technology, Tarleton State University, Texas A&M University System, Stephenville, TX, USA. She has one US patent on polysaccharide agents and methods of their use for removing solids from water; a part of the technology has been licensed by a Company Pristana LLC, California.

Prof. Anuradha has more than 100 research publications in reputed journals, book chapters, and conference proceedings to her credit. She has also authored/edited five books on polymers and green and sustainable chemistry. She has guided 14 Ph.D. and 20 master's research students. She has been principal investigator in many sponsored research projects. She has been a member of many academic/research bodies of government organizations in India and abroad.

**Kuldeep Singh** is Associate Professor at Maharishi Markandeshwar (Deemed to be University), Mullana, India. He obtained his M.Sc. (2001) at the Indian Institute of Technology, Roorkee, and Ph.D. in 2007 at the Indian Institute of Technology Bombay, Mumbai (India). In 2007, he worked as a postdoctoral scientist at Laboratoire de Synthèse Bioorganique, Illkirch (France). In 2009, he joined Sai Advantium Pharma Ltd, Pune, as a research scientist. He then joined the Jaypee University of Information Technology as an assistant professor. His research activities focus on (a) the development of new synthetic methodologies using metathesis, coupling reactions, and C–H activation, (b) the design and synthesis of new molecules against antibacterial drug resistance, and natural products, nanotechnology, and biofuels. He has supervised two Ph.D. theses and 18 postgraduate theses. Currently, two research scholars are working with him. He is a member of the Chemical Research Society of India, the Royal Society of Chemistry, and the American Chemical Society. He is also a recipient of the NTSEAG, CSIR-JRF, SRF, GATE, and DST/SERB Start-up grant for Young Scientists (YSS).

# Contributors

**Md. Aftab Alam**
Galgotias University
Greater Noida

**Syed Salman Ali**
Department of Pharmaceutical
    Technology
Meerut Institute of Engineering
    and Technology
Meerut-India

**C. K. Ashok Kumar**
School of Medical and Allied Sciences
GD Goenka University
Gurugram (NCR New Delhi)
Sohna Road, Haryana, India

**Shvetank Bhatt**
Amity Institute of Pharmacy
Amity University Madhya Pradesh
Gwalior, MP

**Shikha Chaudhary**
Department of Pharmacy
School of Medical & Allied
    Sciences
Galgotias University
Greater Noida, India

**Akash Chauhan**
Department of Pharmacy
School of Medical & Allied
    Sciences
Galgotias University
Greater Noida, India

**Anindita De**
Department of Chemistry and
    Biochemistry
School of Basic Sciences
    and Research
Sharda University

**Ayush Dubey**
Department of Pharmacy
School of Medical & Allied Sciences
Galgotias University
Greater Noida, India

**Umar Farooq**
Galgotias University
Greater Noida

**Rajat Goyal**
College of Pharmacy
M.M University Sadopur
Ambala, Haryana, India

**Asheesh Kumar Gupta**
Division of Physics
School of Basic and Applied
    Sciences
Galgotias University,
Greater Noida

**Deepak Gupta**
Division of Physics
School of Basic and Applied Sciences
Galgotias University
Greater Noida

**Chitra K. Iyer**
Koninklijke Philips N. V. Philips
    India Limited

**Priya Jadoun**
Division of Physics
School of Basic and Applied Sciences
Galgotias University,
Greater Noida

**Preeti Jain**
Department of Chemistry and
    Biochemistry
Sharda University
Greater Noida, India

**Jyoti Joshi**
Department of Chemistry and
  Biochemistry Sharda University
Greater Noida, India

**Awaneet Kaur**
School of Medical and Allied Sciences
  Galgotias University,
Greater Noida

**Niranjan Kaushik**
School of Medical and Allied Sciences
Galgotias University,
Greater Noida

**Krishna**
Department of Pharmacy
School of Medical & Allied Sciences
Galgotias University
Greater Noida, India

**Bijoy K. Kuanr**
JNU, Delhi

**Dharmaraj Senthil Kumar**
Galgotias University
Greater Noida

**Bhavna Kumari**
Department of Biosciences
Galgotias University
Greater Noida, Uttar Pradesh 201310,
  India

**Chethana M. V.**
SJB Institute of Technology Department
  of Chemistry Bangalore-60

**S. Mohana Lakshmi**
Amity Institute of Pharmacy
Amity University, M.P.

**Aakash Mathur**
Division of Physics
School of Basic and Applied
  Sciences
Galgotias University,
Greater Noida

**Anupam Prakash**
Galgotias University
Greater Noida

**Manu Sharma**
M.M. College of Pharmacy
M.M. Deemed to be University
Mullana

**Monika Sharma**
Special Centre for Nanoscience
Jawaharlal Nehru University
New Delhi, India and
Department of Physics
Deshbandhu College
University of Delhi, India.

**Pooja Sharma**
Department of Pharmacy
Dr. Bhimrao Ambedkar
  University
Agra, India

**Ravi Shekhar**
Department of Pharmacy
Dr. Bhimrao Ambedkar
  University
Agra, India

**Kuldeep Singh**
Department of Applied
  Chemistry
Amity University Madhya Pradesh,
  Gwalior, M.P.

**Pratibha Singh**
JSS Academy of Technical Education
Noida

**Milan Singh**
Division of Physics
School of Basic and
    Applied Sciences
Galgotias University,
Greater Noida

**Vijay Kumar Singh**
Department of Pharmacy
School of Medical & Allied Sciences
Galgotias University
Greater Noida, India

**Kiran T.**
SJB Institute of Technology,
    Department of Chemistry
Bangalore

# 1 Nanomaterials as Drug Delivery Systems

*Akash Chauhan, Ayush Dubey, Awaneet Kaur,
Shikha Chaudhary, Krishna, Md. Aftab Alam,
Niranjan Kaushik, and Vijay Kumar Singh*

## 1.1 INTRODUCTION

Traditional medicine offers advantages over nanomaterial-based therapies. Scavenger cell medication clearance may be slowed by increasing nanoparticle (NP) size. Capsulizing and transferring drugs that are weakly water soluble provides pharmacological advantages. Controlled drug release alters pharmacokinetics and minimizes side effects. Medication effectiveness is enhanced by biodistribution and intracellular penetration [1]. NPs help cancer chemotherapy treatments work better by maintaining drug levels in tumor tissue and minimizing drug concerns in healthy tissues. Because of their improved permeability and retention in the tumor vasculature, optimized NPs may be able to penetrate the capillary vascular wall when used for passive tumor targeting. Angiogenesis and endothelial cell division in the tumor are separated by a large margin neovasculature that promotes NP extravasation. NP accumulation is aided by impaired lymphatic outflow in tumors. However, NP-targeting ligands might be made using tumor-cell membrane receptors. Endocytosis may be aided by vitamins, lectins, carbohydrates, antibodies, and antibody fragments, which can increase drug delivery selectivity [2]. The size, shape, and surface characteristics of NPs influence their biodistribution. This has been done using nanovesicles, nanomicelles, nanoliposomes, nanogels, and nanodendrimers. Polymeric, mesoporous, magnetic, gold, and other NPs are among them. Design, modification, synthesis, and active functionalization of organic and natural polymers are all diverse. With the application of NP-dependent support, poorly water-soluble tablets and other medications may be given more efficiently. Polymer materials include polycaprolactone, polyglycolic acid, polyethylene glycol, polylactic acid, polysaccharides, and polypeptides. They are all biocompatible polymers. Drug delivery vehicles are also made from polysaccharides. Drug delivery vectors include lipids and synthetic polymers [3]. Nanocarriers for tumor therapy have appeared in recent decades and have shown promise in preclinical studies. Amphiphilic NPs are made using a variety of chemical and physical methods, including carbohydrate- and derivative-based NPs. Hydrophilic and hydrophobic components coexist in amphiphilic medicine administration. Hydration and swelling are caused by hydrophilic parts, while hydrophobic portions minimize water contact since it is dynamically unfavorable. Because polysaccharides are hydrophilic, amphiphilic systems require grafting hydrophobic

DOI: 10.1201/9781003316398-1

1

segments onto them. Polysaccharides and modified polysaccharides have physical and chemical features that may be exploited to create a system that is amphiphilic [4]. Carbohydrates may generate amphiphilic compounds by reacting with hydrophobic moieties through ester, amide, urea, and hydrazone connections. Fatty acids, bile acids, cholesterol, polyester, polyanhydride, and paclitaxel are all hydrophobic moieties. Enzyme cleavability, pH sensitivity, and reductiveness are all examples of labile covalent chemical connections between polysaccharide and the hydrophobic moiety in cancer cells. Hydrophobic interactions allow amphiphilic polymers to self-assemble into NPs in water [5]. Metal ligands, electrostatic contact between hosts and guests, charge transfer, hydrogen bonding, and other noncovalent interactions are examples. The resulting amphiphilic compounds are known as supramolecular or super amphiphiles because noncovalent interactions are weaker than covalent connections. Supramolecular amphiphiles show promise as stimulus-responsive nanocarriers. Carbohydrate-dependent amphiphilic drug delivery systems are made via electrostatic interactions, host-guest recognition, pep stacking, or a mix of these. When cationic and anionic molecules come into contact with one another, polyelectrolytes are formed (charged macromolecules) [6]. Self-assembling polyelectrolytic compounds might produce an NP. Because compressed networks include fewer stable ionic crosslinkers and weaker ionic connections, the resultant NPs are labile in the external environment. Understanding a polyelectrolyte requires an understanding of both its internal and its exterior properties. These include polyelectrolytes' charge density, chain flexibility, and molecular weight; their molecular weight and their charge distribution normally regulate the generation and stability of complexes. Polyelectrolytes need polysaccharides to function. Polysaccharides containing positive and negative charges, such as chitosan and hyaluronan, may form complexes with polyelectrolytes. Chemical changes may cause neutral polysaccharides to become charged polyelectrolytes. Chemical changes are often made by adding an amino group to positively charged polysaccharides and a carboxylic or sulphate group to negatively charged polysaccharides [7].

## 1.2 THE FUNCTION OF NANOPARTICLES IN MEDICATION DELIVERY

Noncovalent and ionic bonding are used in host-guest chemistry to identify molecules. Supramolecular amphiphilic delivery devices are made using this technology. Cyclodextrins, cucurbiturils, calixarenes, and porphyrins are all common host chemicals. Hydrophobic molecules may enter the hydrophobic hollow void of cyclodextrins (cavity). Through hydrophobic interactions, dihydrate cavities hold the hydrophobic head sections of molecules that come to visit [8]. Polypseudorotaxanes are formed when linear polymers enter cyclodextrin cavities. Because of the existence of host cyclodextrins and guest molecules, the complex's amphiphilicity may be modified. Hydrophobic moieties may be included in cyclodextrins or guest compounds by functionalization and modification. Noncovalent connections between aromatic systems are referred to as "pep stacking." Catalysis, molecular identification, and protein-ligand interactions all rely on these interactions. Medication distribution is increasingly using intermolecular pep stacking. Graphene, carbon nanotubes, fullerenes, and carbon black nanoparticles are used to make effective medication distribution [9].

Carbon nanoparticles cannot penetrate water without causing surface changes, limiting their use. Many biocompatible polysaccharides are being developed for use with aromatic moieties in order to transport aromatic medications through intermolecular pep stacking attraction. Aromatic moieties may be added to nonaromatic drugs through cleavable chemical bonds, making them pep stacking possibilities. For pharmacological encapsulation and self-assembly of vesicles or micelles, hydrophobic interactions have a synergistic effect [10]. The features of several nanomaterials are detailed next.

1. Polar and fat-loving structural units coexist in amphiphilic or amphipathic polymers. In contrast, these architectural designs, which are frequently used in nanoscience research, may be created via flash solvent or nonsolvent nanoprecipitation procedures, both of which allow for dimension control. When functional groups are added to the surface of NPs, this leads to a wide range of activities and effective drug delivery [11].
2. Systematized self-assembly in aqueous solution with bifunctional proteins creates microemulsion polymers, yielding amphiphilic multiblock copolymers with a core-shell nanostructure made up of hydrophilic and hydrophobic units. An inner core of hydrophobic units is surrounded by a hydrophilic corona. Micelle size is determined by changing the composition, which includes both hydrophobic and hydrophilic molecules. Freight molecules are covalently linked to polymers or encapsulated in micelles [12].
3. Vesicles are formed of organic, naturally modified, or synthetic polymers. Vesicles are generated by aggregating these molecules and have a solvent-filled inside and an encapsulating outside that mimics a cell membrane. Both hydrophilic and hydrophobic medicines may be found in the same vesicles [13].
4. Several drug-delivery nanogels are made from polymeric particles that are covalently coupled or self-assembled to approximate a three-dimensional molecular structure. The porous interior of the nanogel might be exploited to insert medicinal ingredients. Sasaki et al. 2011 discovered a simple method for processing nanogels. Under the correct circumstances, nanogel swelling may be controlled for controlled medicine release [14].
5. Dendrimers are created by combining constituent units with a tree-like initiator molecule. To preserve pharmaceuticals in branching structures, noncovalent bonding is employed. Nanomaterials may have multiple functions for the same purposes. Lipids, phospholipids, lactic acid, dextran, and chitosan are examples of biological nanoparticles. Phospholipids, for example, interact with cells differently from nonbiological components [15].

## 1.3 AN OVERVIEW OF NANOCARRIERS FOR MEDICATION DELIVERY

This section looks at some of the current NP trends in smart drug administration systems (Figure 1.1). They are distinguished by their capacity to react to physical, chemical, and biological stimuli (including certain substances and enzymes). We also look

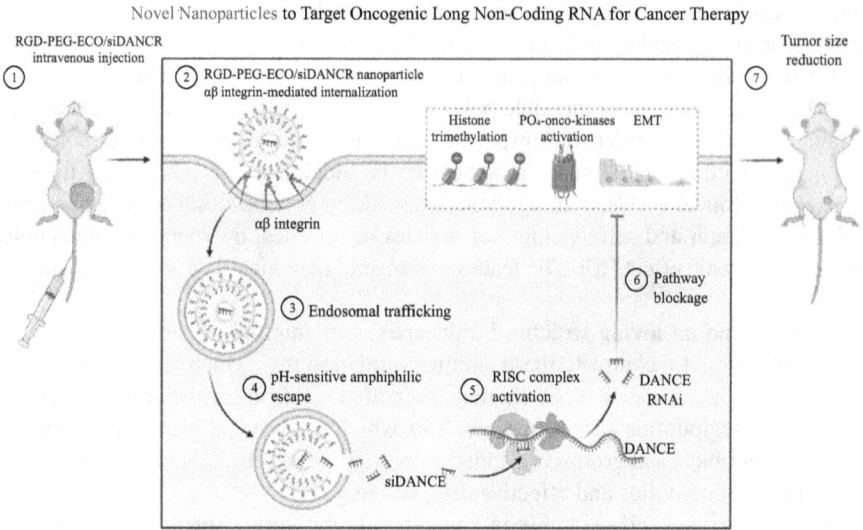

Novel Nanoparticles to Target Oncogenic Long Non-Coding RNA for Cancer Therapy

**FIGURE 1.1**    Nanocarriers in drug delivery.

at the limits of these niosomes, their potential for use in biological systems, and the toxicity of these solid lipid nanoparticles to the cellular sensing effects of protein corona. The history and working procedures of each class are also discussed [16–17].

## 1.3.1 NANOCARRIER STUDIES

Colloidal nanoparticles called nanocarriers are utilized to transport substances to a specified location. According to studies, therapeutic nanocarriers must be less than 200 nm in diameter due to the 200 nm diameter of body microcapillaries. These latent nanocarriers are biocompatible and harmless. By avoiding the endosome–lysosome pathway, these nanocarriers will have a long circulation period. Surface, composition, and shape changes might boost nanocarrier activity while reducing negative effects [18–20]. This has a number of implications for pharmaceutical delivery. Only a few nanocarriers are capable of delivering medications to the desired location over time, improved biodistribution, pharmacokinetics, stability, solubility, toxicity, and drug administration. In nanotransporters, the surface-to-volume ratios of organic, inorganic, and hybrid nanocarriers are all high. Organic nanocarriers that may be employed include lipids, dendrimers, polymers, and micelles. Organic nanocarriers have a wide range of applications, are low toxic, and can transport medicines and ligands simultaneously [21–23].

## 1.3.2 NANOPARTICLES WITH THERMORESISTANCE

In stimulus-responsive, spatiotemporally regulated medicine administration, heat and temperature are the most prevalent triggers for drug release [24, 25]. The majority of infected swollen tissues and tumors are hyperthermic, indicating that they are being stimulated inside. The use of appropriate ambient temperature changes to switch on thermoresponsive NPs is also a viable alternative for a stimuli-responsive medication delivery system. Another advantage is the ability to implant contaminated tissue and cells. A drug delivery system must be able to hold its load at inconvenient locations while still delivering medicine in order to be effective in target cells and tissues, such as tumors, with regulated kinetics [26–28]. Temperature-sensitive polymers, which expand as the temperature increases, are the most prevalent materials utilized in drug-delivery systems. These polymers alter the release of encapsulated drugs because their solubility fluctuates with temperature. A phase shift occurs when the temperature of the UCST or LCST is changed. Nano valve devices control drug release by opening gates in mesopores of mesoporous silica nanoparticles, which are light, pH, and temperature sensitive [29]. Thermoresponsive polymers function as an "off switch." For the first time, the scientists linked N-isopropylacrylamide (NIPAAm) to a flexible nylon membrane. At temperatures above 35 °C, NIPAAm shrank and collapsed, the nano valve closed, and drug delivery was reduced. The valve opens and medication release is boosted at temperatures below LCST (35 °C) due to polymer swelling [30]. Liu et al. 2014 employed NIPAAm and the NIPAAm-co-AAc brush proved successful in polycarbonate "brush hydrogels" grafted with acrylic acid (AAc) copolymers, which shrank at temperatures over the LCST, enhancing drug release from the pores. Brush swelling hindered the drug delivery route by lowering the temperature below LCST. Temperature-sensitive medication delivery systems and smart nano bombs undergo a reversible phase change as the temperature lowers. This temperature reduction is referred to as "cold shock" or "cryotherapy." At 37 °C, Lee et al. 2008 created pluronic/polyethyleneimine (PEI) nanogels that were used to infect cells. When the temperature was reduced to 20 °C after cell absorption, the nanogel expanded 800-fold to 1.3 mm. The increased dimension "blew up" the tumor cells, resulting in necrotic cell death as a result of the high hydrostatic pressure created within the cells [31]. One potential is the formation of active stimulus-responsive nanocarriers that are very sensitive to internal temperature variations, such as those found in cancer. Advances in thermosensitive drug-delivery systems include tissue engineering and bone tissue regeneration using protein delivery systems and drug and gene code delivery. If uterine sickness can be effectively treated with distribution to the uterus, effective moderate hyperthermia with improved tumor retention may be accomplished (e.g., bovine reproductive issues). Strong selectivity towards cancer and angiogenic cells and other uses have been demonstrated.

Magnetism is one of the best external stimuli because, unlike light and electrical fields, it scarcely interacts with the body. Magnetically controlled and established NPs may be made and investigated as delivery and biological target vehicles by researchers. In their 1960 publication, they (the researchers who developed magnetic nanoparticles in 1960) advocated employing magnetic fields to accelerate medication absorption, and their characteristics have fewer biomedical application prospects than ferromagnetic NPs due to limited colloidal stability and homogeneity [32]. Magnetic hyperthermia is the ability of magnetic nanoparticles to generate heat in a high-frequency alternating magnetic field (AMF). Because of the changing nature of magnetic forces, magnetic NPs may tolerate internal stresses and strains. Magnetic NPs in an AMF are thought to generate heat by a combination of intrinsic rotational motion, most likely a Brownian type of movement, and extrinsic motion, which is primarily caused by the thermal rotation of the particles' magnetic critical moment and relaxation through diffusion. Magnetic saturation and hysteresis coefficient, both of which are linked to NP size, are required for inductive hyperthermia. Hyperthermal therapy, controlled medicine release, targeting, imaging, and guiding are all possible using magnetic fields. Based on their UCST and LCST, polymeric NPs may undergo thermodynamic phase transitions. Magnetic heating, which accelerates drug-release structural disruption, or a pumping action, might produce this singularity. By adsorbing polymer chains to their surfaces, magnetic NPs may be confined inside polymer particles like hydrogels. An external magnetic field might also help NP absorb more. Louguet et al. 2014 used thermosensitive block-copolymer brushes to make silica-coated magnetic NPs. By changing the conformation of the brushes, an AC magnetic field enabled the magnetic core to develop heat, which increased the activated release of the medication. Magnetic hyperthermia is controlled by varying field strengths [33]. In polymer-based medication delivery induced by magnetic fields over a distance of several centimeters, magnetic-responsive materials are more effective than temperature or pH triggers. Extracorporeal magnetic teaching may help overcome a variety of drug-delivery system problems, such as physiological obstacles or a lack of tissue selectivity. After a nanocarrier is injected intravenously, magnets can be used to stop blood flow in arteries and capillaries.

### 1.3.3 Electro-NPs

Nanomaterial-based drug delivery may employ external electric fields to deliver medicine in a continuous, pulsed, or on-demand fashion. Electro responsive materials, compound-loaded nanostructures, or a combination of the two, temperature and magnetic field-responsive vehicles are used in electro responsive drug release. Polyelectrolytes containing ionizable species taper or expand in response to an electrical input. An electric field enhanced the ionization of the structure due to polyelectrolyte poly (sodium 4-vinylbenzene sulfonate). The swelling ratio and particle size may be affected by an external electric field. Researchers at Nagoya University devised a method for producing stimulus-responsive polymers in a predictable manner. When carbon nanorings and iodine were combined, researchers created an electrically conductive material that generated white light when charged. The current method could help put together a good set of materials that respond to stimuli [34].

### 1.3.4 Nanoparticles That Respond to Light

Irradiation aids drug release. Light is a uniquely powerful tool for controlling molecular events in biology. No other external input (e.g., heat, ultrasound, magnetic field) can be so tightly focused or so highly regulated as a clinical laser. Drug delivery vehicles that can be photonically activated have been developed across many platforms, from the simplest "caging" of therapeutics in a prodrug form, to more complex micelles and circulating liposomes that improve drug uptake and efficacy, to large-scale hydrogel platforms that can be used to protect and deliver macromolecular agents including full-length proteins. UV (10–300 nm), visible, or near-infrared (650–900 nm) radiation zones are frequently used to liberate drugs or genes from nanocarriers. Despite the fact that UV light can ionize and break covalent bonds with energies of100 kcal per mole which leads to risk of tissue ham and usage. Pharmaceuticals and tissues may be preserved using long-wave UV lasers (200–300 nm). UV light is more harmful than other light wavelengths, and it cannot penetrate deep into tissue due to endogenous chromophores (oxy- and deoxyhemoglobin, lipids, and water). In other words, the lower the wavelength, the better for stimulating drug release in the skin for the treatment of pathological disorders that affect both the skin and the mucosa. Supramolecular aggregates triggered by visible light are uncommon. Because hemoglobin, water, and lipids do not absorb near-infrared light, it has a higher tissue transmission and causes less cell and tissue damage than visible light. Two low-energy photons may be absorbed by some organic chromophores [35].

### 1.3.5 NANOPARTICLES THAT ARE MECHANICALLY SENSITIVE

Mechanoresponsive nanostructures for medicine administration have been the subject of a few papers. To modify volume, mechanoreceptive molecules are formed and deconstructed using pressure and other physical inputs. This proposal provides a biomimetic and simple technique for delivering regulated medicine. Mechanical compression destabilizes inclusion complexity in cyclodextrin alginate hydrogels for patient-regulated ondansetron release, suggesting that pressure may influence cyclodextrin addition complexes. Stress on an implanted gel reduces b-inclusion cyclodextrin's capacity, resulting in quicker drug release. Chemotherapy-induced nausea is increasingly being treated with this strategy. Shear stress might be used to create smart medication.

### Mechanisms of delivery

Biological or chemical triggers for the release of drugs in arterial blood vessels may not be as effective as this physical or mechanical trigger. Stenosis and atherosclerosis may benefit from an endogenous mechanical stimulation response. Take, for example, a study found that organized lenticular liposomes were stable in a controlled setting but not during atherosclerosis. The lenticular form caused a preferred rupture along the equator of the nanostructures, resulting in greater drug release. A second study revealed microscale NP aggregation for thrombus therapy [36].

### 1.3.6 NANOPARTICLES THAT RESPOND TO ULTRASOUND

Ultrasound is a kind of external mechanical and thermal stimulation. Ultrasound is a dominant force in medical research and development, with applications in diagnostic imaging and therapy of aberrant tumors. The notion of ultrasound-responsive nanocarriers needs two major characteristics: effective medicine release in response to particular ultrasound waves, and the ability to detect drug release for imaging and therapeutic applications. Increased drug diffusion may be achievable as a result of cavitation bubbles and temperature rises produced by drug release from carriers. Biomedical ultrasonic waves vary from 0.1 to 50 MHz. Unidirectional water flow may cause tissue disruption. By causing fluctuating pressure between the biological membrane layers, these waves create holes in the cell membrane. Ultrasound is absorbed differently in various organs and tissues. Soft tissues have lower coefficients than hard tissues. The ultrasonic pressure, frequency, and compressed length minimum thresholds may affect tissue disruption and brain injury caused by head ultrasonography.

Microbubbles' contrast-enhancing characteristics are being used by scientists to design a modern medicine delivery system. Ultrasound operators may use this method to detect tumors based on their location and tumor-specific blood vessels. When a tumor is identified, the microbubbles are shattered by a high-energy ultrasonic pulse. When the shells of the microbubbles break, the contents leak out into the environment. This lets drugs reach a tumor without going through the bloodstream [37].

Microbes might be to blame for the pH differences between healing and non-healing wounds. Tumors have a lower pH when compared to normal tissues. When cancer cells outnumber normal cells and oxygen and nutrients are insufficient, glycolysis creates lactic acid instead of oxidative phosphorylation. As a consequence of the various pH conditions in biological systems, pH-sensitive drug delivery devices were created. Nanomaterials that are pH responsive, such as sensors and wettable nanoparticles, have been used. Biphasic colloidal particles that are "Janus particles" are mixed in pH-responsive nanocarriers that can change shape as the pH changes. These nanocarriers have improved properties. The pH responsiveness of nano systems has influenced recent innovations in acute medicine delivery systems significantly and may have led to remarkable breakthroughs in the detection and treatment of a variety of ailments, including malignancies and infections. They have increased anticancer activity, decreased chemotherapy's bad effects, and supplied nucleic acids, proteins, peptides, and other compounds. The pH-sensitive drug-delivery devices may increase the oral bioavailability of anticancer drugs, decreasing tumor growth, reducing toxicity, and improving tumor cell selectivity while reducing normal cell cytotoxicity. Recently, the production, properties, and applications of pH-sensitive nanoparticles used in drug-delivery systems were investigated. Nanocarriers that are pH responsive require a deep understanding of their advantages and uses. Peptides are recognized as efficient and secure nanocarriers for nonviral gene transfer to cells, compared to bacterial gene delivery methods, despite their structure being impacted by bacterial dimensions. Peptide sizes help with things like moving peptides across cell membranes, fusing endosomes together (at low pH), and sending them to the nucleus.

## 1.3.7  Redox Nanoparticles

Redox-responsive drug-delivery devices are effective for treating stimulated-responsive cancers and genes. Redox active nanoparticles containg antioxidants create a new therapeutic perspetive over pH. These advantages include a good response to high intracellular blood glutathione levels; direct drug release into the nucleus and cytoplasm; and a consistent extracellular environment with low blood glutathione levels [38]. NADP+/NADPH influences the redox environment. The cytoplasm contains constant metabolic oxidases, which impact nitric oxide synthases and NADPH oxidases. Each cell's mitochondria contain reducing components since they're oxidation sensitive. Mitochondria have more electron transit than other cells. Nuclei are oxidation resistant yet have modest redox potentials. Through oxidizing mechanisms and enzymes, the secretory pathway creates disulfide bonds in exportable proteins. The way a cell works affects its redox potential. This includes things like death and growth.

### 1.3.8 Biomolecular-Responsive Nanoparticles

Glucose, adenosine triphosphate, reactive oxygen species, and deoxyribonucleic acid under certain physiological or pathological situations prompted the development of biomolecule-sensitive sensors. Most smart drug delivery systems for diabetics are glucose-responsive controlled-release devices. These smart systems rely on glucose oxidase to convert glucose to gluconic acid (GOx). Glucose-responsive microgels stimulate glucose enzymatic transformation, protonation of the pH-responsive chitosan matrix, and insulin release in a self-regulating nano valve system for type 1 diabetes therapy [39].

### 1.3.9 Enzyme-Responsive Nanoparticles

Enzymes are crucial due to their biorecognition and catalytic properties. Their remarkable selectivity and enzyme-catalyzed reaction benefits are hallmarks of nanomedicine. Evaluating or discovering enzyme activity is also important for medical diagnostics, such as if its dysregulation is a disease symptom. A drug delivery device may use physical or chemical modes of action. If, for example, the enzyme can control the nanomaterials, they could be made to release their drugs by breaking down their different structures, which could be anything from polymers to mesoporous silicate. If enzyme transformation or breakdown of nanocarriers releases the therapies further, multimodal nanomedicines with mutually beneficial consequences may be added. Enzyme-responsive NPs are designed to modify their macroscale structure and govern their release. Adding chemicals affected by enzymatic responses to the nanomaterial's surface changes its physical characteristics. Physical and chemical techniques are advised in this example to make the system functional. Dysregulation of enzyme activity in many pathological circumstances has led to innovative ideas for medication administration in *in vivo* and ultrasensitive disease sensors. This course of action does not require adjusting to accessible medications since they aren't chemically linked; nonetheless, they may be physically imprisoned. The enzyme eats the carrier parameters involved in a rational drug delivery system when their concentrations are adequate. This method is especially helpful for chemotherapy because it lessens the side effects of some dangerous drugs. Several papers address enzyme-degradable nanocarrier improvements. These enzymes are essential for mobilization, immunological response, necrosis, and apoptosis, among other functions. Reduced doses will improve treatment outcomes while reducing side effects [40].

## 1.4 TARGETING NANOCARRIERS

It is important to discover a specific disease target, a suitable treatment to treat it, and the best method for delivering it. In 1906, Paul Ehrlich proposed that focused medicine distribution be essential for illness treatment. Treatment efficacy can be improvd by boosting drug targeting at site of action which helps in non-specific medication distribution and lesser adverse effects. Drug delivery relies on the target area's physiology. Major tactics for targeting nanocarriers include:

- Target-site nanocarrier administration
- Using magnetic fields to target magnetic nanocarriers

## 1.4.1 Active Targeting

Directly administering therapeutic nanocarriers to the target organ seems difficult. This is because sickness spreads systematically across the body's cells and tissues. Sickness is rarely localized. Noninvasive magnetic fields are used to direct magnetic nanocarriers to a specific location, including increased magnetic nanocarrier absorption at the sick region, but this method has limitations. Drug distribution efficiency relies on field strength, particle physicochemical qualities, vascular supply, target tissue depth, and blood flow rate. This affects blood vessel flow. Active and passive targeting are commonly used nowadays. The active targeting technique uses nanocarriers with tiny ligand molecules that actively attach to a specific receptor, remain in the target region, and are actively taken up by sick cells. This well-known method binds the target location with great selectivity. The binding ligand should be specific to sick cells' overexpressed protein receptors. Cancer cells contain surface proteins that are overexpressed. Active targeting increases sick cell medication absorption. High surface-to-volume ratio nanocarriers may target many moieties. Active targeting avoids off-site medication delivery and reduces multi-drug resistance. The ligand–receptor interaction reduces tumor accumulation, attracting active targeting in cancer therapy due to cancer-specific receptors [41]. Epidermal Growth Factor Receptor (EGFR), folate receptors in ovarian and lung cancer cells, aptamer, and the transferrin receptor in breast cancer cells are a few examples. Folate ligands are smaller than antibodies, which are 160,000 kDa. The smaller molecule may reach the target spot more easily. Endothelial cells on the blood–brain barrier contain transferrin receptors, allowing transferrin ligands to be used as a targeting agent for anticancer medication delivery. Active targeting may also target leaky tumor vascular vasculature. This would cut off the tumor's nutrition and oxygen supply, eliminating its cells. Active vascular targeting may target vascular cell adhesion molecules, vascular endothelial growth factors, v3 integrins, and matrix metalloproteinases. This vascular targeting reduces tissue penetration and treatment resistance since endothelial indicators are more persistent than tumor markers. Active targeting of caveolar markers in endothelial cells is another method. Actively targeted nanocarriers may target bodily locations without increasing permeability and retention. A decrease in the cost of actively targeting nanocarriers is required for commercially relevant medicine to be developed for the treatment of illness [42].

## 1.4.2 Passive Targeting

The fundamental approach to passive targeting is increased permeability and retention effect. Cancer causes a leaky vasculature with endothelial cell dimensions that are several times (50–70 times) larger than healthy blood vessels. This has lymphatic drainage as well as interior vascular development, resulting in uneven angiogenesis. As a result, nanocarriers with molecular weights of more than 40 kDa may extravasate via the inflamed area's leaky vasculature, tumor tissue, or ischemia tissue. Passive tar accumulation in this leaky vasculature may allow the nanocarrier to pass through the interstitial gap in the endothelial barrier, allowing the drug-conjugated nanocarrier to aggregate at the disease site. They described this effect as the improved permeability and retention effect, two examples. This increased permeability and retention effect minimizes unwanted side effects caused by medication buildup at the sick location.

The disadvantage of targeting nanocarriers with low molecular weight drugs is that they diffuse away from the target region and re-enter the bloodstream. The charge of the nanocarrier owing to the functional group and its size determine its accumulation at the target location. Because their blood circulation length is longer, nanocarriers with hydrophilic surfaces and sizes smaller than 200 nm are reported to have improved permeability and retention. Thus, the foregoing disadvantage is avoided by targeting tumors based on their immunochemical and pathophysiological features [43]. To construct nanocarriers with greater effectiveness and target-specific therapeutic effects, researchers must investigate increased permeability and retention effects in various tumors. Only 15 nanocarriers have been effective in preclinical research and are indicated for clinical usage in passive targeting (Doxil™, Marqibo™, Abraxane™, Onivyde™, and DaunoXome™—US; Genexol-PM™—Korea; Mepact™ and Myocet™—Europe; and SMANCS™—Japan) [201]. However, no active targeting nanocarriers have been authorized for clinical usage. This is owing to the challenges of actively targeting nanocarriers in terms of regulatory and scale-up processes. The DOXIL™, which contains doxorubicin in a PEGylated liposome, is the first passively targeted nanocarrier to enter clinical usage. A liposomal nanocarrier carrying the medicines cytarabine and daunorubicin (CPX-351 [Vyxeos™]) has been shown to be effective in the treatment of acute myeloid leukemia. Specifically, passively targeted nanocarriers rely primarily on increased permeability and retention, although the previously mentioned disadvantages restrict their application in diagnostic and therapeutic processes. To overcome the restrictions, new methodologies and tools for evaluating nanocarrier interactions have been developed. For example, multilayered passively targeting nanocarriers for diagnostics and therapies contain ionizable lipids in the event of endosomal escape [44].

## TABLE 1.1
### Nanocarrier-Based Drug-Delivery Systems

| Drug-Delivery System | Drug (Loading Capacity) | Description | References |
|---|---|---|---|
| GCPQ, a micellar formulation based on quaternary ammonium palmitoyleglycole chitosan | Celecoxib | Micelles of amphiphilic GCPQ self-assembled by vortexing and celecoxib (size range 200–300 nm) | [45] |
| Hollow chitosan particles | Doxorubicin (<19% w/w) | Layer-by-layer chitosan/heparin hollow nanocapsule system (200 nm in size) | [46] |
| Cholanic acid-modified chitosan nanoparticles containing paclitaxel (PX). | Paclitaxel (<10% w/w) | Nanoparticles (about 260 nm in diameter) created using the dialysis technique | [47] |
| Chitosan nanoparticles | Cyclosporin A | Chitosan nanoparticles (size range 104–148 nm) produced via emulsification solvent diffusion | [48] |

(*Continued*)

**TABLE 1.1** (*Continued*)
**Nanocarrier based Drug-Delivery Systems**

| Drug-Delivery System | Drug (Loading Capacity) | Description | References |
|---|---|---|---|
| PLGA nanoparticles coated with chitosan | Albendazole (<54% w/w) | PLGA nanoparticles coated with chitosan and synthesized by emulsion (with a size range of 260-480 nm) | [49] |
| Chitosan nanoparticles | Enalaprilat (<31% w/w) | Ionic gelation produces spherical nanoparticles (size around 200 nm) | [50] |
| Chitosan nanoparticles derived from thiolated glycol | Calcitonin (<64% w/w) | Liquid microsprayer-prepared thiolated chitosan nanoparticles (size range 232–332 nm) | [49] |
| Chitosan nanocapsules | Salmon calcitonin (<51% w/w) | Resin nanocapsules made via solvent displacement in the range of 266–333 nm | [51] |
| Plasmid DNA-loaded chitosan nanoparticles | Hepatitis B virus surface protein-encoding DNA vaccination | DNA-loaded chitosan nanoparticles (size range 300–400 nm) were synthesized utilizing a complicated co-acervation technique | [52] |
| Chitosan-coated PLGA nanoparticles | Bevacizumab (<69% w/w) | Chitosan-coated PLGA nanoparticles (approximately 222 nm in diameter) were produced by double emulsion solvent extraction | [53] |
| Chitosan nanoparticles | Insulin (<55% w/w) | Cross-linked TPP particles (sizes between 300 and 400 nm) | [54] |
| Alginate-based beads coated with Poly(Lys) | Vancomycin | Beads made from alginate and calcium solution are mixed and added dropwise (size around 3000 mm) | [55] |
| Beads of calcium/ alginate | VEGF | By using extrusion/external gelation, you may make beads (particle size range 500–600 mm) | [56] |
| Alginate hydrogel with PLGA microspheres infused with VEGF | VEGF | Combination of PLGA microspheres with hydrogels loaded with VEGF and made by ionic crosslinking of an alginate solution with PLGA microsphere suspension using calcium sulphate | [57] |
| PVA-g-PAAm and PVA polymeric blend beads with alginate | Diclofenac sodium | Beads made by crosslinking glutaraldehyde with (particle size range 510–1320 mm) | [58] |
| Alginate/gelatin blend films | Ciprofloxacin hydrochloride | Casting/solvent evaporation-prepared drug-loaded films | [59] |
| Nanoparticles based on dextran that are loaded with siRNA | Macrophage marker protein CD68 targeted by a siRNA | Prepared using the double emulsion process, nanoparticles in the size range of 160–250 nm | [60] |

# REFERENCES

[1] Ibarra, A. M. F. P. R. (2007). MR Santamaría J. *Nano Today*, *2*, 22–32.

[2] Weiser, J. R., & Saltzman, W. M. (2014). Controlled release for local delivery of drugs: Barriers and models. *Journal of Controlled Release*, *190*, 664–673.

[3] Greish, K. (2010). Cancer nanotechnology. *Edited Anonymous Springer*, 25–37.

[4] Kerbel, R. S. (2008). Tumor angiogenesis. *New England Journal of Medicine*, *358*(19), 2039–2049.

[5] Maeda, H., Bharate, G. Y., & Daruwalla, J. (2009). Polymeric drugs for efficient tumor-targeted drug delivery based on EPR-effect. *European Journal of Pharmaceutics and Biopharmaceutics: Official Journal of Arbeitsgemeinschaft Fur Pharmazeutische Verfahrenstechnik e.V, 71*(3), 409–419.

[6] Astruc, D., Lu, F., & Aranzaes, J. R. (2005). Reduced transition metal colloids: A novel family of reusable catalysts. *Angew Chem Int Ed.*, *44*, 7852–7872.

[7] Shaw, B. J., & Handy, R. D. (2011). Physiological effects of nanoparticles on fish: A comparison of nanometals versus metal ions. *Environment International*, *37*(6), 1083–1097.

[8] Shi, H., Magaye, R., Castranova, V., & Zhao, J. (2013). Titanium dioxide nanoparticles: A review of current toxicological data. *Particle and Fibre Toxicology*, *10*(1), 1–33.

[9] Mohapatra, S., Ranjan, S., Dasgupta, N., Mishra, R. K., & Thomas, S. (Eds.). (2018). *Nanocarriers for drug delivery: Nanoscience and nanotechnology in drug delivery*. Elsevier.

[10] Mishra, R. K., Thomas, S., & Kalarikkal, N. (Eds.). (2017). *Micro and nano fibrillar composites (MFCs and NFCs) from polymer blends* (pp. 27–55). Woodhead Publishing.

[11] Raghvendra, K. M., & Sravanthi, L. (2017). Fabrication techniques of micro/nano fibres based nonwoven composites: A review. *Mod Chem Appl.*, *5*(2), 2.

[12] Rinaudo, M. (2006). Non-covalent interactions in polysaccharide systems. *Macromolecular Bioscience*, *6*(8), 590–610.

[13] Beeren, S. R., Meier, S., & Hindsgaul, O. (2013). Probing helical hydrophobic binding sites in branched starch polysaccharides using NMR spectroscopy. *Chemistry: A European Journal*, *19*(48), 16314–16320.

[14] Stirnemann, G., Hynes, J. T., & Laage, D. (2010). Water hydrogen bond dynamics in aqueous solutions of amphiphiles. *The Journal of Physical Chemistry B*, *114*(8), 3052–3059.

[15] Lalatsa, A., Schätzlein, A. G., Mazza, M., Le, T. B. H., & Uchegbu, I. F. (2012). Amphiphilic poly (l-amino acids)—new materials for drug delivery. *Journal of Controlled Release*, *161*(2), 523–536.

[16] Dutta, B., Barick, K. C., & Hassan, P. A. (2021). Recent advances in active targeting of nanomaterials for anticancer drug delivery. *Advances in Colloid and Interface Science*, *296*, 102509.

[17] Patra, J. K. (2018). Gitishree Das, Leonardo Fernandes Fraceto, Estefania Vangelie Ramos Campos, Maria del Pilar Rodriguez-Torres, Laura Susana Acosta-Torres, Luis Armando Diaz-Torres, Renato Grillo, Mallappa Kumara Swamy, Shivesh Sharma, Solomon Habtemar, Nano based drug delivery systems: Recent developments and future prospects". *Journal of Nanobiotechnology*, *16*(1), 71.

[18] Zhang, L., Wang, L., Jiang, Z., & Xie, Z. (2012). Synthesis of size-controlled monodisperse Pd nanoparticles via a non-aqueous seed-mediated growth. *Nanoscale Research Letters*, *7*(1), 1–6.

[19] Walters, G., & Parkin, I. P. (2009). The incorporation of noble metal nanoparticles into host matrix thin films: Synthesis, characterisation and applications. *Journal of Materials Chemistry*, *19*(5), 574–590.

[20] Xiong, Y., & Xia, Y. (2007). Shape-controlled synthesis of metal nanostructures: The case of palladium. *Advanced Materials*, *19*(20), 3385–3391.

[21] Niu, W., & Xu, G. (2011). Crystallographic control of noble metal nanocrystals. *Nano Today*, *6*(3), 265–285.

[22] Soni, K. S., Desale, S. S., & Bronich, T. K. (2016). Nanogels: An overview of properties, biomedical applications and obstacles to clinical translation. *Journal of Controlled Release*, *240*, 109–126.

[23] Rolland, J. P., Maynor, B. W., Euliss, L. E., Exner, A. E., Denison, G. M., & DeSimone, J. M. (2005). Direct fabrication and harvesting of monodisperse, shape-specific nanobiomaterials. *Journal of the American Chemical Society*, *127*(28), 10096–10100.

[24] Kersey, F. R., Merkel, T. J., Perry, J. L., Napier, M. E., & DeSimone, J. M. (2012). Effect of aspect ratio and deformability on nanoparticle extravasation through nanopores. *Langmuir*, *28*(23), 8773–8781.

[25] Kabanov, A. V., & Vinogradov, S. V. (2009). Nanogels as pharmaceutical carriers: Finite networks of infinite capabilities. *Angewandte Chemie International Edition*, *48*(30), 5418–5429.

[26] Torchilin, V. P. (2014). Multifunctional, stimuli-sensitive nanoparticulate systems for drug delivery. *Nature Reviews Drug Discovery*, *13*(11), 813–827.

[27] Zha, L., Banik, B., & Alexis, F. (2011). Stimulus responsive nanogels for drug delivery. *Soft Matter*, *7*(13), 5908–5916.

[28] Mura, S., Nicolas, J., & Couvreur, P. (2013). Stimuli-responsive nanocarriers for drug delivery. *Nature Materials*, *12*(11), 991–1003.

[29] Motornov, M., Roiter, Y., Tokarev, I., & Minko, S. (2010). Stimuli-responsive nanoparticles, nanogels and capsules for integrated multifunctional intelligent systems. *Progress in Polymer Science*, *35*(1–2), 174–211.

[30] Stuart, M. A. C., Huck, W. T., Genzer, J., Müller, M., Ober, C., Stamm, M., . . . & Minko, S. (2010). Emerging applications of stimuli-responsive polymer materials. *Nature Materials*, *9*(2), 101–113.

[31] Oh, J. K., Drumright, R., Siegwart, D. J., & Matyjaszewski, K. (2008). The development of microgels/nanogels for drug delivery applications. *Progress in Polymer Science*, *33*(4), 448–477. Ayame, H., Morimoto, N., & Akiyoshi, K. (2008). Self-assembled cationic nanogels for intracellular protein delivery. *Bioconjugate Chemistry*, *19*(4), 882–890.

[32] Sathali, A. H., Ekambaram, P., & Priyanka, K. (2012). Solid lipid nanoparticles: A review. *Sci Rev Chem Commun.*, *2*(1), 80–102.

[33] Lv, Q., Yu, A., Xi, Y., Li, H., Song, Z., Cui, J., . . . & Zhai, G. (2009). Development and evaluation of penciclovir-loaded solid lipid nanoparticles for topical delivery. *International Journal of Pharmaceutics*, *372*(1–2), 191–198.

[34] Luo, Y., Chen, D., Ren, L., Zhao, X., & Qin, J. (2006). Solid lipid nanoparticles for enhancing vinpocetine's oral bioavailability. *Journal of Controlled Release*, *114*(1), 53–59.

[35] Paliwal, R., Rai, S., Vaidya, B., Khatri, K., Goyal, A. K., Mishra, N., . . . & Vyas, S. P. (2009). Effect of lipid core material on characteristics of solid lipid nanoparticles designed for oral lymphatic delivery. *Nanomedicine: Nanotechnology, Biology and Medicine*, *5*(2), 184–191.

[36] Xu, Z., Chen, L., Gu, W., Gao, Y., Lin, L., Zhang, Z., . . . & Li, Y. (2009). The performance of docetaxel-loaded solid lipid nanoparticles targeted to hepatocellular carcinoma. *Biomaterials*, *30*(2), 226–232.

[37] Asasutjarit, R., Lorenzen, S. I., Sirivichayakul, S., & Ruxrungtham, K. (2007). Uracha Ruktanonchi and Garnpimol C. Ritthidej. *Pharm. Res.*, *24*(6), 1098–1107.

[38] Rudolph, C., Schillinger, U., Ortiz, A., Tabatt, K., Plank, C., Müller, R. H., & Rosenecker, J. (2004). Application of novel solid lipid nanoparticle (SLN)-gene vector formulations based on a dimeric HIV-1 TAT-peptide in vitro and in vivo. *Pharmaceutical Research*, *21*(9), 1662–1669.

[39] Teja, V. C., Chowdary, V. H., Raju, Y. P., Surendra, N., Vardhan, R. V., & Reddy, B. K. (2014). A glimpse on solid lipid nanoparticles as drug delivery systems. *J Glob Trends Pharm Sci.*, *5*(2), 1649–1657.

[40] Suresh, G., Manjunath, K., Venkateswarlu, V., & Satyanarayana, V. (2007). Preparation, characterization, and in vitro and in vivo evaluation of lovastatin solid lipid nanoparticles. *Aaps Pharmscitech.*, *8*(1), E162–E170.

[41] Lennernas, H., & Fager, G. (1997). Pharmacodynamics and pharmacokinetics of the HMG-CoA reductase inhibitors: Similarities and differences. *Clin Pharmacokinet.*, 32, 403–425.

[42] Kaur, S., Nautyal, U., Singh, R., Singh, S., & Devi, A. (2015). Nanostructured lipid carrier (NLC): The new generation of lipid nanoparticles. *Asian Pacific Journal of Health Science*, 2(2), 76–93.

[43] Trotta, M., Debernardi, F., & Caputo, O. (2003). Preparation of solid lipid nanoparticles by a solvent emulsification-diffusion technique. *Int J Pharm.*, 257, 153–160.

[44] Molpeceres, J., Guzman, M., Bustamante, P., & Rosario, M. D. (1996). *Int. J. Pharm.*, 130(1), 75–81.

[45] Mishra, R. K., Ahmad, A., Kumar, A., Vyawahare, A., Raza, S. S., & Khan, R. (2020). Lipid-based nanocarrier-mediated targeted delivery of celecoxib attenuate severity of ulcerative colitis. Materials science & engineering. *C, Materials for Biological Applications*, *116*, 111103. https://doi.org/10.1016/j.msec.2020.111103

[46] Patil, R. R., Guhagarkar, S. A., & Devarajan, P. V. (2008). Engineered nanocarriers of doxorubicin: a current update. *Critical Reviews in Therapeutic Drug Carrier Systems*, *25*(1), 1–61.

[47] Ma, P., & Mumper, R. J. (2013). Paclitaxel Nano-Delivery Systems: A Comprehensive Review. *Journal of Nanomedicine & Nanotechnology, 4*(2), 1000164. https://doi.org/10.4172/2157-7439.1000164

[48] De Campos, A. M., Sánchez, A., & Alonso, M. J. (2001). Chitosan nanoparticles: a new vehicle for the improvement of the delivery of drugs to the ocular surface. Application to cyclosporin A. *International Journal of Pharmaceutics, 224*(1–2), 159–168. https://doi.org/10.1016/s0378-5173(01)00760-8

[49] Makhlof, A., Werle, M., Tozuka, Y., & Takeuchi, H. (2010). Nanoparticles of glycol chitosan and its thiolated derivative significantly improved the pulmonary delivery of calcitonin. *International Journal of Pharmaceutics*, *397*(1–2), 92–95. https://doi.org/10.1016/j.ijpharm.2010.07.001

[50] Racoviceanu, R., Trandafirescu, C., Voicu, M., Ghiulai, R., Borcan, F., Dehelean, C., Watz, C., Aigner, Z., Ambrus, R., Coricovac, D. E., Cîrcioban, D., Mioc, A., Szuhanek, C. A., & Șoica, C. (2020). Solid Polymeric Nanoparticles of Albendazole: Synthesis, Physico-Chemical Characterization and Biological Activity. *Molecules* (Basel, Switzerland), *25*(21), 5130. https://doi.org/10.3390/molecules25215130

[51] Liu, L., Yang, H., Lou, Y., Wu, J. Y., Miao, J., Lu, X. Y., & Gao, J. Q. (2019). Enhancement of oral bioavailability of salmon calcitonin through chitosan-modified, dual drug-loaded nanoparticles. *International Journal of Pharmaceutics*, *557*, 170–177.

[52] Özbaş-Turan, S., & Akbuǧa, J. (2011). Plasmid DNA-loaded chitosan/TPP nanoparticles for topical gene delivery. *Drug Delivery*, *18*(3), 215–222. https://doi.org/10.3109/10717544.2010.544688

[53] De Negri Atanasio, G., Ferrari, P. F., Baião, A., Perego, P., Sarmento, B., Palombo, D., & Campardelli, R. (2022). Bevacizumab encapsulation into PLGA nanoparticles functionalized with immunouteroglobin-1 as an innovative delivery system for atherosclerosis. *International Journal of Biological Macromolecules*, *221*, 1618–1630. https://doi.org/10.1016/j.ijbiomac.2022.08.063

[54] Ali, A., Saroj, S., Saha, S., Gupta, S. K., Rakshit, T., & Pal, S. (2023). Glucose-Responsive Chitosan Nanoparticle/Poly(vinyl alcohol) Hydrogels for Sustained Insulin Release In Vivo. *ACS Applied Materials & Interfaces*, *15*(27), 32240–32250. https://doi.org/10.1021/acsami.3c05031

[55] Ueng, S. W., Yuan, L. J., Lee, N., Lin, S. S., Chan, E. C., & Weng, J. H. (2004). In vivo study of biodegradable alginate antibiotic beads in rabbits. *Journal of Orthopaedic Research: Official Publication of the Orthopaedic Research Society*, *22*(3), 592–599. https://doi.org/10.1016/j.orthres.2003.09.001

[56] Yang, Y., Gruwel, M. L., Dreessen de Gervai, P., Sun, J., Jilkina, O., Gussakovsky, E., & Kupriyanov, V. (2012). MRI study of cryoinjury infarction in pig hearts: i. Effects of intra-pericardial delivery of bFGF/VEGF embedded in alginate beads. *NMR in Biomedicine*, *25*(1), 177–188. https://doi.org/10.1002/nbm.1736

[57] Yang, Y., Liu, X., Zhang, D., Yu, W., Lv, G., Xie, H., Zheng, J., & Ma, X. (2011). Chitosan/VEGF-sIRNA nanoparticle for gene silencing. *Journal of Controlled Release: Official Journal of the Controlled Release Society*, 152 Suppl 1, e160–e161. https://doi.org/10.1016/j.jconrel.2011.08.062

[58] Nayak, A. K., & Pal, D. (2011). Development of pH-sensitive tamarind seed polysaccharide-alginate composite beads for controlled diclofenac sodium delivery using response surface methodology. *International Journal of Biological Macromolecules*, *49*(4), 784–793. https://doi.org/10.1016/j.ijbiomac.2011.07.013

[59] Abdelazeem S. Eltaweil, Maha S. Ahmed, Gehan M. El-Subruiti, Randa E. Khalifa, Ahmed M. Omer. 2023. Efficient loading and delivery of ciprofloxacin by smart alginate/carboxylated graphene oxide/aminated chitosan composite microbeads: In vitro release and kinetic studies, *Arabian Journal of Chemistry*, *16*(4).

[60] Wasiak, I., Kulikowska, A., Janczewska, M., Michalak, M., Cymerman, I. A., Nagalski, A., Kallinger, P., Szymanski, W. W., & Ciach, T. (2016). Dextran Nanoparticle Synthesis and Properties. *PloS One*, (1), e0146237.

# 2 Nanocarriers Manufacturing Approaches and Drug Delivery Applications

*Akash Chauhan, Ayush Dubey, Awaneet Kaur,
Shikha Chaudhary, Krishna, Md. Aftab Alam,
Niranjan Kaushik, and Vijay Kumar Singh*

## 2.1 INTRODUCTION

Since the beginning of the twenty-first century, extensive research has been conducted to address concerns such as limited areas of biology, inadequate selection, drug solubility, overdose or short distribution duration, and unattractive pharmacokinetics or pharmacodynamics. Hydrophobic drugs have lower pharmacological effectiveness and performance because they are difficult to dissolve in the circulation. As a result, their distribution routes are quite constrained. When administered orally or intravenously, some bioactive chemicals, such as proteins and nucleic acids, are rapidly broken down by the body's metabolism and fail to reach their intended target areas. Local delivery techniques, particularly implanted drug delivery, are being studied as an alternative to standard medication delivery mechanisms and systemic route constraints. In the early phases of research [1, 2], poly (methyl methacrylate) implants, poly (glycolic acid) implants, poly (lactic acid implants), hyaluronan implants, collagen implants, hydroxyapatite implants, silk implants, and injectable cement implants were all employed to generate biodegradable implants. Bioactive chemical distribution methods and implantation procedures have been used in bone repair and tissue regeneration. However, these systems have a number of limitations, including uncontrolled burst drug release and the difficulty of maintaining a targeted dose over an extended period of time. Because of their intrinsic inorganic nature, amorphous or porous biomaterials generate the majority of these issues. Nanotechnology-based techniques have suggested innovative drug delivery implants with ordered nanoporous and nanotubular designs to solve these challenges [3, 4]. Nanomedicine, or the use of nanotechnology in pharmaceuticals, is commonly regarded as a rapidly expanding field with significant potential for the creation of innovative and high-level medicine delivery systems. Nanoporous anodic alumina and titania nanotubes are the most notable examples. These are two novel, promising biomaterials created using a top-down, lithography-free nanoengineering method. Their manufacturing strategy is based on a number of elements, including cost, ease of use, productivity,

DOI: 10.1201/9781003316398-2

scalability, and a suitable time period. In the past, titanium nanotubes have been explored for their potential to be used in drug delivery due to their multifaceted and intrinsic properties, such as excellent biocompatibility, nontoxicity, durability, mechanical stability, chemical resistance, inertness, high surface area, manageable nanotube thickness (1–300 mm) and diameter (10 nm), and simple, low-cost fabrication [5–8]. For drug delivery systems, nanomaterials have been employed in a number of morphologies, including one-dimensional (nanocrystal, dendrimer, micelle, aquasome, and core shell structures) that may be improved by adjusting the monomer ordering and layer-to-layer thicknesses. By alternating the monomer on a regular basis, hydrolysis, degradation, and surface energy may all be changed. Biopolymers having reversibly altered physicochemical characteristics against external stimuli, particularly those that are crosslinked or even covalently attached to surfaces, are a hot research area as light-responsive hydrogels capable of transporting medicines and genes selectively [9, 10].

### 2.1.1 METAL AND METAL OXIDE NANOCARRIERS AND METAL OXIDE DRUGS

Scientists resorted to inorganic materials to create higher quality nanocarriers with specific characteristics. While traditional nanocarriers can encapsulate their loads safely and securely, inorganic nanocarriers have several advantages over them (Figure 2.1), including the ability to include targeting ligands and the most readily available materials, all of which reduce manufacturing costs. Imaging modalities and photothermal characteristics of inorganic nanocarriers are not present in

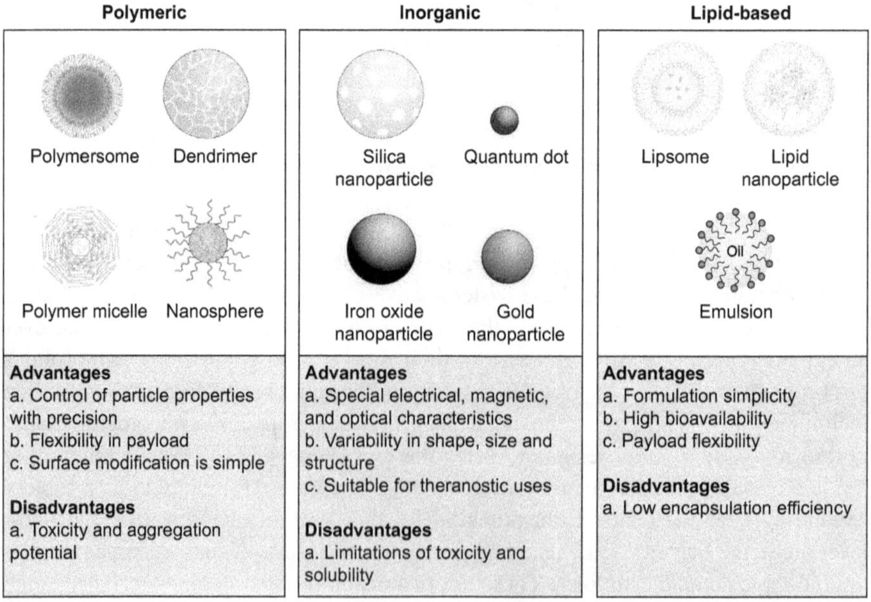

| Polymeric | Inorganic | Lipid-based |
|---|---|---|
| Polymersome     Dendrimer | Silica nanoparticle     Quantum dot | Lipsome     Lipid nanoparticle |
| Polymer micelle   Nanosphere | Iron oxide nanoparticle     Gold nanoparticle | Emulsion |
| **Advantages** <br> a. Control of particle properties with precision <br> b. Flexibility in payload <br> c. Surface modification is simple <br><br> **Disadvantages** <br> a. Toxicity and aggregation potential | **Advantages** <br> a. Special electrical, magnetic, and optical characteristics <br> b. Variability in shape, size and structure <br> c. Suitable for theranostic uses <br><br> **Disadvantages** <br> a. Limitations of toxicity and solubility | **Advantages** <br> a. Formulation simplicity <br> b. High bioavailability <br> c. Payload flexibility <br><br> **Disadvantages** <br> a. Low encapsulation efficiency |

**FIGURE 2.1**   Drug delivery using nanoparticles.

polymer-based NPs and are unique to inorganic nanocarriers. In terms of single-setup efficiency, there are presently no better inorganic nanotransporters than valuable gold metal NPs [11, 12]. Because of their intrinsic imaging characteristics, Quantum Dots (QDs) containing iron oxide are preferred to conventional nanocarriers for diagnostic and drug administration tracking. Because of the optical properties of QDs, fluorescent probes may conduct imaging and analysis at the same time. The paramagnetic properties of iron oxide nanoparticles (NPs) have aided magnetic resonance imaging (MRI), making them more powerful contrast agents.

Despite their wide range of uses, the biodegradability and biocompatibility of inorganic nanocarriers should be thoroughly investigated before they are used in medical settings. The dissolution of their heavy metals, which are frequently composed of toxic heavy metals, is a constraint of QDs in health-related practical applications (e.g., cadmium). According to analytical investigations, heavy metals accumulate in these organs for up to 3 months following exposure. This shows that QD death and removal are both delayed, potentially resulting in systemic toxicity through bioaccumulation [13]. Another example is gold nanoparticles. Because of the ease with which they can be manufactured, the benefits of their biocompatibility, the clarity of their surface chemistry, and the ease with which they can be imaged using fluorescence resonance energy transfer, gold metal NPs are being studied for their potential as multifunctional gene carriers [14, 15]. There is a positive charge that allows for DNA binding or even RNA silencing for the formation of complexes that cells can grip. Additional techniques for producing gene nanocarriers using PEI and silencing RNA involve gold-thiol chemistry. One significant disadvantage of these techniques is the need for surface functionalization and cationic moieties [16, 17], which necessitates several stages in the artificial process. Another method for avoiding this outcome has recently been proposed: the development of very efficient controllable catechol-conjugated PEI was used to regulate the production of PEI-coated gold NPs for RNA silencing delivery. In aqueous solution, reductive catechol active groups form spherical multicore micelles. These templates are often used to create gold nanoparticles with precise sizes and surface charges. NPs coated with PEI have low toxicity and significant gene silencing effects in cancer cells [18]. Gold NPs functionalized with double-stranded DNA or single-stranded DNA were generated faster than gold NPs shielded by a monolayer or polymer-gold NP complex. For dependable gene delivery, gold nanoparticles functionalized with single-stranded DNA were produced. When compared to antisense DNA, gene knockdown efficacy was higher with these NP forms, while cellular toxin levels were lower. Improved DNA binding affinity and nuclease resistance were observed [19].

## 2.1.2 MAGNETIC NANOPARTICLES

Magnetic materials are now being carefully explored in industrial diagnostics and biosensors, as well as for other biological applications. Control of the material's inorganic–organic interface is critical, and mediation plays a key role in this. This is a key aspect in achieving this aim. The organic phase is modulated to change biomolecular appreciation and biocompatibility, whilst the inorganic phase generates physical features such as magnetism. In medical applications, magnetic spheres with diameters ranging from 100 nm to less than 100 nm are used [20, 21]. Magnetic NPs that are commercially available have previously been used as gene-delivery nanocarriers. Due to their distinct size, magnetic NPs are able to infiltrate cells. Magnetic nanoparticles formed from magnetic ferrite NPs are known as ferrofluids. Ferrofluids are widely employed in a variety of applications. Magnetic area sensors and insulated electromagnetic circuits are featured. Electromagnets have been suggested as NP carriers for bioseparation and immobilization of enzymes and proteins [22]. In addition, a reliable and biocompatible substance, such as silica, surrounds the NPs ($SiO_2$) to protect cells from hazardous effects. One of the most significant applications for magnetic NPs is biomedical imaging.

### 2.1.3 NANOPARTICLES WITH SOLID LIPIDS

Lipids contain fats, waxes, sterols, and phospholipids. They perform specialized duties and may maintain organ immunity. These supplements help absorb fat-soluble vitamins, nutraceuticals, and pharmaceutical molecules. They create and receive energy. Dietary lipids may impart texture, consistency, rheological characteristics, taste, flavour, colour, smell, and pleasant sensations as food additives. Lipid-based delivery systems are gaining popularity because of their eligibility, nontoxicity, and convenience of use in culinary items. Lipid-based delivery strategies include liposomal. Next [23–25], we will discuss solid lipid NPs. Due to their intrinsic functionality and features, which enable endurance in complex systems, and their capacity to segregate bioactive cells into the critical core of NPs, they are a rapidly increasing food carrier system. Liposomes and emulsions have fewer of these features. The high expense of hazardous-level reporting renders artificial polymeric NPs less suited for food delivery than pharmaceutical distribution techniques [26, 27]. These submicrometer-sized, 50 to 500 nm wide solid lipid NPs were first produced in 1991 for use as drug carriers and have high ambient and physiological temperatures. As a result, they were effectively managed by a biodegradable, biocompatible, and nontoxic surfactant. Scientists have synthesized and analysed para-acyl-calixarenes, which are used to create solid lipid NPs for medication delivery [28–30]. They may be used to adorn objects as well as transport both hydrophilic and hydrophobic molecules, with hydrophobic molecules benefiting the most. Hydrophobic molecules are more attracted to the hydrophobic matrix of solid lipid NPs, while hydrophilic molecules are less attracted to the hydrophobic matrix. This technique has previously been used to deliver nutrition, medicine, and proteins and is continued in use today. Solid lipid NPs for commercial application, high-pressure homogenization, and microfluidization, all of which are easily scalable, are utilized. Lipid-based NP medicines for medical and cosmetic delivery have been available for quite some time. The following steps can be used to make solid lipid nanoparticles that are full of bioactive compounds [31–33].

- microfluidization
- homogenization under high pressure
- contactor membrane technique
- bottom-up innovation

- top-down approach
- emulsification evaporation solvent

Bioavailability is the amount of bioactive compound that can be used in the area of a biological system's physiological process. Curcumin, quercetin, and other water-soluble bioactive compounds, as well as bioactive compounds with low cell permeability, may have their oral bioavailability improved by utilizing lipid nanoparticles. The particle size of solid lipid NPs influences aspects. Phospholipid presence influences the physical and chemical characteristics of solid lipid NPs, such as size, surface charge, stability, polymorphism, and many more. Until recently, a variety of surfactants were used and have been permitted for use in foods and other items, either alone or in combination, such as solid lipid NPs, which seem to have been used to stabilize lipid NPs. The selection of an adequate food-grade surfactant is one of the most difficult aspects of effectively stabilizing in food-based matrices. Solid lipid NPs are used. For solid lipid NPs needing matrix and suspension stability to be achieved in pharmaceutical formulations, mixtures of ionic and non-ionic surfactants are used. Ionic surfactants, on the other hand, are not suggested for food applications due to their high emulsifying capacity, which causes gelling during storage. In principle, the method of choosing is determined by the excipient's physicochemical properties, as well as its bioactivity, output volume, and other factors [34–36]. Microfluidization and high-pressure homogenization are the most frequently employed methods of the aforementioned techniques because of their ease of scaling and lack of toxicity. Solid lipid NPs, such as lipids, surfactants, cosurfactants, cryoprotectants, and others, are selected based on their encapsulation capability as well as loading performance, target ability, stability, and other factors. Candidates for adequate and well-suited excipients will be in high demand in the future. The following are some characteristics of solid lipid NPs:

1. Most solid lipid nanoparticles are made without using chemical solvents, which cannot be used in food [37, 38].
2. Sterilization of solid lipid NPs is usually done with methods like UV and heat sterilization [39, 40], which are also used to sterilize food.
3. Lipid NPs in solid form are present in both fluid and solid forms. Because of this, they can be used in both solid and liquid food [41, 42].

### 2.1.3.1 Phospholipids
Liposomes are spherical vesicles made up of lipid bilayers around an aqueous core that may transport both lipophilic and hydrophilic medications. Vesicles formed by the bilayer might be uniflagellar or multilamellar (more than one bilayer). Physiologically active chemicals are transported by this vesicle. These compounds have a shorter systemic half-life.

### 2.1.3.2 Graphene, Fullerenes, and Nanodiamond
According to studies, carbon nanoparticles inhibit bacterial development in a variety of ways. Graphene oxide is made from graphite with the use of oxidizing acids [43]. According to Liu et al., the antibacterial properties of graphite, graphene oxide, and

reduced graphene oxide are all similar. Reactive oxygen species (ROS), or photocata-lytic activity generated by fullerenes (such as the soccer ball shaped C60 buckmin-sterfullerene), is effective in killing bacteria. Without the presence of ROS, diamond nanoparticles may generate oxidative stress [44]. Graphene Oxide is composed of three nanoparticles: the DNA aptamer, the graphene oxide nanoparticle, and the semicon-ductor quantum dots. In 2017 Debopam et.al validated that the DNA-based optical nano sensor is discovered to be capable of detecting potassium concentrations across a wide range from 1 pM to 1 mM by employing the FRET principle when used for physi-ological potassium sensing. This device might be used to identify multidrug-resistant infections using certain multidrug-resistant genes [45].

Carbon-based QDs are fluorescent, as are graphene QDs. There are two types of graphene QDs: those that depend on faults in carbon nanostructures and those that rely on isolated p-domains on the graphene layer that are physically separated. When there are flaws in graphene QD, they cause it to fluoresce more brightly than when there are flaws in carbon dots [46, 47]. In the next few lines, we will discuss carbon dots and their mechanically equivalent graphene QDs as fluorescent bioimaging agents that outperform standard semiconductor QDs. Carbon dots are NPs with heavily modi-fied or functionalized surfaces. Since the 2006 paper on carbon dots, several organi-zations across the globe have conducted studies into the structural and mechanical understanding of carbon dots, as well as their potential uses. Studies on functionalized carbon nanotubes inspired the discovery of carbon dots. Chemically functionalizing single-walled and multiwalled carbon nanotubes resulted in brilliant and multicoloured fluorescence emission. When chemically functionalized, they must be equally or more luminous across equivalent spectral zones. In the visible and near-infrared spectrum, surface-functionalized tiny carbon NPs, known as carbon dots, create bright and colourful fluorescence emissions. Carbon dots with multiphoton fluorescence [48–50] were discovered. The two-photon cross sections of carbon dots excited by a pulsed laser beam (800–900 nm) for visible fluorescence emission were considerable. Because of their high fluorescence emission and their linear and nonlinear optical characteris-tics, carbon dots have been employed in bioimaging, photodynamic treatment, fluores-cence probing and analysis, and photocatalysis applications. This is a viable alternative because carbon dots are safe and do not break down when exposed to light, and their advantages for fluorescence imaging may be very important [51].

### 2.1.4 CARBON NANOTUBES

Carbon nanotubes, fullerenes, carbon nanoparticles, graphene, nanodiamonds, car-bon nanocaps, and carbon nanohorns exhibit exceptional electrical, mechanical, optical, and magnetic properties. UV-visible near-infrared light absorption, photolu-minescence, novel Raman signals, outstanding photothermal response, photosensi-tized singlet oxygen generation, and a large surface area for covalent and noncovalent conjugation of contrast agents, drugs, DNA, and RNA are all appealing for biological applications [52–57]. Carbon nanoparticles are used in bioimaging, biosensing, gene and medication delivery, and photothermal and photodynamic cancer and infection therapy. Carbon nanostructures are hydrophobic and incompatible as prepared, neces-sitating conjugation with hydrophilic compounds [58–60]. CNTs have been studied for biological applications since 1991. Carbon Nanotubes (CNTs) are rolled-up graphene sheets that range in length from hundreds to tens of micrometres. Carbon nanotubes have a graphene covering of 0.4 nm. Multiwalled carbon nanotubes are made from 100 nm single-walled carbon nanotubes [61, 62]. CNTs are functionalized by biomolecules, imaging agents, and medicines. The CNT surface is usually functionalized by acidic oxidation followed by a 1,3-dipolar cycloaddition reaction, resulting in a covalently bonded modification, or through a hydrophobic process between the CNTs and another nonpolar ring, such as DNA, resulting in noncovalent conjugates. Water-soluble carbon nanotubes might be used for gene delivery. Electrostatically, ammonium-functionalized nanotubes bind plasmid DNA. Microscopy reveals that the bioconjugate can pass through the cell membrane through nanoneedles, catatonically glyco-polymerized single-walled carbon nanotubes. They are used in gene delivery. Single-walled biocompatible copolymer-functionalized nanotubes are good transfection agents. Because carbon nanotubes absorb themselves, they are perfect for gene and medication delivery. Single- and multiwalled carbon nanotubes interact more consistently with plasmid DNA, allowing gene transfer. To investigate cellular uptake, fluorescent markers and biomolecules are bonded to CNT walls. Biotin avidin is created by covalently attaching carbon nanotubes to fluorescein or biotin [63].

### 2.1.4.1   Graphene and Devices

Graphene-based composites can be exploited to transfect two distinct cell lines with gene delivery. Plasma treatment adds oxygen groups to CNT surfaces. Composites and SFP-conjugated plasmids were coupled with other plasmids and used for gene transfection with little toxicity and great performance [64]. It is because of the conjugated graphene basal plane that organic–inorganic hybrids may be made and may link and absorb water-insoluble medications. Amidated chitosan was used to synthesize graphene oxide. Graphene oxide is made of graphene oxide, which is biocompatible and water soluble. Using chitosan-grafted graphene, plasmid DNA may be compressed into nanosized complexes. Graphene-oleate polyamidoamine dendrimer hybrids were evaluated for gene transfection [65–68].

### 2.1.4.2   Dendrimers

Dendrimers are branching macromolecules with an initiator core and active terminal arms [69]. Dendrimers are made from nucleotides, sugar, and amino acids. Multivalent, highly branching, and with different peripheral groups, they constitute a unique drug delivery source. Stepwise dendrimer synthesis can result in a dendrimer with irregular branches [70]. Adding a generation to the core offers a bulk external group. Hydrophobic connections, chemical contact, or hydrogen bonds with increased surface function may enclose drug molecules in core cavities. Covalent bonding may link medicinal compounds to active groups [71]. Single-generation dendrites may dissociate molecules [72]. Physical and chemical bond formation is key in drug–dendrimer interactions. This dendrimer is used for MRI, gene therapy, medication, antiviral, and vaccine administration [73]. It helps connect prodrugs. Different anticancer medications like cisplatin and doxorubicin are coupled to dendrimers to boost anticancer action [74–78].

### 2.1.5  POLYMERIC NANOCARRIERS

It is possible to decompose polymeric nanoparticles [79]. Drug molecules may be dissolved or dispersed in a polymer core or encased in a polymer matrix (nanospheres) and held in place by the matrix. Both types have the ability to chemically conjugate or adsorb drugs [80]. Biodegradation of polymeric nanocarriers produces monomers readily destroyed by metabolic processes [81]. Polymeric nanocarriers may be made from a number of natural and synthetic polymers. Chitosan, gelatine, albumin, collagen, alginate, and other natural polymers are examples [82]. Polymer-based nanocarriers target malignant cells with anticancer drugs like doxorubicin. Changing the polymer's physicochemical characteristics may improve medication release. Multifunctional polymeric nanocarriers allow multiple drug incorporations [83]. The synthesis of stimuli-sensitive polymeric nanoparticles for targeted medication delivery has advanced (smart polymers). Smart polymers release medication in response to both internal and external stimuli. Scalability, toxicity/biocompatibility, and stimulus sensitivity are smart polymer design issues. Intrinsic stimuli vary across clinical and preclinical models, whereas extrinsic stimuli provide compliance, tissue penetration, and localization [84]. Polymeric nanocarriers provide tailored medication delivery despite these obstacles. When McBain invented it in 1913, colloidal aggregates created when detergent and water are mixed together were dubbed "micelles." They are amphiphilic molecules because they have a hydrophobic tail and a hydrophilic head. Amphiphilic or detergent molecules produce inverted micelles in a nonpolar solvent, with the head pointing toward the centre and the tail pointing away from it. Temperature and pH, as well as amphiphilic molecular type and ionic strength, influence micelle nanoparticle size and form. Surfactant concentration influences micelle production. Micelle formation fails below the threshold concentration [85]. Two copolymers generate polymeric micelles with amphiphilic molecules in specific solvents. One copolymer is solvent-soluble while another is not. Insoluble copolymer creates the chain's centre while soluble copolymer produces the shell [86]. Polymeric micelles have industrial and medicine delivery uses [87, 88]. Polymeric micellar nanocarriers target the pilosebaceous unit in hair follicle disorders. Effectiveness was increased by 4.5% and 3.3%, respectively, when micelles were used as nanocarriers for adapalene [89, 90].

Gold, magnetic nanocarriers, quantum dots, mesoporous silica, and other inorganic nanocarriers are examples. Inorganic materials are used to create traceable nanocarriers. Biosensing, cell labelling, targeting, imaging, and diagnostics all make use of inorganic nanocarriers. Synergistic therapeutic effects of inorganic nanocarriers can be generated by changing their composition or size, nanocarriers' magnetic, plasmonic, and optical properties may be improved [91]. Heavy metal inorganic nanocarriers cause long-term health problems [92].

### 2.1.6  NANOGOLD AND NANOSILVER

With their size, biocompatibility, and controllable release of drugs, metal nanoparticles alone or in combination with plant extracts or antibiotic chemicals are fascinating candidates for the administration of drugs. They are both cost effective and environmentally friendly [93–95]. Biogenic nanoparticles have been studied using absorption spectroscopy, scanning electron microscopy, X-ray diffraction, and Fourier transform infrared analysis. These silver and gold nanoparticles have been demonstrated to be "magic bullets in the fight against disease" [96]. When given to mouse pancreatic islets, polyvalent gold–DNA conjugates ameliorate diabetes [97]. Gold NP-22 reduces Ab fibril cytotoxicity and peroxidase activity [98]. Gold nanoparticles also detect mycobacterium complexes *in vivo* [99].

Researchers have developed gold nanoparticles for a range of purposes, including photoacoustic imaging [100], chemotherapy, and surface-enhanced resonance spectroscopy, using the noble metal gold as a paradigm for nanotechnology [101–104]. Gold nanoparticles may be synthesized top-down or bottom-up. Nanostar, nanorod, nanocage, nanoshell, nanoprism, and so on are gold anisotropies. Gold nanocarriers' optical properties interest biological researchers. Enzymes, sugars, fluorophores, peptides, proteins, and DNA may adhere to gold nanoparticles. Gold nanocarriers efficiently image malignant cells [105, 106]. The nanoshell with Optical Coherence Tomography (OTC) agents enables 3D tissue imaging [107, 108]. Positron Emission Tomography (PET), Single Photon Emission Tomography (SPECT), and Computed Tomography Angiography (CTA) employ gold nanocarriers.

### 2.1.7 SILICA MESOPORES

Mesoporous silica's honeycomb structure allows for drug integration. Its simplicity and availability make it useful in biomedicine [109]. It can encapsulate both drugs that do not mix well with water and drugs that do. It is formed from silica that has large pores, a high surface area, and excellent biocompatibility as well as high drug-loading capability. Active and passive cancer targeting may be done using mesoporous silica [110]. Camptothecin and methotrexate are effectively administered by mesoporous silica [111, 112].

Hybrid nanocarriers combine organic and inorganic nanocarriers. Multi-components, organic–inorganic, inorganic–organic hybrid nanocarriers include lipid-polymer, ceramic-polymer, and so on. Combining two nanoparticles significantly improves their properties [113]. Organic nanocarriers like liposomes are leaky and unstable. The leaky and unstable nature of organic nanocarriers helps eliminate organic nanocarrier from the bloodstream. Additionally this combination of two nanoparticles makes the product more stable and effective. This is countered by a hybrid nanocarrier system [114]. Selecting optimal nanocarriers for medication delivery aims to maximize bioavailability with minimal negative effects. Hybrid nanocarriers are used for research. The mesoporous silica nanoparticle lipid bilayer hybrid nanocarrier technology revealed intracellular transport of zoledronic acid in high-retention breast cancer [115, 116]. This mechanism allows stimuli-response medication release, avoiding premature body release. Novel albumin hybrid nanocapsules encapsulate hydrophilic peptides or tiny medicinal molecules to target cancer cells. Lower toxicity and better uniformity may be achieved by using this method [117]. Encapsulation of medicinal compounds with ferritin and stimuli-response drug release provides a long-term release of drugs into the target area [118]. Polyethylene has been used in research to deliver small interfering RNA throughout the body. Theragnostic gene silencing can be done with this method because it is stable, gets into cells well, and works [119].

## 2.2 NANOCARRIER FUNCTIONALIZATION

Functionalization adds functional groups to the nanocarrier surface. Medication delivery and targeting must be controlled during medication delivery and targeting [120]. Functionalization improves the intracellular drug delivery nanoparticle payload, binding capability, and cytotoxicity [121]. Using various processes, nanoparticles may be functionalized with a range of small molecules and biomolecules [122]. It is said that bioactive compounds and biological macromolecules can be covalently or noncovalently attached to a multivalent surface in order to interact with the target and be biocompatible.

### 2.2.1 POLYMER-COATED NANOCARRIERS

The polymer coating on nanoparticles' macromolecular characteristics makes them target specific. Polymer coating allows for long-term pharmaceutical release as well as passive tumour targeting [123]. Polyethylene glycol, for example, improves nanoparticle permeability and retention [124, 125]. The stealth nanocarriers used in this work increase particle penetration, circulation time, and serum protein adsorption [126–128]. Targeting compounds coupled with stealth nanoparticles are used to target cancerous cells [129]. PEGylation enhances nanoparticle stability over unmodified nanoparticles. A functionalized polymer nanocarrier is employed to passively target mouse tumour tissues. The cooperative nanocarrier system consists of polyethylene glycol-coated gold nanorods and doxorubicin-loaded liposomes or a magnetic nanoworm. The gold nanorod functions as a photothermal antenna to heat tumour tissue when irradiated by infrared. This second component, which targets cyclic peptide species, raises P32 stress protein levels in tumour cells. The tumour shrank with heat [130–132].

### 2.2.2 FUNCTIONALIZING WITH LIGANDS

Nanocarriers attach to ligands like green fluorescent protein for biosensing and disease diagnostics [133–135]. Electrostatically complexed nanocarriers with green fluorescent protein distinguish metastatic tumours [136]. Green fluorescent protein-gold nanocarrier sensing is regulated by ligand headgroup modification and cell-nanoparticle affinity [137]. Negatively charged cells may readily ingest positively charged functionalized nanocarriers. The guanidine head group functionalizes magnetic iron oxide nanoparticles. To link with the cell membrane through negatively charged functional groups on the nanocarrier, diffusion or pinocytosis may be employed (carboxylate on iron oxide) [138, 139]. The researchers found that the surface charge of neutral nanocarriers could give information about how cells interact with each other.

### 2.2.3 NANOCARRIER BIOFUNCTIONALIZATION

Nanoparticles may be functionalized using monoclonal antibodies, oligonucleotides, proteins, and siRNA. These biomolecule-conjugated nanocarriers minimize cytotoxicity and allow targeted medication administration by binding to cell surface receptors [140]. Alkyl thiol groups or glycol spacer molecules may covalently attach biomolecules to nanocarriers [141, 142]. This bio-inspired nanocarrier's cell-mimicking property allows for long-term biosystem circulation. It can overcome many disorders. The chemotherapy drug doxorubicin is in the erythrocyte bioinspired membrane nanoprobe so that it can be used to treat disease and find out what is wrong [143].

## 2.3 NANOCARRIER DRUG LOADING AND PROPERTIES

Surface functional groups facilitate medication loading and release. There are three main ways to load therapeutic agents onto nanocarriers [144–146].

### 2.3.1 ENCAPSULATION

#### 2.3.1.1 Electron Irradiation (ESD) Bonding with Covalent

Functional groups on a nanocarrier's surface allow it to carry a high concentration of medicinal substances. Due to the nanocarrier's many functional groups, the medication binds covalently to it [363]. After conjugation, hydrolytic bonds may be broken via enzymatic cleavage or chemical decomposition. Due to the slow diffusion rate of the drug-carrier nanoparticle mixture across the cell membrane, precise drug delivery is possible. This covalent coupling allows stable nanocarrier medication delivery [147].

##### 2.3.1.1.1 Encapsulation

Encapsulating therapeutic drugs is another nanocarrier loading technique. The nanocarrier's hollow area allows medication encapsulation. Polymeric nanocarriers, nanocapsules, dendrimers, and the like may encapsulate drugs efficiently [148, 149]. Hydrophobic inner cavities allow for hydrophobic medication incorporation through hydrophobic contact or hydrogen bonding. Physical interactions may cause encapsulation. Active and passive drug loading encapsulate liposomes. Hydrolysis, thiolysis, and thermolysis release the medication [150].

##### 2.3.1.1.2 Electrostatics

Functional nanocarriers like carboxyl and amine promote hydrophobic medication solubility. These high-density functional groups interact electrostatically with the nanocarrier system. Nonsteroidal anti-inflammatory medications such as indomethacin, ciprofloxacin, diflunisal, ibuprofen, and others are effectively incorporated by electrostatic interaction [151]. Liposomes' double electrical layer gives them electrostatic charges.

### 2.3.2 CHARACTERIZATION OF NANOMATERIALS

Nano pharmaceutical characterization is key to FDA guideline publications. As indicated, nanosizing may drastically affect the physicochemical characteristics of active ingredients, and a comprehensive assessment of nanosized formulations is required for effectiveness and safety approval. An Investigational New Drug (IND) application may be submitted after identifying all of these features. As with ordinary drug development, this identification cycle comprises four distinct phases:

- Phase I: healthy people
- Patients in Phase II have a known medical problem.
- Phase III: multicenter placebo-controlled randomized clinical trials
- Phase IV: After a product has been approved for sale, these studies are often asked for to answer questions from Phase III [152].

Analytical approaches can characterize nanoparticles. Particle size is crucial. Photon correlation spectroscopy (PCS), or dynamic light scattering, is used to determine nano pharmaceutical size distribution. PCS measures the hydrodynamic diameter of Brownian nanoparticles to determine their size. This approach measures nanoparticle size distribution down to 1 nm. Nano pharmaceutical fate is also influenced by molecular weight, density, z potential, and hydrophobicity. Scanning Electron Microscopy (SEM) or Atomic Force Microscopy (AFM) is used to examine nano pharmaceutical surfaces. Differential scanning calorimetry and thermogravimetric analysis may be used to characterize nano pharmaceuticals [153]. Therapeutic or diagnostic nano pharmaceuticals might be inorganic or organic. Some inorganic particles have been approved for imaging applications and anaemia treatment [154, 155], including intraoperative sentinel lymph node imaging and thermal ablation of cancers in clinical studies. Biodegradable organic nanoparticles are often used to make vaccines, and controlled-release formulations are often used to treat skin problems locally or over a long period of time [156, 157].

### 2.3.3 DISPERSION OF SHAPE AND SIZE

Particle size influences several physicochemical aspects and indicates quality and performance. Cytotoxicity is caused by nanomaterial surface interactions with biological components. Increasing particle diameter increases surface area exponentially [158]. Thus, particles with the same composition and components might vary in cytotoxicity depending on size and surface reactivity. Size impacts drug release, aggregation, and transport. As Dissolving smaller particles is more effective than bigger particles, and also increases the suspension viscosity. Milling, crushing, homogenization, and emulsification are all processes that reduce particle sizes to the nanoscale. Before administering NPs *in vivo*, it is important to describe and manage their size and size distribution *in vitro* [159]. Microscopy can now determine NP size. Later, we will cover NP characterization approaches. Particle size distribution determines the sample's mean, median, and mode size. The mode is the highest-frequency peak. Mode is the most common particle size. Breadth is another significant characteristic. Standard deviation or variance may be used to calculate distribution breadth [160].

## 2.3.4 Surfaces

Nanomaterials' surface composition and structure influence how they react to inter-actions. This must be considered before administering a nanomaterial physiologically. Nanomedicine requires the ability to study nanomaterial surfaces. Surface charge, energy, and wettability are essential surface qualities. Each parameter affects nanomate-rial functionality. Surface charge may alter binding to receptors or ligands and penetra-tion through physiological barriers. Solubility, stability, and aggregation are all affected by surface charge and energy [161, 162]. Surface charge is estimated using Z potential.

## 2.3.5 Shape/Morphology

Like size, NP shape affects flow and compaction. Spheres, which have a length several times their width, flow more readily than high-aspect-ratio particles. For these and other reasons, we measure and regulate particle morphology. Shape affects medication distribution, transport, targeting, and degradation. The nano-material's form impacts cellular absorption, retention, and biocompatibility [163].

## 2.3.6 Purity, Composition

A nanocomposite is made from nanomaterials. Nanomaterial composition affects deliv-ery, transport, and degradation. Nanomaterial toxicology depends on their composition [164]. Therefore, nanocomposites are formed of two or more nanomaterials, and analys-ing their composition and biological impact may be difficult. Polymeric NPs, liposomes, micelles, carbon nanotubes, dendrimers, and hydrogels may be used in nanocomposites. Metals, metal oxides, proteins, lipids, and nucleic acids are all constituents [165].

## 2.3.7 Aggregation/Agglomeration

Particle agglomeration and aggregation are critical in many processes and applica-tions. Aggregation, or agglomeration, refers to particle assembling. These names have distinct meanings, yet they are occasionally used interchangeably. Aggregation, unlike agglomeration, is reversible. Only particle clusters may be detected using dynamic light scattering (DLS), NP tracking analysis, or electron microscope imaging. Particles in a sample clump together in a variety of ways. When randomly moving particles collide and cling together, Brownian agglomeration occurs. The size and terminal velocity of the particles affect gravitational aggregation. Clusters arise when faster-settling par-ticles grab slower-settling particles. Ionic strength and pH are affected by clustering. Agglomeration and aggregation of NPs must be determined [166].

## 2.4 MANUFACTURING OPPORTUNITIES FOR NANODRUGS, NANOADDITIVES, AND NANOCARRIERS

Nanoscale materials and electronics may be created using both "bottom-up" and "top-down" manufacturing processes. Bottom-up approaches, as their basic principle, focus on the self-assembly of nanomaterials from atoms or molecules. Top-down micro technologies such as photolithography, nano moulding, dip-pen lithography, and nanofluidics, on the other hand, might be used to create nanomaterials [167, 168]. By covering nanomaterials with a polymer and letting the polymer diffuse, it is possible to control the rate at which the core material is released. The nanoscale components, functions, and efficiency of therapeutic medications are all tough to comprehend and analyse. Six elements determine the performance of a nanoparticulate drug delivery system: When making a product, it is important to think about the active component concentration (small or large molecular weight), the surface characteristics of the drug, additive, or carrier, and the chemical composition of the drug, additive, or carrier. When creating a design space, bear in mind the goal product's quality characteristics such as composition and structure. There is a list of the most important quality factors that must be studied carefully to meet the product profile and performance criteria.

To prolong bronchodilation in an animal model the Nanomaterial Registry has suggested data structure that would allow for trend analysis and the discovery of data that was not easily accessible prior to data integration [169]. This method should improve decision-making and risk management as more data is filtered. Nanoparticles are abundant and found in a variety of environments. Data collection and expression are affected by (1) the physicochemical property being studied; (2) the way the property is recorded; (3) the measurement technique and equipment; and (4) the record of how samples were collected and prepared [170].

### 2.4.1    Drug Loading Issues in Nanotechnology

In nanoparticle delivery systems, encapsulation efficiency is a measure of medicine loading efficiency. In comparison to traditional medication delivery systems, the drug reservoir is small. Because of their small size, drug molecules interact with polymeric structures. The maximum amount of medicine encapsulated in micellar nanoparticles is 20%–30% of the polymer quantity. The interaction between hydrophobic medicine and polymers is modest. Hydrotropic polymers may be useful [171].

### 2.4.2    Stability and Storage

The polymeric structure of the drug-delivery mechanism may cause it to absorb water during storage. Polymer breakdown may occur as a result of absorption. This cascade might result in physicochemical and *in vivo* changes. The shelf life of a medicine may be affected by residual monomers, solvents, and polymerization catalysts. Storage conditions, like those for other medications, must be set carefully to ensure shelf life [172]. Schroeder et al., 2000 investigated the stability of poly (butyl cyanoacrylate) nanosuspensions over time and under various circumstances. The storage conditions were hydrochloric acid, Phosphate-Buffered Saline (PBS), and human blood serum. Because of agglomeration, PBS storage resulted in increased polydispersity index values, whereas human blood serum storage resulted in stable particle size distribution for at least 5 days and acidic medium storage resulted in

**TABLE 2.1**

Characteristics of Different Nanocarriers. Different nanocarriers are being prepared with different physicochemical methods. Such nanocarriers possess unique properties, most sought for drug delivery such as drug-loading capacity, compatibility, stability, and degradability. Nanocarriers have broader applications such as diagnosis of disease, targeted drug delivery, drug release, and biosensing [178–184].

| Nanocarrier | Mode of Synthesis | Properties of Nanocarriers | Applications | Reference |
|---|---|---|---|---|
| Solid lipid nanocarriers | Ultrasonication, solvent emulsification, microemulsion, spray drying, high shear homogenization, hot homogenization, cold homogenization, and ultrasonication | Biodegradable, more stable, and easier to update using a colloidal carrier. The burst release of the medication has a low loading capacity. | Topical application, *in vivo*, and *in vitro* medication delivery to solid tumours, antitubercular treatment, gene vector carriers | [175] |
| Liposome | Methods of dispersion include mechanical, solvent, and detergent removal | Non-toxic, biodegradable, biocompatible phospholipid bilayer vesicle | Pharmaceutics that capture and distribute biologically active agents in the most efficient way possible | [176] |
| Dendrimer | Self-assembly, convergent or divergent strategy, cascade response | Hyperbranched molecules that are radially symmetric, homogenous, highly defined, and monodisperse | Photodynamic treatment, neutron capture therapy, imaging, and gene delivery are all methods for delivering drugs to specific organs | [177] |
| Polymeric nanocarriers | Polymerization, emulsification/solvent diffusion, nanoprecipitation, salting-out, dialysis, and supercritical fluid technologies | Membrane-to-membrane penetration that is stable in the bloodstream and biodegradable | The stability of a volatile pharmacological agent is maintained by a high concentration of drug delivery, both active and passive | [178] |
| Micelle | Solution/mechanical dispersion of supramolecular self-assembly | Dynamic system, amphiphilic molecule-based colloidal aggregation | Encapsulates either a water-soluble or water-insoluble medication | [179] |
| Carbon nanotubes | Laser ablation, chemical vapour deposition, and carbon arc discharge are all examples of this | Crystalline third allotropic carbon sheet in a hexagonal pattern with dynamic strength as well as unique electrical and elastic properties | Artificial implants, tissue engineering, and cancer cell detection are some of the techniques that may be used to deliver genes and drugs | [180] |

*(Continued)*

**TABLE 2.1** (*Continued*)

**Characteristics of Different Nanocarriers. Different nanocarriers are being prepared with different physicochemical methods. Such nanocarriers possess unique properties, most sought for drug delivery such as drug-loading capacity, compatibility, stability, and degradability. Nanocarriers have broader applications such as diagnosis of disease, targeted drug release, drug delivery, and biosensing [178–184].**

| Nanocarrier | Mode of Synthesis | Properties of Nanocarriers | Applications | Reference |
|---|---|---|---|---|
| Gold | A biphasic reduction, two-phase synthesis | Surface plasmon resonance property, fluorescence resonance energy transfer phenomena, multi-surface functionality, versatility, great biocompatibility, low toxicity | Drug distribution and imaging are improved by biosensing and functionalized AuNP | [181] |
| Magnetic nanocarriers | Coprecipitation in microemulsion hydroxide coprecipitation, glycothermal synthesis, citrate gel process and glass crystallization are all examples of processes using metal alkoxide hydrolysis. Coprecipitation in microemulsion hydroxide coprecipitation, glycothermal synthesis, citrate gel process, and glass crystallization are all examples of processes using metal alkoxide hydrolysis. | It has superparamagnetism, chemical stability, excellent colloidal stability, and magnetic moment | Magnetic resonance technique can be used for imaging | [182] |

stable nanosuspensions for at least 2 months. According to this research, adopting acidic storage instead of lyophilization may help with reconstitution concerns [173].

### 2.4.3 THE NANOCARRIERS COMPLEXITY

They were compounded by research into multifunctional, targeted, and more effective nanoparticle delivery systems. Multidisciplinary collaboration is required for nanoscale medication delivery technologies. Because of their complexity and low manufacturing yield, nano systems are more costly than standard equivalents. Because the physical and chemical properties of complex systems are more variable, it is harder to predict what will happen to them and how they will work in the body [174].

### 2.4.4 TOXICITY

Nanomaterials have the potential to be platforms, but their qualities have the ability to damage the environment and human health [175]. Their peculiar characteristics cause concern since their effects on biological systems are unclear. Nano pharmaceutical goods, as well as their creation and disposal, have raised concerns. Researchers are becoming more worried about the environmental and toxicological dangers of nanomedicines. Although nanomedicines have obvious advantages, altering their physicochemical properties may result in unknown limitations. One example is the increased clearing time in relation to the increased circulation time, the effect of which is uncertain. As shown by Singh et al., 2007 [176], fullerenes, or fibrous nanomaterials may induce DNA strand breakage, point mutations, oxidative DNA adducts, and changes in gene expression. Because of the unique features of nano pharmaceuticals, nanotoxicology and nanotechnology are inextricably linked. It is critical that their fate, safety, and toxicity profiles be correctly defined (also shown in Table 2.1) [177].

### REFERENCES

[1] Nandi, S. K., Mukherjee, P., Roy, S., Kundu, B., De, D. K., & Basu, D. (2009). Local antibiotic delivery systems for the treatment of osteomyelitis—A review. *Materials Science and Engineering: C*, 29(8), 2478–2485.

[2] Shojaei, A. H. (1998). Buccal mucosa as a route for systemic drug delivery: A review. *J Pharm Pharm Sci*, 1(1), 15–30.

[3] Brannon-Peppas, L. (1995). Recent advances on the use of biodegradable microparticles and nanoparticles in controlled drug delivery. *International Journal of Pharmaceutics*, 116(1), 1–9.

[4] Lee, S. H., & Shin, H. (2007). Matrices and scaffolds for delivery of bioactive molecules in bone and cartilage tissue engineering. *Advanced Drug Delivery Reviews*, 59(4–5), 339–359.

[5] Monteiro, N., Martins, A., Reis, R. L., & Neves, N. M. (2015). Nanoparticle-based bioactive agent release systems for bone and cartilage tissue engineering. *Regenerative Therapy*, 1, 109–118.

[6] Nel, A., Xia, T., Madler, L., & Lin, N. (2006). Ti is an abundant element in soil and appears in a variety of forms including primary and secondary minerals as well as organically bound and amorphous compounds. *Science*, 311, 622–627.

[7] Losic, D., & Simovic, S. (2009). Self-ordered nanopore and nanotube platforms for drug delivery applications. *Expert Opinion on Drug Delivery*, 6(12), 1363–1381.

[8] Sharma, V. K., Filip, J., Zboril, R., & Varma, R. S. (2015). Natural inorganic nanoparticles—formation, fate, and toxicity in the environment. *Chem Soc Rev*, 44, 8410–8423.

[9] Mandel, K., & Hutter, F. (2012). The magnetic nanoparticle separation problem. *Nano Today*, 7(6), 485–487.

[10] Pavlukhina, S., & Sukhishvili, S. (2011). Polymer assemblies for controlled delivery of bioactive molecules from surfaces. *Advanced Drug Delivery Reviews*, 63(9), 822–836.

[11] Kobayashi, K., Wei, J., Iida, R., Ijiro, K., & Niikura, K. (2014). Surface engineering of nanoparticles for therapeutic applications. *Polymer Journal*, 46(8), 460–468.

[12] Cheng, C., Nie, S., Li, S., Peng, H., Yang, H., Ma, L., . . . & Zhao, C. (2013). Biopolymer functionalized reduced graphene oxide with enhanced biocompatibility via mussel inspired coatings/anchors. *Journal of Materials Chemistry B*, 1(3), 265–275.

[13] Bhattacharyya, A., Datta, P. S., Chaudhuri, P., & Barik, B. R. (2011, March). Nanotechnology-A new frontier for food security in socio economic development. In *Disaster risk vulnerablity conference* (pp. 116–120).

[14] Dumestre, F., Chaudret, B., Amiens, C., Renaud, P., & Fejes, P. (2004). Superlattices of iron nanocubes synthesized from Fe [N (SiMe3) 2] 2. *Science*, 303(5659), 821–823.

[15] Emami, S., Siahi-Shadbad, M., Adibkia, K., & Barzegar-Jalali, M. (2018). Recent advances in improving oral drug bioavailability by cocrystals. *BioImpacts: BI*, 8(4), 305.

[16] Zhao, Z., Zhou, Z., Bao, J., Wang, Z., Hu, J., Chi, X., . . . & Gao, J. (2013). Octapod iron oxide nanoparticles as high-performance T2 contrast agents for magnetic resonance imaging. *Nature Communications*, 4(1), 1–7.

[17] Tchounwon, P. B., Yedjou, C. G., Patlolla, A. K., & Sutton, D. J. (2012). Heavy metal toxicity and the environment. *Molecular, Clinical and Environmental Toxicology, Experentia Supplementum*, 101, 133–164.

[18] Murphy, C. J., Gole, A. M., Stone, J. W., Sisco, P. N., Alkilany, A. M., Goldsmith, E. C., & Baxter, S. C. (2008). Gold nanoparticles in biology: Beyond toxicity to cellular imaging. *Accounts of Chemical Research*, 41(12), 1721–1730.

[19] Shenoy, D., Fu, W., Li, J., Crasto, C., Jones, G., DiMarzio, C., . . . & Amiji, M. (2006). Surface functionalization of gold nanoparticles using hetero-bifunctional poly (ethylene glycol) spacer for intracellular tracking and delivery. *International Journal of Nanomedicine*, 1(1), 51.

[20] Conde, J., Ambrosone, A., Sanz, V., Hernandez, Y., Marchesano, V., Tian, F., . . . & de la Fuente, J. M. (2012). Design of multifunctional gold nanoparticles for in vitro and in vivo gene silencing. *ACS nano*, 6(9), 8316–8324.

[21] Ghosh, P. S., Kim, C. K., Han, G., Forbes, N. S., & Rotello, V. M. (2009). Efficient gene delivery vectors by tuning the surface charge density of amino acid-functionalized gold nanoparticles. *ACS Nano*, 2, 2213–2218.

[22] Gu, Y. J., Cheng, J., Lin, C. C., Lam, Y. W., Cheng, S. H., & Wong, W. T. (2009). Nuclear penetration of surface functionalized gold nanoparticles. *Toxicology and Applied Pharmacology*, 237(2), 196–204.

[23] Agarwal, S., Greiner, A., & Wendorff, J. H. (2013). Functional materials by electrospinning of polymers. *Progress in Polymer Science*, 38(6), 963–991.

[24] Sattel, T. F., Knopp, T., Biederer, S., Gleich, B., Weizenecker, J., Borgert, J., & Buzug, T. M. (2008). Single-sided device for magnetic particle imaging. *Journal of Physics D: Applied Physics*, 42(2), 022001.

[25] Bernal, J., & Mendiola, J. (2011). E. Ibá nez, A. Cifuentes. *A Journal of Pharmaceutical and Biomedical Analysis*, 55, 758–774.

[26] Nichols, D. S., & Sanderson, K. (2002). *The nomenclature, structure, and properties of food lipids*. CRC Press, 29–59.

[27] Lason, E., & Ogonowski, J. (2011). Solid Lipid Nanoparticles—characteristics, application and obtaining. *Chemik*, 65(10), 960–967.

[28] Pardeike, J., Hommoss, A., & Müller, R. H. (2009). Lipid nanoparticles (SLN, NLC) in cosmetic and pharmaceutical dermal products. *International Journal of Pharmaceutics*, 366(1–2), 170–184.

[29] Weber, S. (2014). Zimmer a., Pardeike J. Solid Lipid Nanoparticles (SLN) and Nanostructured Lipid Carriers (NLC) for pulmonary application: A review of the state of the art. *Eur J Pharm Biopharm [Internet]*, 86(1), 7–22.

[30] Salminen, H., Gömmel, C., Leuenberger, B. H., & Weiss, J. (2016). Influence of encapsulated functional lipids on crystal structure and chemical stability in solid lipid nanoparticles: Towards bioactive-based design of delivery systems. *Food Chemistry*, 190, 928–937.

[31] Fonseca-Santos, B., Gremião, M. P. D., & Chorilli, M. (2015). Nanotechnology-based drug delivery systems for the treatment of Alzheimer's disease. *International Journal of Nanomedicine*, 10, 4981.

[32] Das, S., & Chaudhury, A. (2011). Recent advances in lipid nanoparticle formulations with solid matrix for oral drug delivery. *Aaps Pharmscitech*, 12(1), 62–76.

[33] Porada, C. D., & Almeida-Porada, G. (2010). Mesenchymal stem cells as therapeutics and vehicles for gene and drug delivery. *Advanced Drug Delivery Reviews*, 62(12), 1156–1166.

[34] Shi, Z., Shao, L., Jones, T. P., & Lu, S. J. (2005). Microscopy and mineralogy of airborne particles collected during severe dust storm episodes in Beijing, China. *Geophys Res: Atmos* 110, D01303.

[35] McClements, D. J. (2013). Edible lipid nanoparticles: Digestion, absorption, and potential toxicity. *Progress in Lipid Research*, 52(4), 409–423.

[36] Battaglia, L., & Gallarate, M. (2012). Lipid nanoparticles: State of the art, new preparation methods and challenges in drug delivery. *Expert Opinion on Drug Delivery*, 9(5), 497–508.

[37] Kovačević, A. B., Müller, R. H., Savić, S. D., Vuleta, G. M., & Keck, C. M. (2014). Solid lipid nanoparticles (SLN) stabilized with polyhydroxy surfactants: Preparation, characterization and physical stability investigation. *Colloids and Surfaces A: Physicochemical and Engineering Aspects*, 444, 15–25.

[38] Tripathy, D. B., Gupta, A., Jain, A. K., & Mishra, A. (Eds.). (2021). *Surfactants from Renewable Raw Materials*. CRC Press.

[39] Lémery, E., Briançon, S., Chevalier, Y., Bordes, C., Oddos, T., Gohier, A., & Bolzinger, M. A. (2015). Skin toxicity of surfactants: Structure/toxicity relationships. *Colloids and Surfaces A: Physicochemical and Engineering Aspects*, 469, 166–179.

[40] Peres, L. B., Peres, L. B., de Araújo, P. H. H., & Sayer, C. (2016). Solid lipid nanoparticles for encapsulation of hydrophilic drugs by an organic solvent free double emulsion technique. Colloids and Surfaces B: Biointerfaces, 140, 317–323.

[41] Peres, L. B., Peres, L. B., de Araújo, P. H. H., & Sayer, C. (2016). Solid lipid nanoparticles for encapsulation of hydrophilic drugs by an organic solvent free double emulsion technique. *Colloids and Surfaces B: Biointerfaces*, 140, 317–323.

[42] Muchow, M., Maincent, P., & Müller, R. H. (2008). Lipid nanoparticles with a solid matrix (SLN®, NLC®, LDC®) for oral drug delivery. *Drug Development and Industrial Pharmacy*, 34(12), 1394–1405.

[43] Cavalli, R., Caputo, O., Carlotti, M. E., Trotta, M., Scarnecchia, C., & Gasco, M. R. (1997). Sterilization and freeze-drying of drug-free and drug-loaded solid lipid nanoparticles. *International Journal of Pharmaceutics*, 148(1), 47–54.

[44] Hu, T., Yang, Y., Tan, L., Yin, T., Wang, Y., & Wang, G. (2014). Effects of gamma irradiation and moist heat for sterilization on sodium alginate. *Bio-Medical Materials and Engineering*, 24(5), 1837–1849.

[45] Blasi, P., Giovagnoli, S., Schoubben, A., Ricci, M., & Rossi, C. (2007). Solid lipid nanoparticles for targeted brain drug delivery. *Advanced Drug Delivery Reviews*, 59(6), 454–477.

[46] Almeida, A. J., & Souto, E. (2007). Solid lipid nanoparticles as a drug delivery system for peptides and proteins. *Advanced Drug Delivery Reviews*, 59(6), 478–490.

[47] Liu, S., Zeng, T. H., Hofmann, M., Burcombe, E., Wei, J., Jiang, R., . . . & Chen, Y. (2011). Antibacterial activity of graphite, graphite oxide, graphene oxide, and reduced graphene oxide: Membrane and oxidative stress. *ACS Nano*, 5(9), 6971–6980.

[48] Niu, K. Y., Yang, J., Kulinich, S. A., Sun, J., & Du, X. W. (2010). Hollow nanoparticles of metal oxides and sulfides: Fast preparation via laser ablation in liquid. *Langmuir*, 26(22), 16652–16657.

[49] Datta, D., Sarkar, K., Mukherjee, S., Meshik, X., Stroscio, M. A., & Dutta, M. (2017). Graphene oxide and DNA aptamer based sub-nanomolar potassium detecting optical nanosensor. *Nanotechnology*, 28(32), 325502.

[50] Fernando, K. S., Sahu, S., Liu, Y., Lewis, W. K., Guliants, E. A., Jafariyan, A., . . . & Sun, Y. P. (2015). Carbon quantum dots and applications in photocatalytic energy conversion. *ACS Applied Materials & Interfaces*, 7(16), 8363–8376.

[51] Zheng, X. T., Ananthanarayanan, A., Luo, K. Q., & Chen, P. (2015). Glowing graphene quantum dots and carbon dots: Properties, syntheses, and biological applications. *Small*, 11(14), 1620–1636.

[52] Bazylinski, D. A., Frankel, R. B., Heywood, B. R., Mann, S., King, J. W., Donaghay, P. L., & Hanson, A. K. (1995). Controlled Biomineralization of Magnetite (Fe (inf3) O (inf4)) and Greigite (Fe (inf3) S (inf4)) in a Magnetotactic Bacterium. *Applied and Environmental Microbiology*, 61(9), 3232–3239.

[53] Chandra, S., Patra, P., Pathan, S. H., Roy, S., Mitra, S., Layek, A., . . . & Goswami, A. (2013). Luminescent S-doped carbon dots: An emergent architecture for multimodal applications. *Journal of Materials Chemistry B*, 1(18), 2375–2382.

[54] Tan, M., Zhang, L., Tang, R., Song, X., Li, Y., Wu, H., . . . & Ma, X. (2013). Enhanced photoluminescence and characterization of multicolor carbon dots using plant soot as a carbon source. *Talanta*, 115, 950–956.

[55] Thomas, S., Thomas, R., Zachariah, A. K., & Mishra, R. K. (Eds.). (2017). *Thermal and rheological measurement techniques for nanomaterials characterization* (Vol. 3). Elsevier.

[56] Thomas, S., Thomas, R., Zachariah, A. K., & Mishra, R. K. (Eds.). (2017). *Microscopy methods in nanomaterials characterization* (Vol. 1). Elsevier.

[57] Bansal, V., Rautaray, D., Ahmad, A., & Sastry, M. (2004). Biosynthesis of zirconia nanoparticles using the fungus Fusarium oxysporum. *Journal of Materials Chemistry*, 14(22), 3303–3305.

[58] Abraham, J., Sharika, T., Mishra, R. K., & Thomas, S. (2017). Rheological characteristics of nanomaterials and nanocomposites. In *Micro and nano fibrillar composites (MFCs and NFCs) from polymer blends* (pp. 327–350). Woodhead Publishing.

[59] Mishra, R. K., Zachariah, A. K., & Thomas, S. (2017). Energy-dispersive X-ray spectroscopy techniques for nanomaterial. In *Microscopy methods in nanomaterials characterization* (pp. 383–405). Elsevier.

[60] Bargel, H., Koch, K., Cerman, Z., & Neinhuis, C. (2006). Evans Review No. 3: Structure-function relationships of the plant cuticle and cuticular waxes—a smart material? *Functional Plant Biology*, 33(10), 893–910.

[61] Mishra, R. K., Abraham, J., Kalarikkal, N., Jayanarayanan, K., Joseph, K., & Thomas, S. (2017). Conducting polyurethane Blends: Recent advances and perspectives. *Polyurethane Polymers*, 203–231.

[62] Yang, W., Ratinac, K. R., Ringer, S. P., Thordarson, P., Gooding, J. J., & Braet, F. (2010). Carbon nanomaterials in biosensors: Should you use nanotubes or graphene? *Angewandte Chemie International Edition*, 49(12), 2114–2138.

[63] Pandey, S., Thakur, M., Mewada, A., Anjarlekar, D., Mishra, N., & Sharon, M. (2013). Carbon dots functionalized gold nanorod mediated delivery of doxorubicin: Tri-functional nano-worms for drug delivery, photothermal therapy and bioimaging. *Journal of Materials Chemistry B*, 1(38), 4972–4982.

[64] Kam, N. W. S., O'Connell, M., Wisdom, J. A., & Dai, H. (2005). Carbon nanotubes as multifunctional biological transporters and near-infrared agents for selective cancer cell destruction. *Proceedings of the National Academy of Sciences*, 102(33), 11600–11605.

[65] Mohapatra S, Ranjan S, Dasgupta N, Kumar R, Thomas S, editors. (2018 Oct 5). *Nanocarriers for drug delivery: Nanoscience and nanotechnology in drug delivery*. Elsevier.

[66] Popov, V. N. (2004). Carbon nanotubes: Properties and application. *Materials Science and Engineering: R: Reports*, 43(3), 61–102.

[67] O'Connell, M. J., Boul, P., Ericson, L. M., Huffman, C., Wang, Y., Haroz, E., . . . & Smalley, R. E. (2001). Reversible water-solubilization of single-walled carbon nanotubes by polymer wrapping. *Chemical Physics Letters*, 342(3–4), 265–271.

[68] Banerjee, S., Hemraj-Benny, T., & Wong, S. S. (2005). Covalent surface chemistry of single-walled carbon nanotubes. *Advanced Materials*, 17(1), 17–29.

[69] Banerjee, S., Kahn, M. G., & Wong, S. S. (2003). Rational chemical strategies for carbon nanotube functionalization. *Chemistry: A European Journal*, 9(9), 1898–1908.

[70] Beyou, E., Akbar, S., Chaumont, P., & Cassagnau, P. (2013). Polymer nanocomposites containing functionalised multiwalled carbon nanoTubes: A particular attention to polyolefin-based materials. *Syntheses and Applications of Carbon Nanotubes and Their Composites*, 1, 77–115.

[71] Nakashima, N., Tomonari, Y., & Murakami, H. (2002). Water-soluble single-walled carbon nanotubes via noncovalent sidewall-functionalization with a pyrene-carrying ammonium ion. *Chemistry Letters*, 31(6), 638–639.

[72] Wu, H. C., Chang, X., Liu, L., Zhao, F., & Zhao, Y. (2010). Chemistry of carbon nanotubes in biomedical applications. *Journal of Materials Chemistry*, 20(6), 1036–1052.

[73] Moshammer, K., Hennrich, F., & Kappes, M. M. (2009). Selective suspension in aqueous sodium dodecyl sulfate according to electronic structure type allows simple separation of metallic from semiconducting single-walled carbon nanotubes. *Nano Research*, 2(8), 599–606.

[74] Hu, S., Wang, T., Pei, X., Cai, H., Chen, J., Zhang, X., . . . & Wang, J. (2016). Synergistic enhancement of antitumor efficacy by PEGylated Multi-walled carbon nanotubes modified with cell-penetrating peptide TAT. *Nanoscale Research Letters*, 11(1), 1–14.

[75] Fan, H., Wang, L., Zhao, K., Li, N., Shi, Z., Ge, Z., & Jin, Z. (2010). Fabrication, mechanical properties, and biocompatibility of graphene-reinforced chitosan composites. *Biomacromolecules*, 11(9), 2345–2351.

[76] Barthlott, W., & Neinhuis, C. (1997). Purity of the sacred lotus, or escape from contamination in biological surfaces. *Planta*, 202(1), 1–8.

[77] Newkome, G. R., Moorefield, C. N., Vögtle, F., Vögtle, F., Vögtle, F., & Chemist, G. (2001). *Dendrimers and dendrons: Concepts, syntheses, applications* (Vol. 623). Wiley-VCH.

[78] Basu S, Sandanaraj BS, Thayumanavan S. (2002 Jul 15). Molecular recognition in dendrimers. *Encyclopedia of polymer science and technology*.

[79] Mignani, S., Tripathi, R. P., Chen, L., Caminade, A. M., Shi, X., & Majoral, J. P. (2018). New ways to treat tuberculosis using dendrimers as nanocarriers. *Pharmaceutics*, 10(3), 105.

[80] Chamundeeswari, M., Jeslin, J., & Verma, M. L. (2019). Nanocarriers for drug delivery applications. *Environmental Chemistry Letters*, 17(2), 849–865.

[81] Stiriba, S. E., Frey, H., & Haag, R. (2002). Dendritic polymers in biomedical applications: From potential to clinical use in diagnostics and therapy. *Angewandte Chemie International Edition*, 41(8), 1329–1334.

[82] Lai, P. S., Lou, P. J., Peng, C. L., Pai, C. L., Yen, W. N., Huang, M. Y., . . . & Shieh, M. J. (2007). Doxorubicin delivery by polyamidoamine dendrimer conjugation and photochemical internalization for cancer therapy. *Journal of Controlled Release*, 122(1), 39–46.

[83] Zhuo, R. X., Du, B., & Lu, Z. R. (1999). In vitro release of 5-fluorouracil with cyclic core dendritic polymer. *Journal of Controlled Release*, 57(3), 249–257.

[84] Malik, N., Evagorou, E. G., & Duncan, R. (1999). Dendrimer-platinate: A novel approach to cancer chemotherapy. *Anti-Cancer Drugs*, 10(8), 767–776.

[85] Bhadra, D., Bhadra, S., Jain, S., & Jain, N. K. (2003). A PEGylated dendritic nanoparticulate carrier of fluorouracil. *International Journal of Pharmaceutics*, 257(1–2), 111–124.

[86] Lee, C. C., Gillies, E. R., Fox, M. E., Guillaudeu, S. J., Fréchet, J. M., Dy, E. E., & Szoka, F. C. (2006). A single dose of doxorubicin-functionalized bow-tie dendrimer cures mice bearing C-26 colon carcinomas. *Proceedings of the National Academy of Sciences*, 103(45), 16649–16654.

[87] Bamrungsap, S., Zhao, Z., Chen, T., Wang, L., Li, C., Fu, T., & Tan, W. (2012). Nanotechnology in therapeutics: A focus on nanoparticles as a drug delivery system. *Nanomedicine*, 7(8), 1253–1271.

[88] Prabhu, R. H., Patravale, V. B., & Joshi, M. D. (2015). Polymeric nanoparticles for targeted treatment in oncology: Current insights. *International Journal of Nanomedicine*, 10, 1001.

[89] Slocik, J. M., Knecht, M. R., & Wright, D. W. (2004). Biogenic nanoparticles. In Nalwa, H. S. (ed.), *Encyclopedia of nanoscience and nanotechnology* (pp. 293–308). American Scientific Publishers.

[90] Wang, X., Wang, Y., Chen, Z. G., & Shin, D. M. (2009). Advances of cancer therapy by nanotechnology. *Cancer Research and Treatment: Official Journal of Korean Cancer Association*, 41(1), 1–11.

[91] Zhu, Y., & Liao, L. (2015). Applications of nanoparticles for anticancer drug delivery: A review. *Journal of Nanoscience and Nanotechnology*, 15(7), 4753–4773.

[92] Rosenblum, D., Joshi, N., Tao, W., Karp, J. M., & Peer, D. (2018). Progress and challenges towards targeted delivery of cancer therapeutics. *Nature Communications*, 9(1), 1–12.

[93] Wennerström, H., & Lindman, B. (1979). Micelles: Physical chemistry of surfactant association. *Physics Reports*, 52(1), 1–86.

[94] Chow, T. S. (2007). Nanoscale surface roughness and particle adhesion on structured substrates. *Nanotechnology*, 18(11), 115713.

[95] Zhang, J., Dubay, M. R., Houtman, C. J., & Severtson, S. J. (2009). Sulfonated amphiphilic block copolymers: Synthesis, self-assembly in water, and application as stabilizer in emulsion polymerization. *Macromolecules*, 42(14), 5080–5090.

[96] Al-Sabagh, A. M., Maysour, N. E., & Noor El-Din, M. R. (2007). Investigate the demulsification efficiency of some novel demulsifiers in relation to their surface-active properties. *Journal of Dispersion Science and Technology*, 28(4), 547–555.

[97] Kandekar, S. G., del Río-Sancho, S., Lapteva, M., & Kalia, Y. N. (2018). Selective delivery of adapalene to the human hair follicle under finite dose conditions using polymeric micelle nanocarriers. *Nanoscale*, 10(3), 1099–1110.

[98] Shen, Y., Zhang, J., Hao, W., Wang, T., Liu, J., Xie, Y., . . . & Liu, H. (2018). Copolymer micelles function as pH-responsive nanocarriers to enhance the cytotoxicity of a HER2 aptamer in HER2-positive breast cancer cells. *International Journal of Nanomedicine*, 13, 537.

[99] Santos, H. A., Bimbo, L. M., Peltonen, L., & Hirvonen, J. (2015). Inorganic nanoparticles in targeted drug delivery and imaging. In *Targeted drug delivery: Concepts and design* (pp. 571–613). Springer.

[100] Xiao, H., Li, C., Dai, Y., Cheng, Z., Hou, Z., & Lin, J. (2015). Inorganic nanocarriers for platinum drug delivery. *Materials Today*, 18(10), 554–564.

[101] Baker, C., Pradhan, A., Pakstis, L., Pochan, D. J., & Shah, S. I. (2005). Synthesis and antibacterial properties of silver nanoparticles. *Journal of Nanoscience and Nanotechnology*, 5(2), 244–249.

[102] Sharma, J., Chhabra, R., Cheng, A., Brownell, J., Liu, Y., & Yan, H. (2009). Control of self-assembly of DNA tubules through integration of gold nanoparticles. *Science*, 323(5910), 112–116.

[103] Pinheiro, A. V., Han, D., Shih, W. M., & Yan, H. (2011). Challenges and opportunities for structural DNA nanotechnology. *Nature Nanotechnology*, 6(12), 763–772.

[104] Zhang, J., Liu, Y., Ke, Y., & Yan, H. (2006). Periodic square-like gold nanoparticle arrays templated by self-assembled 2D DNA nanogrids on a surface. *Nano Letters*, 6(2), 248–251.

[105] Rink, J. S., McMahon, K. M., Chen, X., Mirkin, C. A., Thaxton, C. S., & Kaufman, D. B. (2010). Transfection of pancreatic islets using polyvalent DNA-functionalized gold nanoparticles. *Surgery*, 148(2), 335–345.

[106] Gao, S., Xu, Y., Asghar, S., Chen, M., Zou, L., Eltayeb, S., . . . & Xiao, Y. (2015). Polybutylcyanoacrylate nanocarriers as promising targeted drug delivery systems. *Journal of Drug Targeting*, 23(6), 481–496.

[107] Hussain, M. M., Samir, T. M., & Azzazy, H. M. (2013). Unmodified gold nanoparticles for direct and rapid detection of Mycobacterium tuberculosis complex. *Clinical Biochemistry*, 46(7–8), 633–637.

[108] Wang, Y., Xie, X., Wang, X., Ku, G., Gill, K. L., O'Neal, D. P., . . . & Wang, L. V. (2004). Photoacoustic tomography of a nanoshell contrast agent in the in vivo rat brain. *Nano Letters*, 4(9), 1689–1692.

[109] Qian, X., Peng, X. H., Ansari, D. O., Yin-Goen, Q., Chen, G. Z., Shin, D. M., . . . & Nie, S. (2008). In vivo tumor targeting and spectroscopic detection with surface-enhanced Raman nanoparticle tags. *Nature Biotechnology*, 26(1), 83–90.

[110] García, M. A. (2011). Surface plasmons in metallic nanoparticles: Fundamentals and applications. *Journal of Physics D: Applied Physics*, 44(28), 283001.

[111] Cao, Y. C., Jin, R., & Mirkin, C. A. (2002). Nanoparticles with Raman spectroscopic fingerprints for DNA and RNA detection. *Science*, 297(5586), 1536–1540.

[112] Lu, W., Huang, Q., Ku, G., Wen, X., Zhou, M., Guzatov, D., . . . & Li, C. (2010). Photoacoustic imaging of living mouse brain vasculature using hollow gold nanospheres. *Biomaterials*, 31(9), 2617–2626.

[113] Loo, C., Lowery, A., Halas, N., West, J., & Drezek, R. (2005). Immunotargeted nanoshells for integrated cancer imaging and therapy. *Nano Letters*, 5(4), 709–711.

[114] Huang, X., El-Sayed, I. H., Qian, W., & El-Sayed, M. A. (2006). Cancer cell imaging and photothermal therapy in the near-infrared region by using gold nanorods. *Journal of the American Chemical Society*, 128(6), 2115–2120.

[115] Chen, J., Saeki, F., Wiley, B. J., Cang, H., Cobb, M. J., Li, Z. Y., . . . & Xia, Y. (2005). Gold nanocages: Bioconjugation and their potential use as optical imaging contrast agents. *Nano letters*, 5(3), 473–477.

[116] Gobin, A. M., Lee, M. H., Halas, N. J., James, W. D., Drezek, R. A., & West, J. L. (2007). Near-infrared resonant nanoshells for combined optical imaging and photothermal cancer therapy. *Nano Lett*, 7, 1929–1934.

[117] Li, Y., Li, N., Pan, W., Yu, Z., Yang, L., & Tang, B. (2017). Hollow mesoporous silica nanoparticles with tunable structures for controlled drug delivery. *ACS Applied Materials & Interfaces*, 9(3), 2123–2129.

[118] Wang, Y., Zhao, Q., Han, N., Bai, L., Li, J., Liu, J., . . . & Wang, S. (2015). Mesoporous silica nanoparticles in drug delivery and biomedical applications. *Nanomedicine: Nanotechnology, Biology and Medicine*, 11(2), 313–327.

[119] Rosenholm, J. M., Peuhu, E., Bate-Eya, L. T., Eriksson, J. E., Sahlgren, C., & Lindén, M. (2010). Cancer-cell-specific induction of apoptosis using mesoporous silica nanoparticles as drug-delivery vectors. *Small*, 6(11), 1234–1241.

[120] Lebold, T., Jung, C., Michaelis, J., & Brauchle, C. (2009). Nanostructured silica materials as drug-delivery systems for doxorubicin: Single molecule and cellular studies. *Nano Letters*, 9(8), 2877–2883.

[121] Qian, W. Y., Sun, D. M., Zhu, R. R., Du, X. L., Liu, H., & Wang, S. L. (2012). pH-sensitive strontium carbonate nanoparticles as new anticancer vehicles for controlled etoposide release. *International Journal of Nanomedicine*, 7, 5781.

[122] Spring, S., & Schleifer, K.-H. (1995). Syst Appl Microbiol., 18, 147–153.

[123] Han, N., Zhao, Q., Wan, L., Wang, Y., Gao, Y., Wang, P., . . . & Wang, S. (2015). Hybrid lipid-capped mesoporous silica for stimuli-responsive drug release and overcoming multidrug resistance. *ACS Applied Materials & Interfaces*, 7(5), 3342–3351.

[124] Desai, D., Zhang, J., Sandholm, J., Lehtimaki, J., Gronroos, T., Tuomela, J., & Rosenholm, J. M. (2017). Lipid bilayer-gated mesoporous silica nanocarriers for tumor-targeted delivery of zoledronic acid in vivo. *Molecular Pharmaceutics*, 14(9), 3218–3227.

[125] Zhou, J., Zhang, X., Li, M., Wu, W., Sun, X., Zhang, L., & Gong, T. (2013). Novel lipid hybrid albumin nanoparticle greatly lowered toxicity of pirarubicin. *Molecular Pharmaceutics*, 10(10), 3832–3841.

[126] Khoshnejad, M., Parhiz, H., Shuvaev, V. V., Dmochowski, I. J., & Muzykantov, V. R. (2018). Ferritin-based drug delivery systems: Hybrid nanocarriers for vascular immunotargeting. *Journal of Controlled Release*, 282, 13–24.

[127] Gao, L. Y., Liu, X. Y., Chen, C. J., Wang, J. C., Feng, Q., Yu, M. Z., . . . & Zhang, Q. (2014). Core-shell type lipid/rPAA-Chol polymer hybrid nanoparticles for in vivo siRNA delivery. *Biomaterials*, 35(6), 2066–2078.

[128] Chou, L. Y., Ming, K., & Chan, W. C. (2011). Strategies for the intracellular delivery of nanoparticles. *Chemical Society Reviews*, 40(1), 233–245.

[129] Saha, K., Bajaj, A., Duncan, B., & Rotello, V. M. (2011). Beauty is skin deep: A surface monolayer perspective on nanoparticle interactions with cells and bio-macromolecules. *Small*, 7(14), 1903–1918.

[130] Kumar, D., & Ahmad, S. (2017). Green intelligent nanomaterials by design (using nanoparticulate/2D-materials building blocks) current developments and future trends. In *Nanoscaled films and layers*. IntechOpen.

[131] Yavuz, M. S., Cheng, Y., Chen, J., Cobley, C. M., Zhang, Q., Rycenga, M., . . . & Xia, Y. (2009). Gold nanocages covered by smart polymers for controlled release with near-infrared light. *Nature Materials*, 8(12), 935–939.

[132] Matsumura, Y., & Maeda, H. (1986). A new concept for macromolecular therapeutics in cancer chemotherapy: Mechanism of tumoritropic accumulation of proteins and the antitumor agent smancs. *Cancer Research*, 46(12_Part_1), 6387–6392.

[133] Petros, R. A., & DeSimone, J. M. (2010). Strategies in the design of nanoparticles for therapeutic applications. *Nature Reviews Drug Discovery*, 9(8), 615–627.

[134] Peer, D., Karp, J. M., Hong, S., Farokhzad, O. C., Margalit, R., & Langer, R. (2020). Nanocarriers as an emerging platform for cancer therapy. *Nano-Enabled Medical Applications*, 61–91.

[135] Kommareddy, S., & Amiji, M. (2004). Targeted drug delivery to tumor cells using colloidal carriers. In *Cellular drug delivery* (pp. 181–215). Humana Press.

[136] Niidome, T., Yamagata, M., Okamoto, Y., Akiyama, Y., Takahashi, H., Kawano, T., . . . & Niidome, Y. (2006). PEG-modified gold nanorods with a stealth character for in vivo applications. *Journal of Controlled Release*, 114(3), 343–347.

[137] Wang, Y., Brown, P., & Xia, Y. (2011). Swarming towards the target. *Nature Materials*, 10(7), 482–483.

[138] Wang, Z., Tiruppathi, C., Cho, J., Minshall, R. D., & Malik, A. B. (2011). Delivery of nanoparticle-complexed drugs across the vascular endothelial barrier via caveolae. *IUBMB Life*, 63(8), 659–667.

[139] Jin, Y., Jia, C., Huang, S. W., O'donnell, M., & Gao, X. (2010). Multifunctional nanoparticles as coupled contrast agents. *Nature Communications*, 1(1), 1–8.

[140] Von Maltzahn, G., Park, J. H., Lin, K. Y., Singh, N., Schwöppe, C., Mesters, R., . . . & Bhatia, S. N. (2011). Nanoparticles that communicate in vivo to amplify tumour targeting. *Nature Materials*, 10(7), 545–552.

[141] You, C. C., Miranda, O. R., Gider, B., Ghosh, P. S., Kim, I. B., Erdogan, B., . . . & Rotello, V. M. (2007). Detection and identification of proteins using nanoparticle—fluorescent polymer 'chemical nose' sensors. *Nature Nanotechnology*, 2(5), 318–323.

[142] Ding, N., Cao, Q., Zhao, H., Yang, Y., Zeng, L., He, Y., . . . & Wang, G. (2010). Colorimetric assay for determination of lead (II) based on its incorporation into gold nanoparticles during their synthesis. *Sensors*, 10(12), 11144–11155.

[143] El-Boubbou, K., Zhu, D. C., Vasileiou, C., Borhan, B., Prosperi, D., Li, W., & Huang, X. (2010). Magnetic glyco-nanoparticles: A tool to detect, differentiate, and unlock the glyco-codes of cancer via magnetic resonance imaging. *Journal of the American Chemical Society*, 132(12), 4490–4499.

[144] Bajaj, A., Rana, S., Miranda, O. R., Yawe, J. C., Jerry, D. J., Bunz, U. H., & Rotello, V. M. (2010). Cell surface-based differentiation of cell types and cancer states using a gold nanoparticle-GFP based sensing array. *Chemical Science*, 1(1), 134–138.

[145] Moyano, D. F., Rana, S., Bunz, U. H., & Rotello, V. M. (2011). Gold nanoparticle-polymer/biopolymer complexes for protein sensing. *Faraday Discussions*, 152, 33–42.

[146] Kim, B., Han, G., Toley, B. J., Kim, C. K., Rotello, V. M., & Forbes, N. S. (2010). Tuning payload delivery in tumour cylindroids using gold nanoparticles. *Nature Nanotechnology*, 5(6), 465–472.

[147] Shi, X., Thomas, T. P., Myc, L. A., Kotlyar, A., & Baker Jr, J. R. (2007). Synthesis, characterization, and intracellular uptake of carboxyl-terminated poly (amidoamine) dendrimer-stabilized iron oxide nanoparticles. *Physical Chemistry Chemical Physics*, 9(42), 5712–5720.

[148] Verma, A., & Stellacci, F. (2010). Effect of surface properties on nanoparticle-cell interactions. *Small*, 6(1), 12–21.

[149] Jiang, W., Kim, B., Rutka, J. T., & Chan, W. C. (2008). Nanoparticle-mediated cellular response is size-dependent. *Nature Nanotechnology*, 3(3), 145–150.

[150] Giljohann, D. A., Seferos, D. S., Prigodich, A. E., Patel, P. C., & Mirkin, C. A. (2009). Gene regulation with polyvalent siRNA– nanoparticle conjugates. *Journal of the American Chemical Society*, 131(6), 2072–2073.

[151] Balmert, S. C., & Little, S. R. (2012). Biomimetic delivery with micro-and nanoparticles. *Advanced Materials*, 24(28), 3757–3778.

[152] Verma, M. L., Chaudhary, R., Tsuzuki, T., Barrow, C. J., & Puri, M. (2013). Immobilization of β-glucosidase on a magnetic nanoparticle improves thermostability: Application in cellobiose hydrolysis. *Bioresource Technology*, 135, 2–6.

[153] Verma, M. L., Naebe, M., Barrow, C. J., & Puri, M. (2013). Enzyme immobilisation on amino-functionalised multi-walled carbon nanotubes: Structural and biocatalytic characterisation. *PloS One*, 8(9), e73642.

[154] Verma, M. L., Barrow, C. J., Kennedy, J. F., & Puri, M. (2012). Immobilization of β-d-galactosidase from Kluyveromyces lactis on functionalized silicon dioxide nanoparticles: Characterization and lactose hydrolysis. *International Journal of Biological Macromolecules*, 50(2), 432–437.

[155] Liu, J. M., Zhang, D. D., Fang, G. Z., & Wang, S. (2018). Erythrocyte membrane bioinspired near-infrared persistent luminescence nanocarriers for in vivo long-circulating bioimaging and drug delivery. *Biomaterials*, 165, 39–47.

[156] Chang, Y., Liu, N., Chen, L., Meng, X., Liu, Y., Li, Y., & Wang, J. (2012). Synthesis and characterization of DOX-conjugated dendrimer-modified magnetic iron oxide conjugates for magnetic resonance imaging, targeting, and drug delivery. *Journal of Materials Chemistry*, 22(19), 9594–9601.

[157] Contri, R. V., Fiel, L. A., Pohlmann, A. R., Guterres, S. S., & Beck, R. C. (2011). Transport of substances and nanoparticles across the skin and in vitro models to evaluate skin permeation and/or penetration. In *Nanocosmetics and nanomedicines* (pp. 3–35). Springer.

[158] Arpicco, S., Battaglia, L., Brusa, P., Cavalli, R., Chirio, D., Dosio, F., . . . & Ceruti, M. (2016). Recent studies on the delivery of hydrophilic drugs in nanoparticulate systems. *Journal of Drug Delivery Science and Technology*, 32, 298–312.

[159] Patil, H., Tiwari, R. V., & Repka, M. A. (2016). Recent advancements in mucoadhesive floating drug delivery systems: A mini-review. *Journal of Drug Delivery Science and Technology*, 31, 65–71.

[160] Kumari, A., Singla, R., Guliani, A., & Yadav, S. K. (2014). Nanoencapsulation for drug delivery. *EXCLI Journal*, 13, 265.

[161] Bobo, D., Robinson, K. J., Islam, J., Thurecht, K. J., & Corrie, S. R. (2016). Nanoparticle-based medicines: A review of FDA-approved materials and clinical trials to date. *Pharmaceutical Research*, 33(10), 2373–2387.

[162] Emeje, M. O., Obidike, I. C., Akpabio, E. I., & Ofoefule, S. I. (2012). Nanotechnology in drug delivery. *Recent Advances in Novel Drug Carrier Systems*, 1(4), 69–106.

[163] Anselmo, A. C., & Mitragotri, S. (2015). A review of clinical translation of inorganic nanoparticles. *The AAPS Journal*, 17(5), 1041–1054.

[164] Huang, H. C., Barua, S., Sharma, G., Dey, S. K., & Rege, K. (2011). Inorganic nanoparticles for cancer imaging and therapy. *Journal of Controlled Release*, 155(3), 344–357.

[165] McCarthy, J. R., Bhaumik, J., Karver, M. R., Erdem, S. S., & Weissleder, R. (2010). Targeted nanoagents for the detection of cancers. *Molecular Oncology*, 4(6), 511–528.

[166] Anselmo, A. C., & Mitragotri, S. (2014). An overview of clinical and commercial impact of drug delivery systems. *Journal of Controlled Release*, 190, 15–28.

[167] Yu, L., Lei, Y., Ma, Y., Liu, M., Zheng, J., Dan, D., & Gao, P. (2021). A comprehensive review of fluorescence correlation spectroscopy. *Frontiers in Physics*, 9, 644450.

[168] Dettmer, K., Aronov, P. A., & Hammock, B. D. (2007). Mass spectrometry-based metabolomics. *Mass Spectrometry Reviews*, 26(1), 51–78.

[169] Ogawa, M., Regino, C. A., Choyke, P. L., & Kobayashi, H. (2009). In vivo target-specific activatable near-infrared optical labeling of humanized monoclonal antibodies. *Molecular Cancer Therapeutics*, 8(1), 232–239.

[170] ZINK, J. MOLECULAR PROBES OF PHYSICAL AND CHEMICAL PROPERTIES OF SOL-GEL FILMS J. I. ZINK*, B. DUNN**, B. DAVE**, F. AKBARIAN** Dept. of Chemistry, University of California, Los Angeles, CA 90095, zink@ chem. ucla. edu"* Dept. of Materials Science and Engineering, University of California, Los Angeles, CA 90095.

[171] Patra, J. K., & Baek, K. H. (2014). Green nanobiotechnology: Factors affecting synthesis and characterization techniques. *Journal of Nanomaterials*, 2014.

[172] Denmark, D., Mukherjee, D., Bradley, J., Witanachchi, S., & Mukherjee, P. (2015). Systematic study on the remote triggering of thermo-responsive hydrogels using RF heating of Fe3O4 nanoparticles. *MRS Online Proceedings Library (OPL)*, 1718, 35–40.

[173] Xu, R. (2008). Progress in nanoparticles characterization: Sizing and zeta potential measurement. *Particuology*, 6(2), 112–115.

[174] Schellekens, H., Stegemann, S., Weinstein, V., de Vlieger, J. S., Flühmann, B., Mühlebach, S., . . . & Crommelin, D. J. (2014). How to regulate nonbiological complex drugs (NBCD) and their follow-on versions: Points to consider. *The AAPS Journal*, 16(1), 15–21.

[175] Borchard, G., Flühmann, B., & Mühlebach, S. (2012). Nanoparticle iron medicinal products—Requirements for approval of intended copies of non-biological complex drugs (NBCD) and the importance of clinical comparative studies. *Regulatory Toxicology and Pharmacology*, 64(2), 324–328.

[176] Pautler, M., & Brenner, S. (2010). Nanomedicine: Promises and challenges for the future of public health. *International Journal of Nanomedicine*, 5, 803.

[177] Peppas, N. A. (2004). Intelligent therapeutics: Biomimetic systems and nanotechnology in drug delivery. *Advanced Drug Delivery Reviews*, 56(11).

[178] Mishra, B. B. T. S., Patel, B. B., & Tiwari, S. (2010). Colloidal nanocarriers: A review on formulation technology, types and applications toward targeted drug delivery. *Nanomedicine: Nanotechnology, Biology and Medicine*, 6(1), 9–24.

[179] How, C. W., Rasedee, A., Manickam, S., & Rosli, R. (2013). Tamoxifen-loaded nanostructured lipid carrier as a drug delivery system: Characterization, stability assessment and cytotoxicity. *Colloids and Surfaces B: Biointerfaces*, 112, 393–399.

[180] Ahmad, S., Tripathy, D. B., & Mishra, A. (2016). Sustainable nanomaterials. *Sustainable Inorganic Chemistry*, 205.

[181] Malam, Y., Loizidou, M., & Seifalian, A. M. (2009). Liposomes and nanoparticles: Nanosized vehicles for drug delivery in cancer. *Trends in Pharmacological Sciences*, 30(11), 592–599.

[182] Müller, R. H., Mäder, K., & Gohla, S. (2000). Solid lipid nanoparticles (SLN) for controlled drug delivery—a review of the state of the art. *European Journal of Pharmaceutics and Biopharmaceutics*, 50(1), 161–177.

[183] Üner, M., & Yener, G. (2007). Importance of solid lipid nanoparticles (SLN) in various administration routes and future perspectives. *International Journal of Nanomedicine*, 2(3), 289.

[184] Hallan, S. S., Kaur, P., Kaur, V., Mishra, N., & Vaidya, B. (2016). Lipid polymer hybrid as emerging tool in nanocarriers for oral drug delivery. *Artificial Cells, Nanomedicine, and Biotechnology*, 44(1), 334–349.

# 3 Magnetization Dynamics of Ferromagnetic Nanostructures for Spintronics and Biomedical Applications

*Monika Sharma and Bijoy K. Kuanr*

## 3.1 INTRODUCTION

Recently, a lot of research has been focused on fifth-generation technology systems, which require high-performance devices for storing enormous amounts of information. Such novel devices can be achieved by replacing charge-based electronics to spintronics where the electron spin plays an important role [1–3]. The utilization of magnetic materials in spintronic devices yields the non-volatility and high speed to store information and fortitude that is unmatched by other memory technologies such as resistive or phase-change memory [4–9]. The discovery of giant magnetoresistance (GMR), which is widely used in magnetic reading heads in hard disk drives, has attracted massive research in the field of spintronics from the last few decades [10–18]. GMR is usually investigated in layered structures (trilayer) consisting of two ferromagnetic layers separated by a nonmagnetic layer. The spin-dependent scattering causes the change in resistance at the ferromagnetic/normal metal interface [19–28]. Depending upon whether the magnetization is parallel or antiparallel in two adjacent ferromagnetic layers, the resistance will be low or high respectively [29–35]. From the last three decades, a lot of research has been done in understanding the fundamental phenomena of spin dynamics and spin relaxation [36–45].

Recently, nanostructured magnetic materials, especially one-dimensional nanowires, have attracted a lot of interest in a variety of biomedical applications such as hyperthermia therapy, drug delivery, cell separation and manipulation, barcoding, nano-swimmers and GMR biosensors [46–52]. In hyperthermia, nanowires provide more frictional reactive areas than nanoparticles resulting in better heating efficiency and reduced cancer treatment time [53]. One-dimensional nanostructures have large magnetic moments and shape anisotropy-enhancing forces compatible for cell culture as they do not disrupt the cellular growth cycle and can be functionalized with biologically active molecules. In addition, nanowires have shown novel properties such

DOI: 10.1201/9781003316398-3

as better dispersibility, less cytotoxicity, better conjugation ability with other bio-active molecules, biocompatibility and enhanced thermal stability. These improved properties make nanowires an ideal candidate to couple biomolecules to nanoscale devices which can be used for biosensing applications.

In 1986, the existence of antiferromagnetic interlayer exchange couplings was revealed by Peter Grunberg and co-workers in Fe/Cr trilayers and multilayers [54]. These multilayer systems were found to be suitable for spintronic devices as it was possible to switch the magnetization orientation from antiparallel to parallel in adjacent magnetic layers by applying a magnetic field. Also, such systems attracted intense interest due to their fundamental properties which open many possible applications for the detection of small magnetic fields. The magnetoresistance can be further enhanced by replacing the metallic spacer layer with an insulating layer. This increase in magnetoresistance is due to the spin-dependent tunnelling effect, which was realized in ferromagnetic tunnel junctions (MTJs) as tunnelling magnetoresistance (TMR) [55–62]. In 1995, about 20% of TMR was first observed in disordered aluminium oxide tunnel barrier MTJs at room temperature [63]. Later research confirmed TMR value today reaching as high as 400% in MTJs having MgO single crystal tunnelling barrier [64]. Currently, many researchers are still very active on TMR by changing the barrier layers. Certainly, these devices are promising candidates for many spintronic applications such as Magneto-resistive random-access memory (MRAM) and Hard Disk Drive (HDD) read heads and highly sensitive field sensors [65–74].

High-speed spintronic devices push the operational frequency higher and higher, which results in the frequency properties of interest to shift in the gigahertz range. This operational frequency is close to the precessional frequency of magnetization in ferromagnetic materials. Usually, the ferromagnetic materials experience ferromagnetic resonance (FMR) in this frequency range, which causes the magnetic moment to precess in the presence of an external time-dependent magnetic field [75]. Many interesting phenomena, such as microwave-assisted magnetization reversal due to nonlinear motion of magnetization and injection of spins in neighbouring layers due to precession magnetization, are observed near ferromagnetic resonance [76–80]. Therefore, it is essential to understand how the magnetization reversal switching happens by the external magnetic field or applied electric (spin) currents. Magnetization dynamics can be an effective tool which provides qualitative as well as quantitative information about the magnetic nanostructures [81–83]. The experimental studies focused on such processes are often challenging. This chapter emphasizes studying the connection between the physical structure and magnetic response of the system, leading to rich and varied dynamic phenomena. The finer details of magnetization dynamics can be probed by various techniques in frequency, time and the wave-vector domain, which includes conventional FMR, time-resolved magneto-optical Kerr effect (TR-MOKE) microscopy and Brillouin light scattering (BLS) respectively [84–86]. In conventional FMR, the sample is excited at a particular frequency and the external magnetic field is varied to investigate the magnetization dynamics through the resonance. To measure the magnetization dynamics in a wave-vector domain, one of the efficient techniques is the BLS technique. The progress of space-resolved and time-resolved BLS approaches have carried the recent measurements to advanced levels. Further, time domain probing of magnetization dynamics is offered by TR-MOKE microscopy having high time resolution of the order of sub-hundred femtoseconds and spatial resolution in the sub-micron regime.

The first section of this chapter focuses on the understanding of magnetization dynamics within ferromagnetic nanostructures. We will next discuss the experimental techniques, namely ferromagnetic resonance and Brillouin light scattering, which can be used to probe the magnetization dynamics in these nanostructures. The static and dynamic measurements can be done by these techniques to study spin wave phenomena. In the BLS process, spin waves of wave vector $\vec{k}$ can be created (Stokes) or annihilated (anti-Stokes). Whereas in FMR, uniform precession modes are considered in which no momentum is transferred and only those modes are excited for which the total momentum is zero, $\vec{k} = 0$. The next section of this chapter considers various nanostructures, namely ferromagnetic trilayer and multilayers separated by a nonmagnetic spacer layer, nanostrips and one-dimensional nanocylinders which were analyzed by both FMR and BLS measurements. The key parameters of the nanostructures, such as saturation magnetization, internal anisotropies, demagnetizing field and exchange coupling energy, were determined. Finally, we will consider various future aspects for development of high-frequency (microwave/mm-wave) spintronic devices.

## 3.2   MAGNETIZATION DYNAMICS IN FERROMAGNETIC NANOSTRUCTURES

For an atom possessing magnetic moment $m$ and orbital angular momentum $J$, the motion of an electron in atomic orbit can be contemplated as a small current loop under the ascendancy of a magnetic field having a magnetic moment

$$m = IA\hat{n} \tag{3.1}$$

where $\hat{n}$ is the unit vector which is normal to the area $A$ enclosed by the loop and $I$ is the current flowing in the loop. The magnetic moment and the total angular momentum can be related to each other and is given by [87, 88]

$$m = \gamma J \tag{3.2}$$

where $\gamma$ is a constant usually referred as gyromagnetic ratio and is given by

$$\gamma = \frac{g|e|}{2m_e} \tag{3.3}$$

g is called Lande's splitting factor and its value depends upon many aspects. The value of g is approximately 2 for free spin and 1 for pure angular momenta. For combinations of **L** and **S**, it can take various other values indicating the protrusion of $m$ along the direction of $J$. The rate of change of total angular momentum $\frac{dJ}{dt}$ is proportional to the torque $m \times B_{eff}$, and the equation of motion for a single magnetic moment can be written as [87, 88]

$$\frac{dm}{dt} = -\gamma \mu_0 m \times H_{eff} \qquad (3.4)$$

where $B_{eff} = \mu_0 H_{eff}$. As the magnetic flux in a magnetic sample may vary by an applied field due to the presence of demagnetizing fields, anisotropy fields, magnetostrictive fields and so on, $H_{eff}$ represents the effective magnetic field. One may obtain the effective magnetic field by taking the negative gradient of the internal energy of the system with respect to the magnetization ($H_{eff} = -\nabla_M U$). The magnetization $M$ is the magnetic moment per unit volume, since any magnetic system generally consists of a large number of atoms in which magnetic moments are coupled to each other by exchange interaction. Magnetization dynamics explains the time evolution of the magnetization vector $M$ out of equilibrium. This leads to dynamical model as given by

$$\frac{dM}{dt} = -\mu_0 \gamma M \times H_{eff} \qquad (3.5)$$

where $\gamma = g \left( \dfrac{e}{2m_e} \right)$ is the gyromagnetic ratio of the charge $e$ and mass $m_e$ of the electron. The constant $g$ is Landé g-factor, also known as the spectroscopic splitting factor, and its value for a free electron is $g = 2 \times 1.001159657$. Eq. (3.5) defines the Larmor precession, with the Larmor precession frequency $\omega_L = \gamma H$. This equation does not describe the relaxation of magnetization to $H_{eff}$. However, a real ferromagnetic system must include a damping term to elucidate the magnetization dynamics. Landau and Lifshitz (LL) amended this behaviour by adding a phenomenological dissipation term [89]:

$$\frac{dM}{dt} = -\mu_0 \gamma M \times H_{eff} - \frac{\mu_0 \lambda}{M_s^2} M \times \left( M \times H_{eff} \right) \qquad (3.6)$$

where $M_s$ is the saturation magnetization and the constant $\lambda$ is phenomenological damping constant, with $\lambda > 0$ in the damping term having the dimension of a frequency. Eq. (3.6) explicates the magnetization precession around the effective magnetic field $H_{eff}$. It includes internal energy of the nanostructures (first term) and dissipation of energy or damping (second term). However, the most often used equation of motion of magnetization is the Landau–Lifshitz–Gilbert (LLG) equation, which consists of the same precession terms of the LL equations Eq. (3.5) in addition to a different damping term given by Gilbert (1955) and hence the name [90]. This dissipation term models a viscous damping, which depends on the time derivative of the magnetization. The modified LLG equation of motion [90, 91] has its form as

$$\frac{dM}{dt} = -\mu_0 \gamma M \times H_{eff} + \frac{\alpha}{M_s} \left( M \times \frac{dM}{dt} \right) \qquad (3.7)$$

where $\alpha$ is the phenomenological dimensionless Gilbert damping parameter. As a result of the damping term, the motion of the magnetization follows a helical trajectory as shown in Figure 3.1.

The equations (3.6) and (3.7) becomes equal for the limit $\alpha \ll 1$ with $\alpha = \dfrac{\lambda}{\gamma M_s}$. On the other hand, when $\alpha \gg 1$, the LL equation predicts the quick loss of energy by the magnetic moments to rapidly reach its low energy state, whereas in the LLG equation the dissipation of energy and the approach to the low energy state will become increasingly slow. The time evolution of magnetization in the LLG equation results in $\dfrac{dM}{dt} \to 0$ for $\alpha \to \infty$, whereas in the LL equation it results in $\dfrac{dM}{dt} \to \infty$ for $\lambda \to \infty$. The LLG equation is more suitable to describe the precession of magnetization in a material than the LL equation, which has infinite change of magnetization. The damping constant $\alpha$ represents all the relaxation mechanisms, and in magnetic materials its observed value is typically small, within the range of 0.01 and 0.1 [92]. It can be considered as a sum of intrinsic and extrinsic damping. The intrinsic damping depends on unavoidable processes which results from the scattering of electrons by phonons and magnons in the lattice via spin-orbit interaction and eddy currents present in conducting electrons, while the extrinsic damping is caused by two-magnon scattering due to structural defects. An experimental access to purely intrinsic damping is given by conventional FMR.

### 3.2.1  MAGNETIC DAMPING

The theory of magnetization damping has mesmerized researchers over the past few years due to its technological pertinence in spintronics [93]. In ferromagnetic nanostructures, magnetization relaxation switching time is one of most important parameters which can be determined by the measurements of resonance linewidth ($\Delta H$).

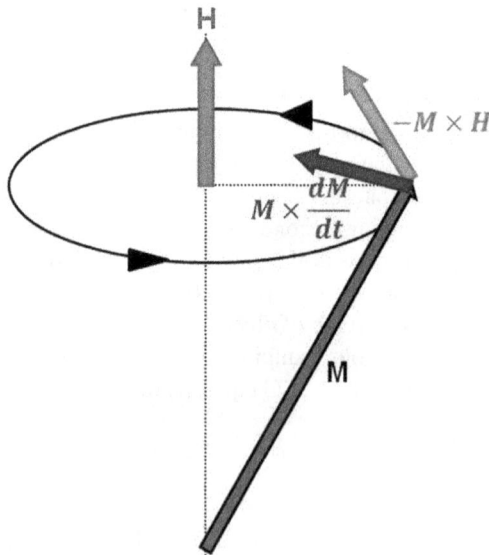

**FIGURE 3.1**  Schematic of magnetic moments dynamics showing the torques due to external magnetic field and damping.

The FMR linewidth can be used to determine the role of magnetic inhomogeneity and to characterize the contribution of the intrinsic loss mechanism. It is found to be linearly dependent on the microwave frequency ω and inversely proportional to the saturation magnetization as [94]

$$\Delta H = \Delta H_0 + 1.16\alpha \left( \frac{\omega}{\gamma} \right) \qquad (3.8)$$

where the first term represents contributions from inhomogeneous broadening, which is frequency independent, and the second term is due to frequency-dependent extrinsic contribution arising from structural inhomogeneity and defects existing in magnetic nanostructures. The linewidth is a measure of the distribution of various parameters such as uniaxial anisotropy and effective anisotropy field of individual nanostructure. The linewidth increases with the Gilbert damping constant ($\alpha$) [95–96].

### 3.2.1.1   Intrinsic Damping Mechanism

The intrinsic Gilbert damping originates basically due to spin-orbit coupling since the mobile electrons first transfer energy to the spin system. It is also possible to explain the intrinsic damping due to scattering of electrons by phonons and magnons. Spin-orbit interaction dominates in the phonon scattering process, while in the magnon scattering angular momentum relaxes through exchange interaction as scattered electrons repopulate the magnetization direction. Afterwards the spin of the conduction electron relaxes to the lattice through spin-orbit interaction. In a metallic ferromagnetic system, any change in the magnetization induces eddy currents, which tend to compensate for this change, and thus provides a damping mechanism.

### 3.2.1.2   Extrinsic Damping Mechanism

Two-magnon scattering is one of the most imperative extrinsic contributions to magnetic damping which arises due to surface defects and roughness, and inhomogeneities in materials due to disorder at atomic and molecular scale. The source for such inhomogeneities is quantized spin waves (magnons) which process as a wave when excited with wave-number $k$. Under uniform precession $k = 0$, these spin waves scatter to degenerate non-uniform states and dissipate energy into the lattice. The total number of magnons remains conserved since one magnon is annihilated and another is created. It also results in an increase of the FMR linewidth and hence damping.

### 3.2.2   Uniform Ferromagnetic Resonance Mode

As we move from bulk to nanoscale materials, the material properties change drastically such as surface-to-volume ratio, thermodynamic fluctuations, ballistic electron transport and into the picture [97]. Magnetic nanomaterials also exhibit distinct magnetic properties due to various anisotropies. Nowadays, magnetic nanostructures such as nanoparticles, thin films and one-dimensional nanocylinders are the area of interest for data storage applications [98], sensors [99], biomedical drugs [100], microwave devices [101] and so on. The FMR technique involves the absorption of microwave energy when the magnetic moments of the system are excited by a spatially uniform microwave (rf) magnetic field. The resonance of the absorption will occur when the Larmor precession frequency coincides with the rf frequency, causing an increase in the energy of the system.

The linearized Landau–Lifshitz–Gilbert equation can be used to solve the equation of motion of magnetization for uniform precessions in the lossless case ($\alpha = 0$). The nontrivial solutions of the linearized equation lead to the precession frequency as a function of the second derivative of the energy [91]:

$$\frac{\omega}{\gamma} = \frac{\gamma}{M_s \sin\theta} \sqrt{\frac{\partial^2 E}{\partial\theta^2} \frac{\partial^2 E}{\partial\varphi^2} - \left(\frac{\partial^2 E}{\partial\theta\partial\varphi}\right)^2} \tag{3.9}$$

where $\theta$ and $\varphi$ represent the angles of magnetization ($M$) with reference to nanostructure axis in spherical coordinates. The transmission response of the magnetic nanostructures can be calculated using the transmission parameters. The microwave absorption power as a function of external magnetic field ($H$) and its derivative with reference to $H$ are used in FMR measurements. The power transmitted per unit area in the coplanar waveguide is calculated from the Poynting vector using the high-frequency field components at all positions around the central signal line. The power absorbed by the nanostructure matrix is given by the relation:

$$P_{abs} = \frac{i\omega}{2} Re(h_X^* M_X + h_Z^* M_Z) \tag{3.10}$$

where $h_X^*$ and $h_Z^*$ are the components of the rf field along the appropriate coordinate axes, $\mathbb{1}$ is the angular frequency. Here

$$M_X = \chi_{eff}^{xx} h_X + \chi_{eff}^{xz} h_Z \tag{3.11}$$

and

$$M_Z = c_{eff}^{zx} h_Z + c_{eff}^{zz} h_X \tag{3.12}$$

$\chi_{eff}^{xx}$, $\chi_{eff}^{xz}$, $\chi_{eff}^{zx}$ and $\chi_{eff}^{zz}$ are the components of the susceptibility tensor of the effective medium which is calculated from Eq. (3.11) and Eq. (3.12). The power transmitted in the dielectric can be calculated from the relation:

$$\langle P_{abs} \rangle = \frac{c}{8\pi\sqrt{\varepsilon_d}} \left(\vec{h_d^*}.\vec{h_d}\right) \tag{3.13}$$

where $\varepsilon_d$ is the relative permittivity of the dielectric medium. Using the law of conservation of power, we can find the transmission coefficients. For thin films, resonance frequency is governed by the demagnetizing field due to the dimensions of magnetic element and is given by

$$f_{res} = \left(\frac{\gamma}{2\pi}\right)\sqrt{H + (N_X - N_Z)4\pi M_s + (N_Y - N_Z)4\pi M_s} \qquad (3.14a)$$

where $N_X$, $N_Y$ and $N_Z$ are the demagnetizing coefficients. The ferromagnetic resonance frequency depends upon various factors such as shape anisotropy and magnetocrystalline anisotropy. The resonance frequency ($f_r$) for one-dimensional nanocylinders in an FMR measurement is written as.

$$\left(\frac{\omega}{\gamma}\right)^2 = \left[H_r cos(\theta - \theta_H) + H_{eff}\theta\right]\left[H_r cos(\theta - \theta_H) + H_{eff} cos2\theta\right] \qquad (3.14b)$$

where $H_{eff}$ is the effective anisotropy field that comes from a combination of effects including shape, magnetocrystalline and magnetoelastic anisotropy. $H_r$ represents the external magnetic field whose orientation is defined by the angles $\theta_H$ and $\varphi_H$. The microwave field is considered to be perpendicular to the external field.

## 3.3  EXPERIMENTAL TECHNIQUES TO PROBE MAGNETIZATION DYNAMICS

### 3.3.1  BRILLOUIN LIGHT SCATTERING

The mapping of spin waves in $f$ and $k$ space can be done by an optical technique known as Brillouin light scattering. In an inelastic scattering process, the spin waves' quasiparticles (magnons) interact with coherent photons. In this process, either the energy is lost by photons to create a magnon (Stokes process) or photons gain energy from an annihilated magnon (anti-Stokes process). This difference between the frequency of the Stokes and anti-Stokes process, also known as the Brillouin shift, gives the frequency shift of scattered photon. One can also determine the wave vector of the magnon by Brillouin light scattering in which the angle of the incident light is varied corresponding to sample surface. The transferred wave vector $k$ can be written in terms of wavelength $\lambda_L$ and incident angle as

$$k = \frac{4\pi}{\lambda_L} sin\theta \qquad (3.15)$$

#### 3.3.1.1  Working Principle

Brillouin light scattering uses multi-pass tandem Fabry–Pérot interferometers to obtain high resolution. The dispersion relations for transparent samples are obtained by measuring phonon frequencies as a function of wave vector $k$ at various scattering angles. For samples which exhibit multiple light scattering such as dry colloidal crystals, the wave vector is not properly defined and hence acoustic-like modes which depend upon $k$ becomes inaccessible. However, the BLS spectrum can record $k$-independent modes and was observed for sub-micron colloidal silica [102] and polymer crystals [103]. BLS has the advantage of measuring numerous thermally

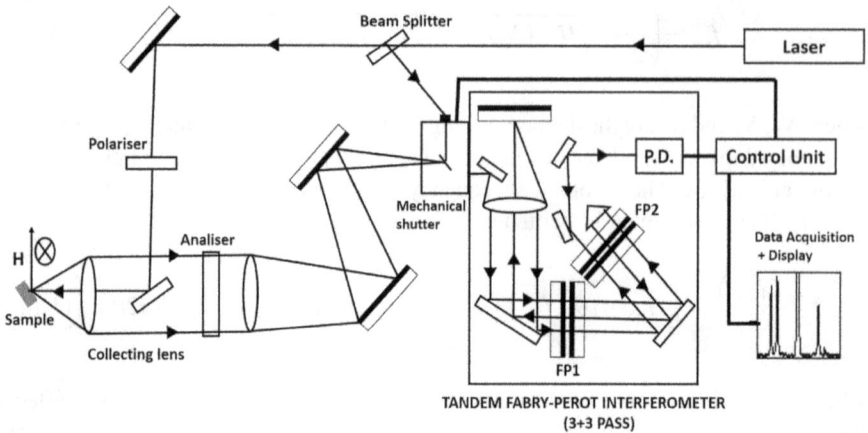

**FIGURE 3.2** Schematic representation of BLS setup used for the measurements, including a (2 × 3) pass tandem Fabry–Pérot interferometer.

excited elastic resonances in a single measurement, resulting in identification of these $k$-independent frequencies at the resonance modes of individual nanostructures. Depending upon the geometrical and elastic parameters of the nanostructures, eigen-frequencies are uniquely defined. Thus, BLS is a powerful tool by which the elastic behaviour of nanostructured materials can be investigated, choosing the suitable combination of $k$-independent and $k$-dependent scattering modes.

Figure 3.2 shows the BLS setup used for the measurements of the trilayer samples, indicating the path of the light inside the Fabry–Pérot interferometer. A (2 × 3) pass tandem Fabry–Pérot interferometer having in-plane wave vector $k_{\parallel} = 1.65 \times 10^5$ cm$^{-1}$ was used to perform BLS experiments at room temperature. The sample was kept in the presence of an externally applied magnetic field, in both parallel and perpendicular directions to the probed magnon wave vector. The surface acoustic and optic spin waves were observed on the Stokes and the anti-Stokes sides of the spectrum. The samples were mounted in the centre of a goniometer in a sample holder at a fixed temperature. A solid state pumped frequency-doubled Nd:YAG laser was fixed on the goniometer and rotated for different scattering angles (0° and 160°) in either transmission or reflection geometry. The light passes through a Glan polarizer with vertical polarization before reaching the sample, which ensures completely polarized incident light. After scattering through the sample, the light was collected by an aperture and focused into the entrance pinhole of the tandem Fabry–Pérot interferometer. The transmitted light from the Fabry–Pérot interferometer was detected by a photo diode and processed with the help of analyzers. Finally, processing was done by computer software. In order to avoid mechanical disruptions, the whole setup was placed on an optical bench with active vibration damping.

### 3.3.2 CONVENTIONAL FERROMAGNETIC RESONANCE

The conventional ferromagnetic resonance is another important technique which can be used to study spin waves in ferromagnetic systems with wave vector $k = 0$. In this technique, a magnetic system is excited by a sinusoidal electromagnetic radiation at a fixed microwave frequency. The sample is kept in a resonant cavity which is coupled to microwave radiation at fixed frequency by a Gunn diode generating a standing microwave field. As discussed in section 3.2.2, the resonance frequency depends upon the effective field which includes internal and external energies. So, by sweeping the external magnetic field the resonance condition can be achieved, usually measured in terms of the maximum power absorption of microwave radiations by the magnetic system. The external magnetic field is adjusted to allow for lock-in detection which enhances the signal to noise ratio (SNR). The measured FMR signal is proportional to the power absorbed or to the field derivative of the imaginary part of high-frequency susceptibility $\frac{\partial \chi''}{\partial H}$ averaged over the sample volume. Conventional FMR is a superlative technique for probing the spin waves in bulk samples demonstrating a single domain magnetization configuration due to its high sensitivity. However, for materials showing complex domain configuration this is not a relevant method. This snag can be dodged by sweeping the frequency instead of the external magnetic field, which can be realized by using a vector network analyzer (VNA).

### 3.3.2.1   Vector Network Analyzer Ferromagnetic Resonance (VNA-FMR)

As mentioned earlier, VNA-FMR is a more advanced technique than conventional FMR, as it exploits the dynamic response of magnetic structures by varying the frequency of the excited rf field at a constant external field (frequency-sweep) and by varying the field at a constant rf frequency (field-sweep). For magnetic systems having a large damping constant and hence large linewidths, the VNA-FMR technique is more advantageous than other mentioned techniques. The transmission and reflection coefficient of FMR signal enables measurement of both the amplitude and phase, which can provide the real and imaginary susceptibility components [104, 105]. High versatility can be achieved in commercially available VNAs due to their broadband operation and sensitivity. One may measure, for example, domain wall resonances in the tens of MHz to FMR in ferromagnetic nanostructures in the 40–100 GHz range. Kalarickal et al. [106] demonstrated that FMR linewidths could be extracted from the frequency linewidth measurements that can provide valuable information on the intrinsic and extrinsic material parameters.

The VNA-FMR technique allows one to follow the changes in the micromagnetic structure induced by changing the applied field. In VNA-FMR, it is also possible that the precession of the magnetization can be excited by sweeping the frequency of the oscillating rf magnetic field at a constant bias field. This offers the conservation of magnetization state which is advantageous for the inspection of complex magnetic nanosystems or which includes effects like exchange bias. This can be achieved by employing a high-bandwidth coplanar waveguide (CPW) as an alternative to a microwave cavity in which a fixed resonance frequency is used. The vector network analyzer sweeps the frequency of outgoing signal over a specified frequency range since it serves as both the source and detector of rf sinusoidal signal. At a particular bias field, maximum microwave power is absorbed when the excitation frequency matches the resonance frequency of the sample so that the magnetization precessional motion is maintained (Figure 3.3). Also at resonance, a destructive interference occurs as the phase of the signal is shifted by $\pi$. Due to the attenuation in

**Vector Network Analyzer**

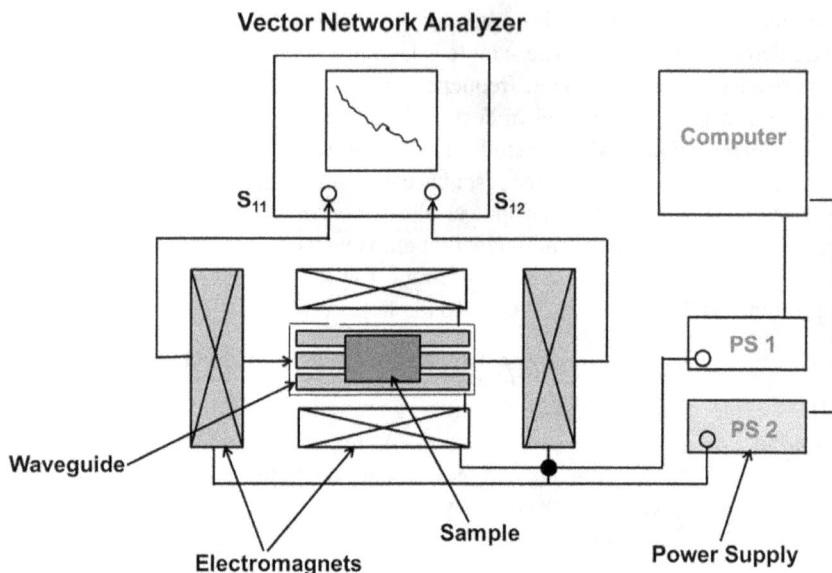

FIGURE 3.3 Block diagram of the broadband VNA-FMR spectrometer. The VNA measures the complex scattering parameters ($S_{11}$ and $S_{21}$) of a sample on a coplanar waveguide (CPW) which is mounted in the middle of electromagnets.

the coaxial cables connected to VNA and coplanar waveguide, the amplitude of the transmitted signals is decreased. The flip-chip measurement is performed using a 50 $\Omega$ characteristic impedance coplanar waveguide and an Agilent vector network analyzer (model no. ZVB 20 Rohde & Schwarz). The VNA was connected to the CPW using coaxial cables with Teflon insulation and SubMiniature Version A (SMA) connectors. Noise is prohibited by using an optical bench with an air cushioning system.

## 3.4 DYNAMIC MEASUREMENTS OF MAGNETIC NANOSTRUCTURES

### 3.4.1 Fe/Al/Fe Trilayer Ultra Thin Films

The trilayer ultrathin films of Fe/Al/Fe with varying interlayer thickness were epitaxially grown by a molecular beam epitaxiy (MBE) technique on (100)/Fe(1 nm)/Ag(150 nm) substrates with deposition parameters as pressure of $\leq 10^{-10}$ mbar and deposition rate of 0.1 Å/s at a temperature of 80 °C. The dynamic measurements of the samples were employed by VNA-FMR technique in the frequency sweep mode. The acoustic and optic resonance frequencies were measured by the complex reflection coefficients $S_{11}$ of the sample at various externally applied magnetic field in the sample plane. A (2 × 3) pass tandem Fabry–Pérot interferometer having in-plane wave vector $k_{\parallel} = 1.65 \times 10^5$ cm$^{-1}$ was used to perform BLS experiments at room temperature. The sample was kept in the presence of an externally applied magnetic field, in both parallel and perpendicular directions to the probed magnon wave vector. The surface acoustic and optic spin waves were observed on the Stokes and the anti-Stokes sides of the spectrum.

The complex reflection coefficient $S_{11}$ as a function of frequency for a (50 Å)/ Al(10 Å)/Fe(50 Å) trilayer sample at a fixed bias field of 0.1 kOe is shown in Figure 3.4. The resonance spectrum shows two peaks correlating to the acoustic and the optic resonance. The solid lines in Figure 3.4 shows the Lorentzian fits to the spectrum to obtain the resonance frequencies and resonance linewidths ($\Delta f$) for acoustic ($f_{ac}$) and optic ($f_{opt}$) modes. The optic resonance was observed at a higher frequency than the acoustic resonance, which can be attributed to the antiparallel or canted alignment of magnetization corresponding to a non-saturated state in coupled layers. The linewidth of optic mode was also observed to be one order of magnitude larger, around 540 MHz, than the acoustic mode linewidth of around 65 MHz for all magnetic fields. The acoustic mode revealed a good magnetic homogeneity and single-domain state corresponding to extremely narrow linewidth (65 MHz) and sharp absorption (225 dB). It was observed that the absorption amplitude for optic mode reduces sharply and becomes practically invisible for very high magnetic fields (>2 kOe). This is due to the fact that the magnetic Fe layers are of equal thickness (50 Å), and when the magnetization of the two layers reaches the saturated state, the out-of-phase amplitude for the optic mode dribbles considerably.

Figure 3.5 collates the experimental resonance frequencies of a (50 Å)/Al(10 Å)/ Fe(50 Å) trilayer for both acoustic (dots) and optic (squares) modes as a function of

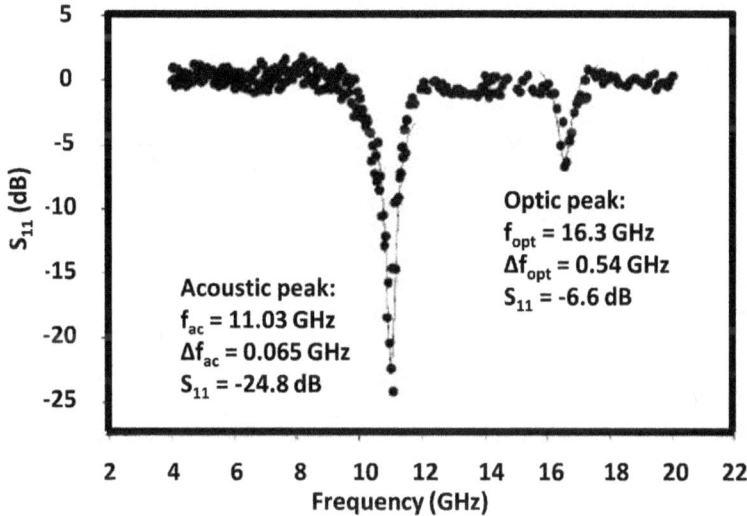

Optic peak:
$f_{opt}$ = 16.3 GHz
$\Delta f_{opt}$ = 0.54 GHz
$S_{11}$ = -6.6 dB

Acoustic peak:
$f_{ac}$ = 11.03 GHz
$\Delta f_{ac}$ = 0.065 GHz
$S_{11}$ = -24.8 dB

**FIGURE 3.4**  The reflection coefficient $S_{11}$ as a function of frequency for a Fe(50 Å)/Al(10 Å)/ Fe(50 Å) trilayer measured at an external field of 0.1 kOe. Solid lines in the spectrum are the Lorentzian fits to the peaks.

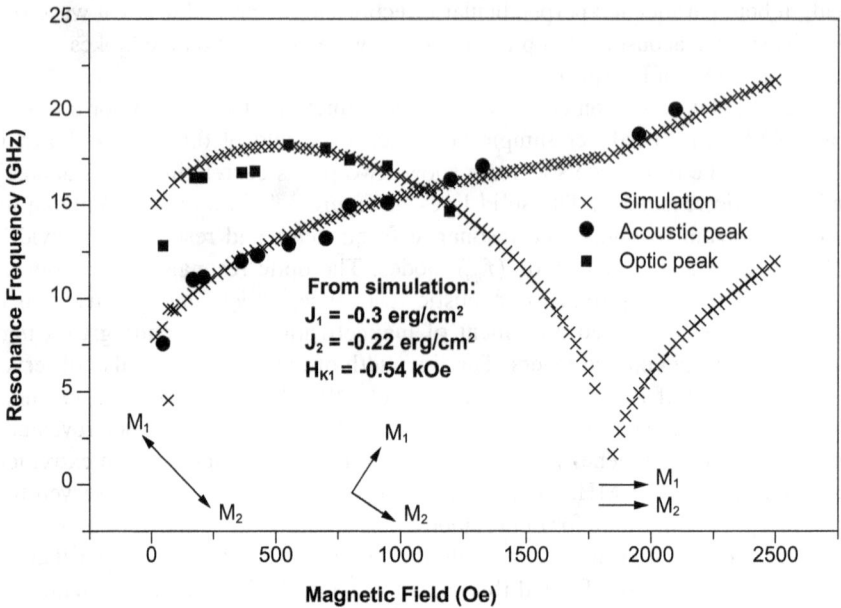

**FIGURE 3.5** VNA-FMR spectrum as a function of applied external magnetic field along hard axis indicating optic (squares) and acoustic (dots) resonance frequencies for a (50 Å)/Al(10 Å)/Fe(50 Å) trilayer. The crosses show the theoretical simulated points from which coupling parameters are obtained.

externally applied magnetic field ($H$) along the hard axis of the sample. The crosses in Figure 3.5 are the theoretical fits from which the various coupling parameters are obtained. The variation of the acoustic and optic mode frequencies with external magnetic field $H$ corresponds to different magnetization states present in the two Fe layers. It is observed that the magnetization is aligned in antiparallel direction for two layers at very low fields ($0 < H < 0.15$ kOe). For the magnetic fields less than 150 Oe, an upward shift of resonance frequencies by 3.5 GHz occurs for both modes, which corresponds to the transition to the spin-flop (SF) phase. In the SF phase for fields greater than 150 Oe, the angle decreases continuously from 180° to 0° between the two magnetizations. The optic frequency increases, reaching a maximum at around 500 Oe and then it again decreases, whereas the acoustic frequency continuously increases starting from 150 Oe, and near the saturation it forms an eccentricity. The acoustic and optic frequencies increase linearly with a field above the saturation field, that is, beyond 1800 Oe, and the difference in these frequencies indicates the coupling strength. The dispersion relation for the trilayer films is obtained by considering Rezende's approach of a thin-film approximation model leading to magnetic excitation frequencies as

$$\left(\frac{\omega}{\gamma}\right)^4 + a\omega^2 + b\omega + c = 0 \qquad (3.16)$$

where parameters $a$, $b$ and $c$ are constants which are derived from the equation of motion ($\omega = 2\pi f$) [13]. Only two modes of $\omega$ are significant for the FMR configuration from the four possible solutions of Eq. (3.16), which results in the acoustic and optic modes. The crosses in Figure 3.5 represents the simulated curves from Eq. (3.16). Since the optic mode is in a beeline susceptible to the coupling strength, the acoustic mode diverges away from it, and the simulation without the $J_2$ term reproduces the acoustic mode frequencies. For the (50 Å)/Al(10 Å)/Fe(50 Å) trilayer samples, the bilinear $J_1$ and biquadratic $J_2$ exchange constants derived from the fits are $J_1 = -0.3$ erg/cm$^2$ and $J_2 = -0.22$ erg/cm$^2$ respectively. The various magnetic parameters derived from the fits to Eq. (3.16) are saturation magnetization as $4\pi M_s = 20.5$ kOe, fourfold cubic anisotropy field as $H_{K1} = 0.5$ kOe, and g-factor as $g = 2.09$.

Figure 3.6(a) shows the BLS spectrum for the Fe(50 Å)/Al(7 Å)/Fe(70 Å) trilayer structure revealing the (Stokes and anti-Stokes) acoustic and optic mode frequencies at various externally applied magnetic fields. The antiferromagnetic coupling can be easily observed from the asymmetry in Stokes and anti-Stokes plots. In the BLS spectrum, the mode frequencies as a function of the magnetic field demonstrates three different magnetic states viz. antiferromagnetic, spin flop and saturation. A perfect antiparallel alignment occurs for the magnetic field below 1 kOe. The magnetization switches to a noncollinear or spin flop phase for fields higher than 1 kOe which may result in a leap of both the acoustic and the optic mode frequency. We derived the exchange coupling constants $J_1$ and $J_2$ from a Levenberg–Marquardt fitting using Eq. (3.16) as $-2.1 \pm 0.006$ and $-0.5 \pm 0.02$ erg/cm$^2$.

To study the nature and strength of coupling for different Al interlayer thickness, we used the focused laser beam embedded within the BLS setup, which efficiently scans the whole Al cram (0–15 Å). Figure 3.6(b) shows the overall coupling strengths of $J_1$ and $J_2$, obtained from the field variation. We observe that for interlayer thicknesses less than 3 Å strong ferromagnetic coupling of about 2.7 erg/cm$^2$ may be originating from pinholes. It is also observed that the ferromagnetic coupling decreases with increasing Al interlayer thickness. The transition from ferromagnetic to antiferromagnetic coupling occurs above 5 Å Al thickness having an antiferromagnetic (AFM) coupling of $J_1 = -0.02$ erg/cm$^2$. The AFM coupling becomes maximum at $t_{Al} = 7.5$ Å, with $J_1$ and $J_2$ as $-2.5$ and $-0.5$ erg/cm$^2$, respectively. However, the biquadratic coupling $J_2$ occurs at the increasing slope of $J_1$ and has a maximum value of $-1.0$ erg/cm$^2$ which is at $t_{Al} = 6$ Å. The reason for this is the strong fluctuation of $\Delta J$ (variation of coupling). As the interlayer thickness approaches 13 Å, the AFM coupling strength decreases to $J_1 = -0.15$ and $J_2 = -0.005$ erg/cm$^2$. There is no further change of the coupling strengths for the higher values of interlayer thickness, which implies that the AFM coupling dominates for Al interlayer thickness until the end of the wedge structure. Neither $J_1$ nor $J_2$ showed any distinctive short-period oscillation.

We observed that the exchange coupling strength strongly depends upon the interlayer thickness. Slonczewski [107] tried to explain this by considering the "loose spin model," which assumes that the loose spins mediate interlayer coupling within the spacer or adjacent to Fe/Al interfaces. The magnetic impurities in the proximity of alloyed/roughened interfaces are the source of loose spins in the case of Fe/Al/ Fe. These thin films can be used for realizing GMR biosensors. The GMR effect can be observed in thin film structures consisting of ferromagnetic layers separated

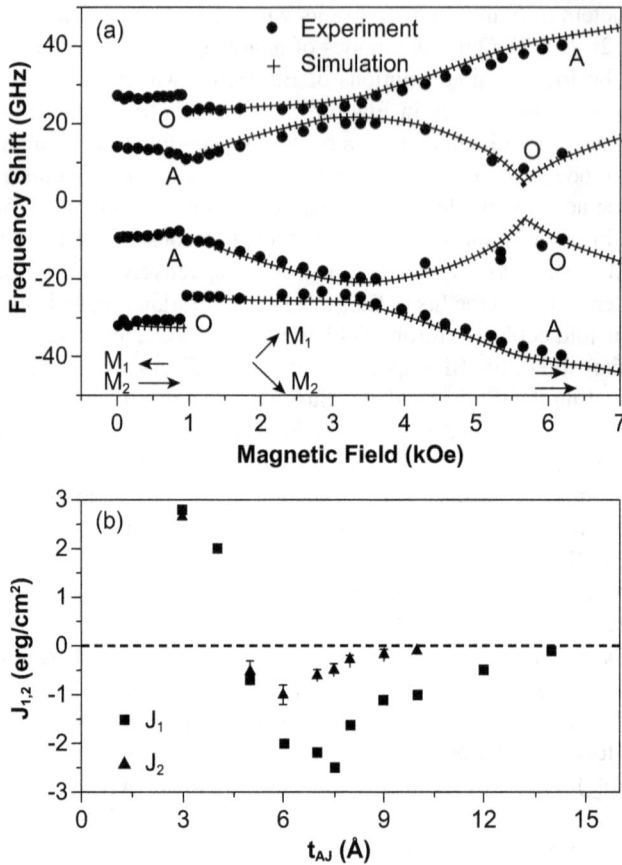

**FIGURE 3.6** (a) BLS spectrum of frequency shift of acoustic and optic modes showing Stokes and anti-Stokes for a Fe(50+)/Al(7+)/Fe(50+) trilayer structure. Solid circles are experimental data and crosses shows the simulated fits of Eq. (3.16). "A" represents the acoustic branches and "O" represents the optic branches. Magnetization orientation relative to $H$ is shown in the bottom by arrows. (b) Coupling strengths bilinear $J_1$ and biquadratic $J_2$ as a function of Al spacer thickness for a Fe(50+)/Al(0–15+)/Fe(50+) structure.

by nonmagnetic layers. The GMR biosensor is based on the detection of biologically functionalized nanosized magnetic labels, using highly sensitive microfabricated magnetic-field sensors. Choi et al., 2016 reported GMR biosensors and their applications in clinical diagnosis [108]. Recently, Su et al. discussed depositing MR stacks onto flexible substrates and transferring the patterned MR devices from rigid substrates to flexible substrates [109].

### 3.4.2 PERMALLOY NANOSTRIPS

In this section, we show the dynamic properties of Permalloy nanostrips at gigahertz frequencies. A Permalloy film was deposited into a resin having emblematic composition of 81% Ni and 19% Fe by rf magnetron sputtering at a rate of 0.2 nm/s and at an Ar pressure of 3 mTorr to obtain a thickness of 100 nm. Electron beam nanolithography was employed for the fabrication of arrays of nanostrips. Arrays of nanosized strips were formed in each patch and had a width of 300 nm separated by a distance of 1 µm. A lift-off process was used to obtain the final structures which involve removal of resin followed by an ultrasonic bath of the overlaying film in acetone. We fabricated a series of six samples consisting of strips length 300 nm, 600 nm, 900 nm, 1.5 µm, 3 µm and 100 µm. The width to length which corresponds to the aspect ratio was given by 1:1, 1:2, 1:3, 1:5, 1:10 and 1:333.

Figure 3.7(a) shows the FMR spectra of Permalloy strips as a continuous film obtained from the vector network analyzer ($S_{21}$ versus frequency) at an externally applied magnetic field of 3 kOe. It is observed that the spectra for Permalloy nanostrips with aspect ratios of 1:1, 1:2, 1:3, 1:5 and 1:10 are shifted to higher frequencies by ~1, 2, 2.15, 2.2 and 2.5 GHz, respectively, when the field is applied along the long axis of the samples as compared to the Permalloy film ($f_r$ = 15 GHz at 3 kOe). Since the arrays have large spacing (1 µm) between the nanosized strips, one can consider the microwave response primarily due to the magnetic properties of the individual strips and the uniform wave vector ($k$=0). For longer strips the rising behaviour of resonance frequency is basically due to the increase in shape anisotropy.

The spins of the magnetic moment precess in uniform mode when the magnetic field is applied along the length of the strip in a microwave experiment. This results in the generation of a demagnetizing field in the "magnetic charges" which are present on the surface. The upshifting of the resonance frequency is predominantly due to these self-demagnetizing fields in Permalloy strips in comparison to a continuous film [110]. All the strips show comparable line shape in the absorption spectra. The 1:1 strip structure shows different frequencies depending on the orientation of the magnetic field. We observed that the 1:1 strip structure is not quite square but slightly elongated in one direction from the AFM images of the arrays [111]. Furthermore, the resonance frequency shifts slightly due to the interactions among different nano-objects. Figure 3.7(b) reveals that for 1:333 strip resonance frequency shifts to higher positions when the magnetic field of 3 kOe is applied along the length ($fr$ = 11.8 GHz) and the width ($fr$ = 17.2 GHz). The reason for this behaviour of resonance frequency shift indicates the presence of uniaxial shape anisotropy which can cause significant changes in the frequency. The difference in resonance frequencies for the two orientations of the sample, that is, along either the length or the width of the strip with the applied magnetic field, gives the magnitude of the shape anisotropy.

The conglomeration of the resonance frequencies done by VNA-FMR for the Permalloy nanostrips and Permalloy film with the applied magnetic field along the length and the width of the strips are shown in Figure 3.8(a). Predictably, we can tune the FMR frequencies with the externally applied magnetic field. It is observed that the resonance frequency increases with the external magnetic field for both the in-plane and out of plane orientation of the samples. However, we noticed that for the 1:10 and 1:333 strips, when the field is applied along the width of the strip there are two frequency-field regimes. The resonance frequency firstly decreases with the increase of magnetic field and then again starts increasing linearly with the increase of magnetic field. In this particular point of the magnetic field, where

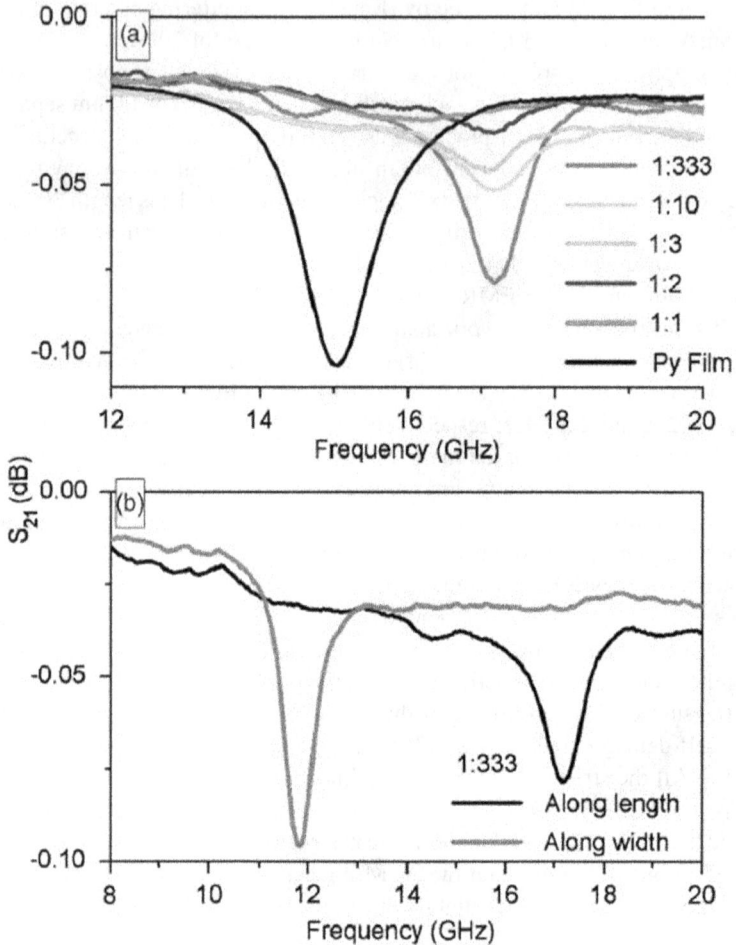

**FIGURE 3.7**   (a) The transmission coefficient evaluated by the network analyzer for various Permalloy strips at an applied field of 3 kOe along the length of the strips. The maximum absorption represents the resonance frequency of the strip. (b) $S_{21}$ versus frequency of a strip structure with aspect ratio 1:333 with the magnetic field applied along the length as well as the width of the strip.

the deviation of frequency-field data occurs corresponds to that below this field, the sample is not in its saturation state. And this field gives the value of shape anisotropy. The magnetic parameters such as g-factor, saturation magnetization and in-plane anisotropy for Py are derived from the best-fit measurements of the frequency-field data of continuous Py film. The saturation magnetization $M_s$ value is found to be low of around 0.612 kG. The uniaxial shape anisotropy in strips of different lengths is calculated from the measured parameters along with the self-demagnetization fields.

These nanostrips can be considered as a segment of small cells having individual magnetization. We used this approach to analyze theoretically the dynamics of

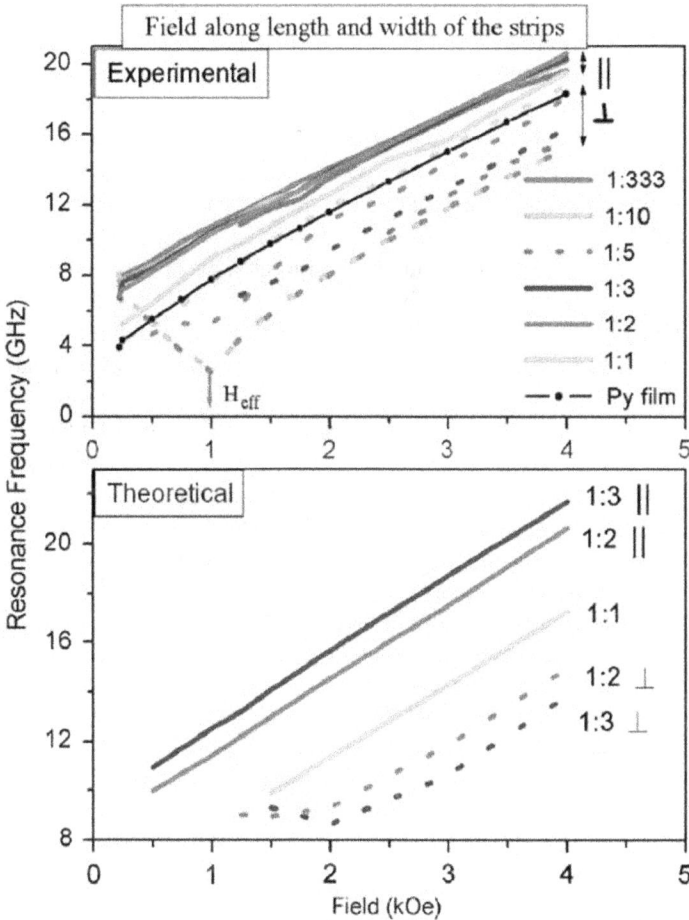

**FIGURE 3.8** (a) Resonance frequency behaviour for nanostrip structures of various aspect ratios and for a continuous Py film with the external magnetic field applied along the width and length of the strips. (b) The bottom portion shows the theoretical results obtained from micromagnetics calculations. The direction of the applied field with regard to the length of the nanostrip is indicated by the parallel and perpendicular symbols.

these structures by micromagnetic simulations for the uniform mode. The dynamical equation of motion of magnetization within an individual cell is obtained from the Landau–Lifshitz equation [90, 91],

$$\frac{\partial M}{\partial t} = -\frac{|\gamma|}{1+\alpha^2}\left[M \times H_{eff} - \frac{\alpha}{M_s}M \times \left(M \times \frac{dM}{dt}\right)\right] \qquad (3.17)$$

where $\alpha$ is the Gilbert damping constant. The internal energies, which are the sum of effective exchange fields, the dipole fields and Zeeman field, result in the effective field ($H_{eff}$) acting on a cell. The normal modes are found by using $M_s = 0.68$ kG, $A = 1.3 \times 10^{-6}$ erg/cm and $\alpha = 0.0001$. The thickness of individual cell size was 100 nm and $10 \times 10$ nm$^2$ in the plane of the sample. For larger structures, we considered a larger cell size for the calculations. Figure 3.8(b) demonstrates the theoretical fits to the experimental data of the resonance frequency. As discussed earlier, the nanostrip structure with 1:1 aspect ratio shows two different frequencies nearly 1 GHz apart with different orientations of the magnetic field, and it is larger than the experimental uncertainty of about of $\pm 0.2$ GHz. The theoretical calculations also revealed the soft mode-like behaviour in the sample undergoing a phase transition, as discussed previously, which has its origin at the edges of the sample when the field is applied perpendicular to the plane. This behaviour is more pronounced in the longer structures than in the shorter structures and is seen both experimentally and theoretically.

Figure 3.9 shows resonance frequencies along the length (easy axis) and width (hard axis) of the strip and frequency linewidth as a function of shape anisotropy ($H_u$). The frequency linewidth ($\Delta f_r$) measured from the ferromagnetic resonance absorption spectrum of the strips shows the signal from the whole sample. As shown in Figure 3.9 (separate star), it is observed that as the length of the strip increases the frequency linewidth of the strips decreases. It has a minimum for a continuous film of the same thickness for $f_r = 1.12$ $GHz$ at 3 kOe. This reduction of the linewidth can be attributed to the shape anisotropy [91, 112].

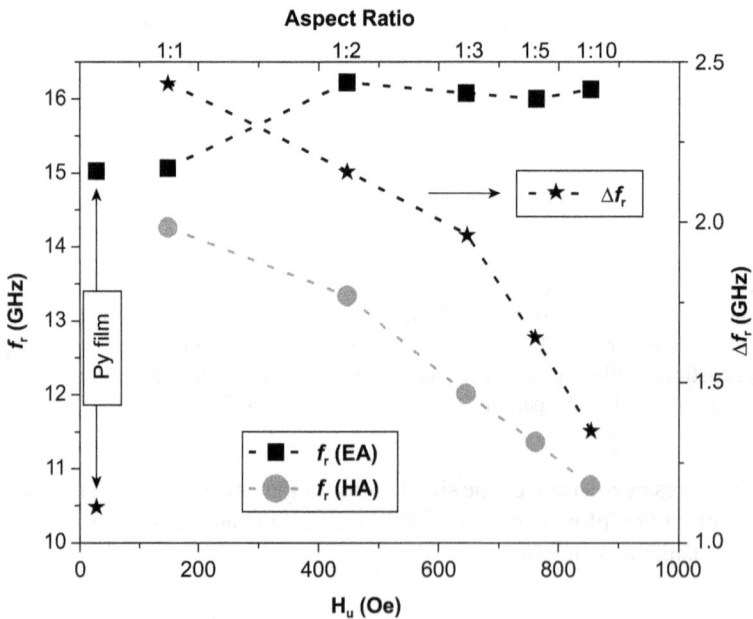

**FIGURE 3.9** VNA-FMR response of the resonance frequency [right side] for easy axis (square) and hard axis (circles) and frequency linewidth ($\Delta f_r$) [left side] as a function of the shape anisotropy ($H_u$) at a magnetic field of 3 kOe. The aspect ratios of the strips (1:1, 1:2, 1:3, 1:5 and 1:10) are shown on the top of the X-axis.

### 3.4.3 ONE-DIMENSIONAL MAGNETIC NANOWIRES

This section focuses on the dynamic properties of arrays of 1-D nickel nanowires (NWs) with high aspect ratio. The arrays of nanowires were fabricated by electrode-position technique in porous anodic alumina template as reported elsewhere [113]. We used six different lengths of NWs: 5, 8.5, 16, 30, 40 and 60 $\mu$ m. A flip-chip technique was used with a CPW to study dynamic parameters of NWs. The sample was flipped on top of the transmission line with the NWs axes parallel to an external static magnetic field. The frequency was swept from 0.05 to 20 GHz at a fixed external magnetic field.

Figure 3.10(a)–(d) represents the transmission response measured for various samples of NW length (8.5, 16, 30 and 40 $\mu$ m) with an externally applied magnetic field parallel to the easy axes of NWs. The resonance frequency ($f_{res}$) and frequency linewidth ($\Delta f_{res}$) are derived from the Lorentzian fits to the experimental data as shown by the solid lines. We observe from the results that the absorption peak varies in position and intensity with the increase in the static magnetic field for the entire samples. For low applied fields specifically, there is a ~0.3 dB absorption dip at the FMR frequency which increases to nearly 1.0 dB for higher fields. The NW system

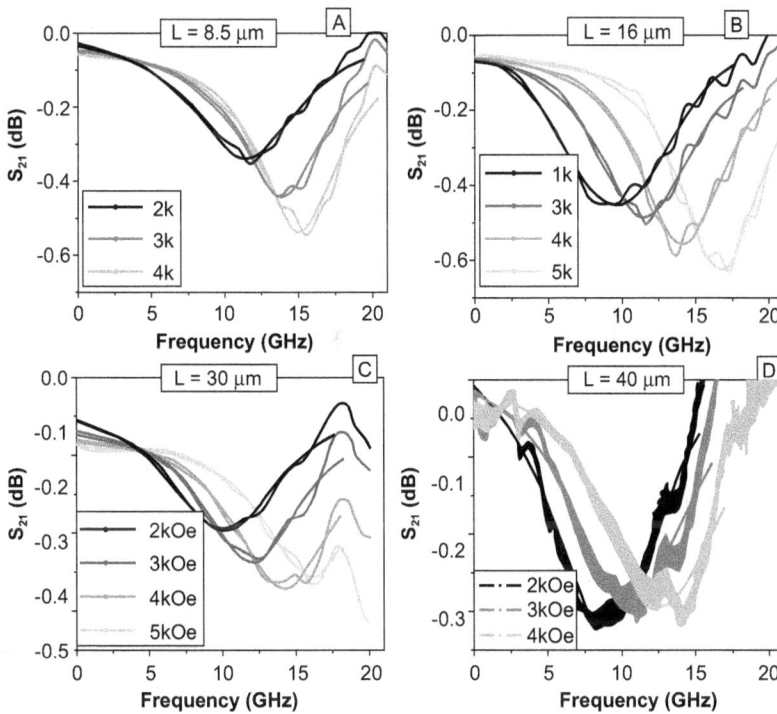

**FIGURE 3.10** (a–d) Transmission response measured by the VNA-FMR as a function of frequency for different Ni NW samples of length 8.5, 16, 30 and 40 μm with fields applied parallel to the easy axis. Solid lines represent the Lorentzian fits to the experimental data.

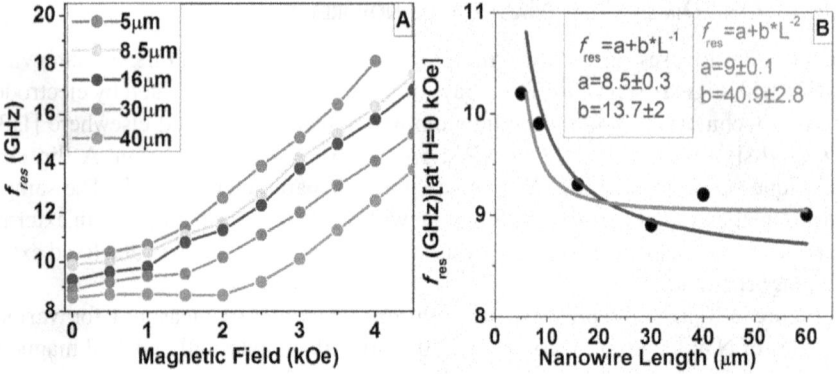

**FIGURE 3.11** (a) Resonance frequency as a function of applied magnetic field for various length samples with field parallel to the easy axis of wires. (b) Zero field $f_{res}$ versus nanowire lengths. The solid dot represents experimental data. The two solid lines are fit to $L^{-1}$ (blue line) and $L^{-2}$ (red line) function.

mainly consists of magnetoelastic, magnetocrystalline and shape anisotropy which contributes to total anisotropy field. Since the one-dimensional nanocylinders have an eminently large aspect ratio, due to their large-shaped anisotropy created along the wire geometry, they behave as unidimensional permanent nanomagnets for low-static magnetic fields. This trademark of the NW system motivates the ferromagnetic resonance to occur even without applying any external magnetic field [114]. Further, along with zero field applications, these NWs can be tuned by the application of magnetic fields.

Figure 3.11(a) reveals the $f_{res}$ $(H_0)$ data for all the fabricated NW samples. The experimental values of zero bias field resonance frequency $f_{res}$ $(H = 0)$ as a function of NW lengths are shown in Figure 3.11(b). It is observed that with the increase of NW length from 5 to 60 $\mu$ m, the $f_{res}$ $(H = 0)$ is decreased from 10.2 to 8.8 GHz. From the results, it can be elucidated that the resonance frequency $f_{res}$ can be enhanced by one and a quarter times with the decrease in the length $(L)$ of the NWs from 40 to 5 $\mu$ m. Thus, by tailoring the NWs length, the $f_{res}$ $(H_0)$ can be adjusted over a small range of values.

The zero-field data (Figure 3.11(b)) show that $H_{eff}$ decreased from 3.6 kOe for Ni1 to 3.15 for Ni6 NW sample. This value of $H_{eff}$ is larger than the demagnetization field $H_d = 2\pi M_s = 3.05 kOe$ for Ni for an infinite cylinder. The reason for large value of $H_{eff}$ may be due to Y-shaped NWs fabricated in the anodic alumina templates which are in unsaturated states as not completely parallel to the applied $H_0$. One-dimensional nanowires give rise to the shape anisotropy which further enhances the effective field. In Figure 3.11(b) solid lines show that the $L^{-1}$ and $L^{-2}$ fit the experimental data. The $L^{-2}$ fits very closely to the experimental data. The resonance frequency increases due to an upward shift of $H_{eff}$ since the applied magnetic field is fixed [see (14)]. The experimental data observed for $H_{eff}$ values as a function of the length of NWs are fitted to a power function. We can express the change of resonance frequency $(\delta f_{res})$ following Eq. (3.14) as $\delta f_{res} \propto H_{eff}$. The effective field varies as an

$L^{-2}$ function, therefore $\delta f_{res}$ also varies as $L^{-2}$ function, given by $\delta f_{res} = a + b \times L^{-2}$ (see Figure 3.11(b)).

The resonance frequency $f_{res}$ measured for Ni NW samples for the external magnetic field applied along the easy axis of the NWs is shown in Figure 3.12(a)–(d). It is observed that at low fields the resonance frequency is almost constant with the increase of $H_0$. After an applied field of ~1 kOe, the resonance frequency increases linearly with the magnetic field. We observed two regimes of frequency field data for parallel configuration of NWs axes which may be due to Y-shaped branches of NWs [71], as these are canted at an angle of 10° with the external magnetic field. A numerical calculation for parallel configuration of NWs shows a constant value of $f_{res}$ $(H_0)$ with the magnetization vector $(\vec{M})$ canted at an angle of 10° with $H_0$ until 0.5 kOe field. Another factor may be the unsaturated state of magnetization due to random orientation of $\vec{M}$ with $H_0$ at very low fields resulting in constant data up to 1–1.5 kOe. Using Eq. (3.14), we fitted the experimental data of $f_{res}$ $(H_0)$ all NWs samples with fitting parameters as $H_{eff}$ and g-factor. From the fitted results we observed that for an increase of NWs lengths from 5 to 60 $\mu$m, the g-factor increases from 2.0 to 2.1, respectively.

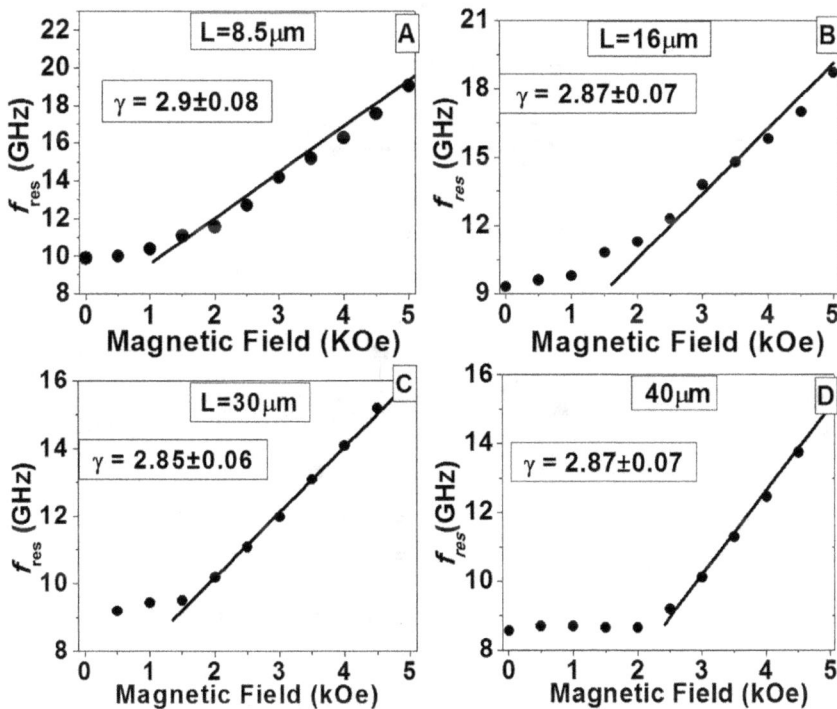

FIGURE 3.12 (a–d) The FMR resonance frequency for NWs samples of different length 8.5, 16, 30 and 40 $\mu$m as a function of magnetic field applied parallel to NWs axes (black symbols). Solid lines are theoretical fits to experimental data.

Figure 3.13(a) shows the frequency linewidth behaviour for different Ni NW samples of various lengths with the magnetic field applied parallel to the wire axis. We derived the frequency linewidths from the Lorentzian fits to the experimental data. In comparison to continuous Ni film, we observed larger frequency linewidth. Various factors contribute for the broadening of $\Delta f_{res}$ such as distributions in NW length, Y-segment growth in AAO template and Y-segment diameters (d, $d_1$ and $d_2$ are branching diameters) as reported in our work [103]. We observed that with the increase of NWs length $\Delta f_{res}$ increases. Frequency linewidth $\Delta f_{res}$ is large at low field for all NW lengths samples. This is because:

1. The sample is in an unsaturated state at zero and very low fields. Therefore, even though the axis of the NWs is parallel to the applied static magnetic field (H), their magnetization is randomly oriented.

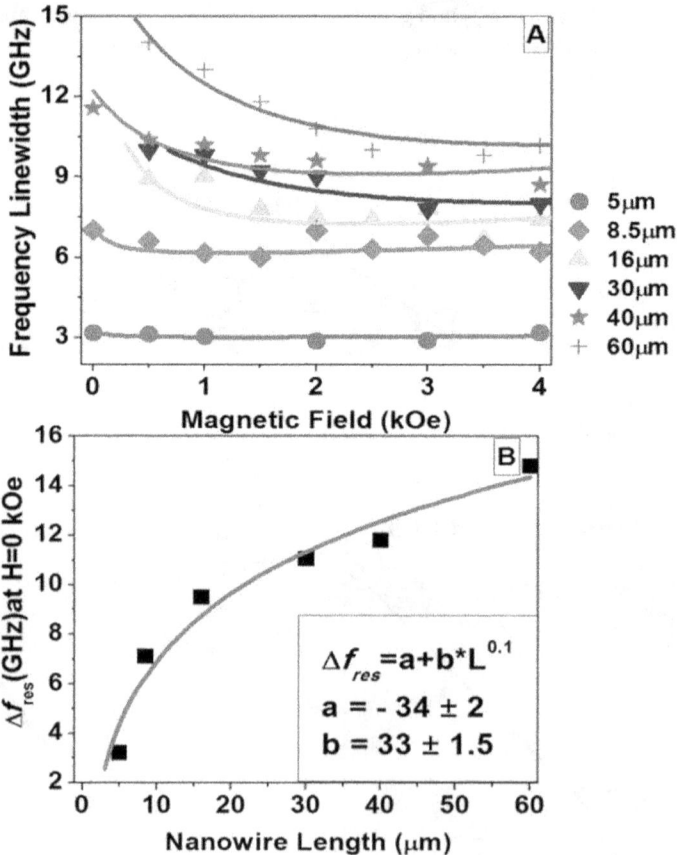

FIGURE 3.13  (a) Frequency linewidth for various length NW samples as a function of magnetic field. (b) Zero field $\Delta f_{res}$ versus nanowire lengths. The solid line is a fit to power function [($\Delta f_{res} = a + b * L^{0.1}$].

2. The magnetization $\vec{M}$ for low or moderate fields (~1 kOe) can be assumed to be along the static magnetic field ($H$) direction in the long arm of the NWs. For short arms ($d_1$ and $d_2$), however, magnetization is aligned at an angle to the static magnetic field. Therefore, at low frequency it corresponds to second resonance very close to main resonance frequency ($f_{res}$).
3. One of the other reasons is that the interwire distance between these Y-segment NWs changes, which causes the resonance frequency to be different in the branches.

The addition of multiple resonances near main resonance is the main reason for the broadening of resonance linewidth in our samples. The resonance linewidth initially decreases with the increase of magnetic field. For higher length samples, it is observed that $\Delta f_{res}$ initially decreases with the field and then subsequently increases at a large field. Figure 3.13(B) shows $\Delta f_{res}$ as a function of NWs length at 0 kOe magnetic field. The solid line is a fit to power function [$\Delta f_{res}= a + b * L^{0.1}$].

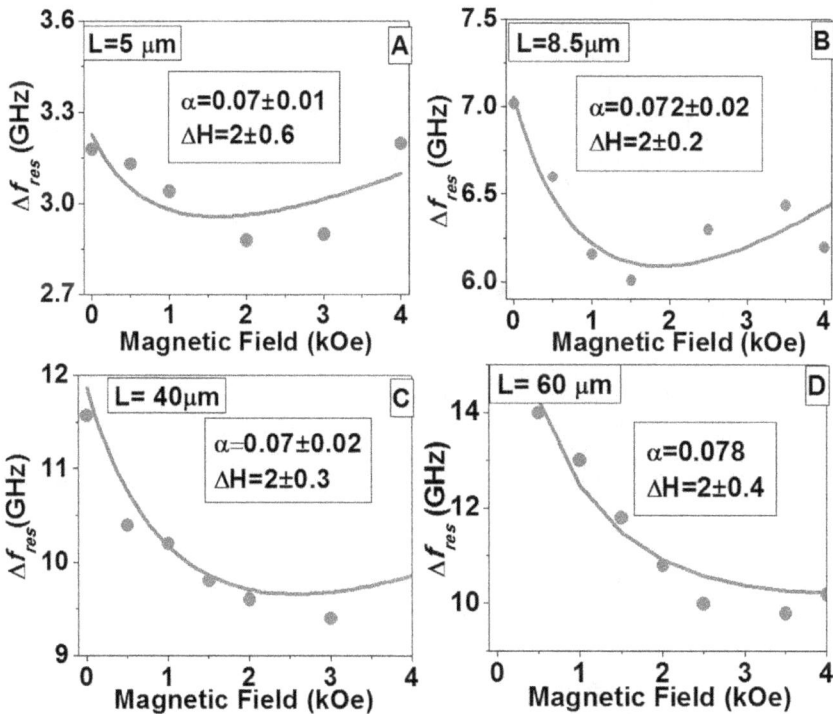

**FIGURE 3.14** (a–d) Frequency linewidth for various NWs of length 5, 8.5, 40 and 60 μm for static magnetic field applied parallel to nanowire axes. The solid lines are theoretical fits from Eq. (3.18).

The frequency linewidth with static magnetic field for various NWs samples is shown in Figure 3.14(a)–(d). Significantly larger linewidths have been observed for longer length NW samples. For all NW samples, at higher frequencies a distinctly narrower frequency linewidth is seen. For a wire-shaped magnetic element, the resonance frequency can be obtained accurately by the FMR condition as given in Eq. (3.14). By differentiating Eq. (3.14), one can obtain the relation between the frequency linewidth and field linewidth as [112]

$$\Delta f_{res} = \frac{\partial f_{res}}{\partial H} \Delta H_0 \qquad (3.18)$$

Thus, the frequency linewidth depends basically on two factors: the slope $\frac{\partial f_{res}}{\partial H}$ and the precise form of $\Delta H_0$. A larger $\Delta \omega$ would be expected if we consider an infinitely long Ni film with $N_y = 0$. In such a case, the slope will be large at $H_0 = 0$ and resonance frequency would be zero, which is the fundamental reason for frequency linewidth to be large at low fields and become smaller at higher fields. This corresponds to a wire-shaped magnetic element, which agrees with the experimental results [113–114]. The theoretical fits for the resonance linewidth are obtained using Eq. (3.18) in conjunction with Eq. (3.14) and are shown by solid lines in Figure 3.14(a)–(d).

Various parameters like the Gilbert damping parameter $\alpha$, zero-frequency offset $\Delta H_0$ and the g-factors are derived from the theoretical fits. Figure 3.15 shows the variation of damping parameter $\alpha$ and g-factor as a function of NW length (L). It

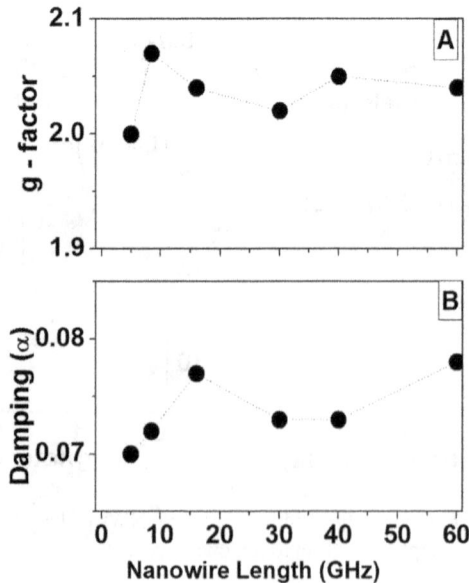

**FIGURE 3.15**  Gyromagnetic ratio ($\gamma$) and damping constant ($\alpha$) as a function of length of nanowires. The lines are guide to the eye.

is observed that with the increase in length of NWs both parameters $\alpha$ and g-factor increases a little.

The extrinsic contribution to linewidth may be the reason for large value of $\alpha$. The frequency linewidth can take a form

$$\Delta f_{res} = \Delta f_{int} + \Delta f_{ext} \tag{3.19}$$

where = $\Delta f_{int}$ and $\Delta f_{ext}$ are intrinsic and extrinsic contributions to linewidth, respectively. Considering bulk value of Ni, $\Delta f_{int}$ is observed to be 0.2 GHz. The various parameters such as g-factor = 2.2, Gilbert damping ($\alpha$ = 0.02) and saturation magnetization $4\pi M_s$ = 6 kOe are derived from the theoretical fittings to experimental data. The dominant contribution to linewidth may be attributed to the extrinsic factors which arises due to the distributions in NWs length and diameters (d, $d_1$ and $d_2$) in Y-shaped sections and variation of $\Delta H_{eff}$ and $\Delta\theta_H$. The variation in effective field $\Delta H_{eff}$ is due to the distribution of dipolar interaction field since both $4\pi M_s$ and shape anisotropy for the wire geometry are constant. Therefore, the variation in interwire distance and in the wire radius in the alumina matrix may contribute to the dipolar interactions. Shore et al. reported that with high saturation magnetization, the specific absorption rate (SAR) value increases [115]. So, we observed that for low-length nanowire sample Ni1 the SAR value is maximum and for Ni6 NW sample SAR value decreases. The dipolar interaction between the nanowires can also affect the magnetization reversal switching. Higher dipolar interaction indicates that the nanowires are close to each other which results in reducing the SAR value.

## 3.5  BIOMEDICAL APPLICATIONS

Cancer, being the leading cause of death, becomes a curse on humanity, with 19.3 million new cancer cases occurring in 2020 as reported by GLOBOCAN [116]. For the treatment of cancer, many conventional techniques have been employed such as chemotherapy, radiation therapy and surgery. Among these techniques, radiotherapy and chemotherapy methods faced severe side effects such as fatigue, alopecia, damage to healthy tissues and multidrug resistance (MDR) [117–119]. However, magnetic hyperthermia technique is a promising non-invasive method to ameliorate cancer by injecting magnetic materials into the tumour tissue which produce by absorbing rf power from oscillating magnetic field [120–121]. The heat produced due to Neel and Brownian relaxation loss processes destroys cancerous cells efficiently without harming the nearby healthy cells [122–123]. For biomedical applications, iron oxide nanoparticles have been explored widely due to their high specific absorption rate and nominal cytotoxicity profile in various applications of magnetic resonance imaging (MRI) contrast enhancement, targeted drug delivery and hyperthermia therapy [124–128]. However, bare iron oxide nanoparticles can accumulate to form bigger clumps preventing immobilization through the system. To overcome the aforementioned issues, iron oxide nanoparticles have to be coated with some suitable surfactants which increase the colloidal stability. Graphene oxide (GO) is one of such surfactants which exhibit remarkable hydrophilic properties, making it suitable for improving the colloidal stability and biocompatibility. We showed in the work an easy and cost-effective electrochemical method for the fabrication of GO-$Fe_3O_4$

nanocomposites. We demonstrated that the SAR, heating efficiency and cytotoxicity have been improved by immobilizing iron oxide on GO sheets. We showed *in vivo* and *in vitro* measurements on A549 cells. The synthesis of nanocomposites has been reported elsewhere [129].

For biomedical applications, the hydrodynamic diameter of particles, electronegativity and UV-visible absorption was determined using dynamic light scattering (DLS), zeta potential and UV-vis spectroscopy. Figure 3.16 (a) shows the hydrodynamic diameter as a function of time of Fe3O4 nanoparticles and composites. It is visible from the graph that the hydrodynamic diameter of composites remains constant up to 48 hours, whereas for $Fe_3O_4$ nanoparticles it increases more than double within 32 hours, hinting at the agglomeration of iron oxide nanoparticles due to magnetic-dipole interaction. We observed that with increase of time $Fe_3O_4$ diameter decreases due to sedimentation of accumulated particles under gravity. The normalized hydrodynamic diameter distribution peak intensity as a function of time for composite and iron oxide nanoparticles is shown in Figure 3.16 (b). We see that the intensity of $Fe_3O_4$ dispersion decreases continuously due to rapid agglomeration, whereas the intensity of nanocomposites does not change much even for 48 hours which depicts uniform distribution and colloidal stability.

The long-term stability of NPs is obtained by using the zeta potential technique. The unstable suspension demonstrates zeta potential values in the range of −30 mV and +30 mV, whereas for stable suspension zeta potential is either higher than +30 mV or lower than −30 mV [130]. For $Fe_3O_4$ nanoparticles, the zeta potential value was observed to be +5 mV, indicating unstable suspension. For GO-$Fe_3O_4$ nanocomposites the zeta potential was obtained to be −40 mV, indicating good stable dispersion. The Prussian blue staining method used to visualize the iron ions inside the cells revealed the internalization of GO-$Fe_3O_4$ nanocomposites. For this, the cells were imposed with nanocomposites with different incubation time (24 hours and 48 hours) and dose concentration. Figure 3.17 showed the intracellular uptake of nanocomposites for A549 cells seen with the help of a light microscope at a magnification of

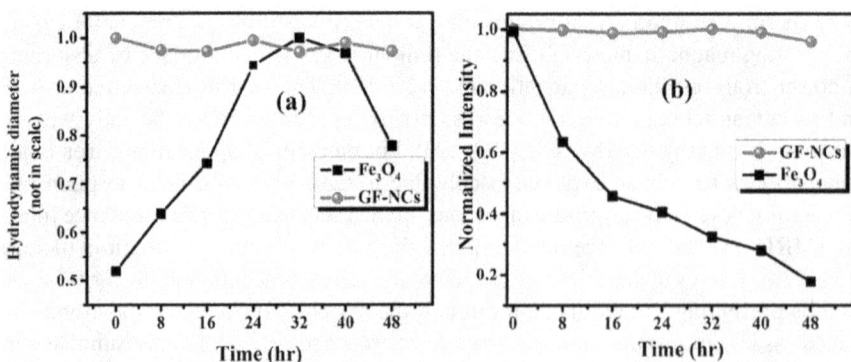

**FIGURE 3.16**    (a) Hydrodynamic diameter as a function of time for $Fe_3O_4$ nanoparticles and GO-$Fe_3O_4$ nanocomposites; (b) normalized intensity as a function of time.

10X. It is observed that as the incubation time is increased, more nanocomposites are internalized by the cells as seen in the formation of clusters of blue granules in the cytoplasm (Figure 3.17 (b and e) and Figure 3.17 (c and f)). Also, more GO-Fe$_3$O$_4$ is internalized by cells for higher concentration as shown in Figure 3.17 (e and f).

The cytotoxicity measurements were done using an MTT (3-(4,5-Dimethylthiazol-2-yl)-2,5-Diphenyltetrazolium Bromide) assay with nanocomposite synthesized on A549 cells as shown in Figure 3.18. We observed low cytotoxicity for different concentrations (25–200 μg/ml) of nanocomposites incubated for 24 hours and 48 hours. The cell viability of nanocomposites at different time period incubation at 200 μg/ml concentration was observed to be around 85% and 75% for most of the samples, indicating good biocompatibility.

For the measurement of SAR values, the heating behaviour of 200 μg/ml nanocomposites subjected to alternating magnetic field (AMF) was studied, as shown in Figure 3.19. From the calorimetric measurements done by the hyperthermia system which generates AMF, we measured the time-dependent gradient increase in temperature for nanocomposite suspensions as shown in Figure 3.19(a). The observed SAR values calculated using the linear slope method at fixed concentration of 200 μg/ml for different samples were 334, 351, 360, 380 and 407 W/gm [Figure 3.19(b)] and measured by the Box–Lucas method were 427, 449, 452, 518 and 543 W/gm [Figure 3.19(c)]. The Box–Lucas method gives higher values as it considers the temperature rise at a higher time scale. The intrinsic power loss (IPL) which defines the heating transformation ability is shown in Figure 3.19(b–c) for all nanocomposites. The IPL values were observed to be 3.56, 3.75, 3.85, 4.1 and 4.34 nHm$^2$Kg$^{-1}$ from linear slope method and 4.7, 4.95, 4.96, 5.71 and 5.98 nHm$^2$Kg$^{-1}$ from the Box–Lucas

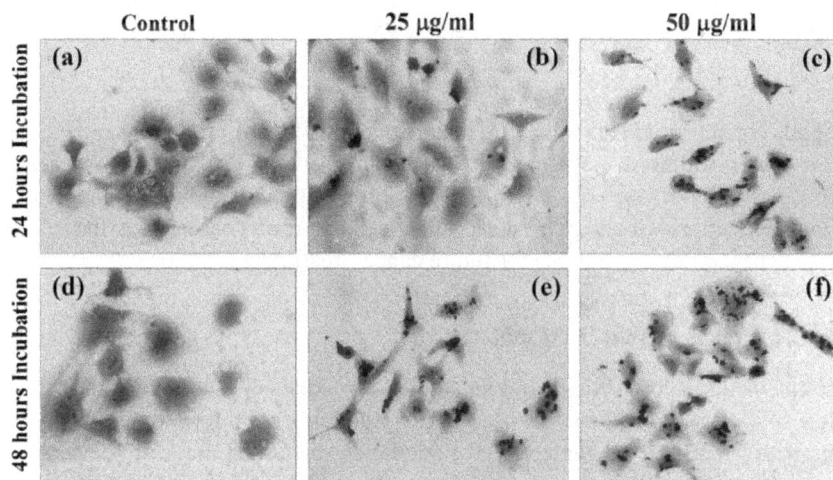

**FIGURE 3.17** Microscope image of Prussian blue staining of A549 cells (a–d) unlabelled cells and (b–e) GO-Fe3O4 nanocomposite incubated cells for 24 hours and 48 hours with 25 μg/ml concentrations, respectively. (c–f) GO-Fe3O4 nanocomposites incubated cells for 24 hours and 48 hours with 50 μg/ml concentrations, respectively.

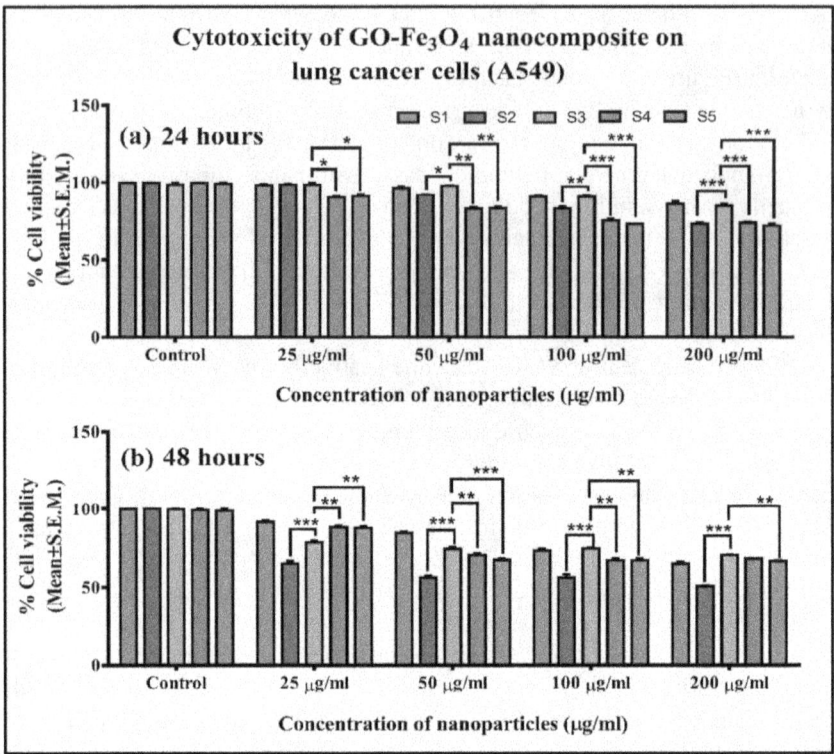

**FIGURE 3.18**   (a) Cell viability measurements of A549 cells imposed with nanocomposites of different concentrations (25, 50, 100, 150, 200 µg/ml) for 24 hours; (b) A549 cells imposed with nanocomposites of different concentrations (25, 50, 100, 150, 200 µg/ml) for 48 hours.

method. The retrieved IPL values are in agreement with FDA protocol. It is observed that as the crystallite size of nanocomposites increased, the IPL value also increased accordingly. Hence, we concluded that with proposed nanocomposites the heating efficiency is improved and up to 80% cancer cell death was observed in the *in vitro* experiments, which are suitable for future cancer treatments.

## 3.6   FUTURE APPLICATIONS

Ferromagnetic nanostructures can be studied for exploring the field of magnonics where spin waves (magnons) are used to carry and process information of spin currents with no dissipative Joule effect. Periodically patterned magnetic devices can open fascinating outlooks for the operation of spin waves at nanoscale, which can lead to the recognition of novel metamaterials. Various microwave devices such as circulators, gyrators, and resonators can be fabricated using these nanostructures. These devices can be useful in communication systems. The circulator ensures the isolation between different parts in transmitting and receiving modules. This

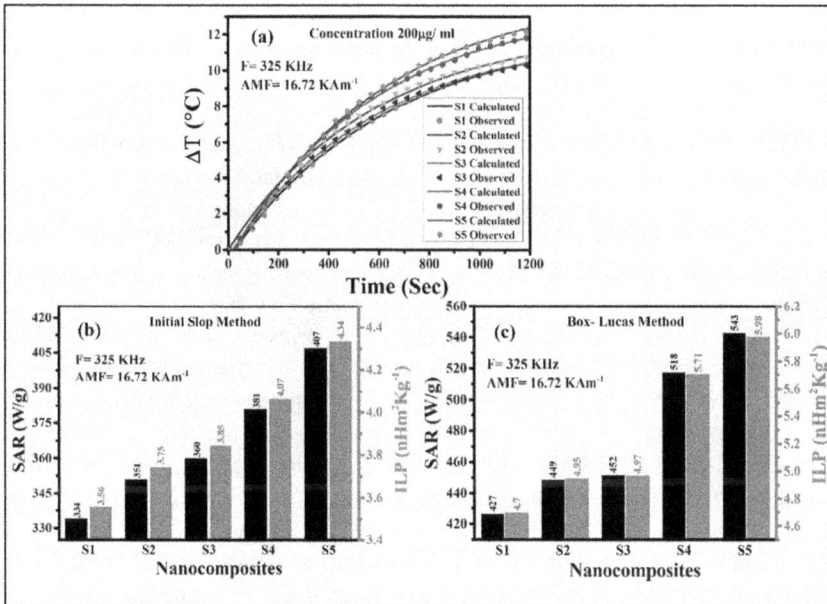

**FIGURE 3.19** Temperature of GO-$Fe_3O_4$ nanocomposites of 200 μg/ml concentration under AC magnetic field (a) AMF-dependent temperature kinetics, (b) SAR and IPL values for GO-$Fe_3O_4$ nanocomposites in suspension using the linear slope method, and (c) SAR and IPL values for GO-$Fe_3O_4$ nanocomposites in suspension using the Box–Lucas method.

circulation operation is achieved by the non-reciprocal behaviour of ferromagnetic materials. Ferromagnetic nanostructures are thus promising candidates for miniaturization of devices in next-generation technology and advantageous for biomedical applications.

## 3.7 CONCLUSION(S)

Recently, a lot of research has focused on the magnetization dynamics of nanosystems due to its potential applications in spintronics, specifically aiming towards design of GMR biosensors and cancer treatment through radiofrequency hyperthermia. These nanostructures can also be useful for signal processing communications and magnetoelectric devices. The development in modern fabrication techniques has opened a broad area for new possibilities in future spintronics applications. The ultrafast magnetization dynamics can be probed by various techniques like conventional FMR, time-resolved magneto-optical Kerr effect microscopy and Brillouin light scattering respectively. These broadband ferromagnetic resonance techniques allow one to follow the changes in the micromagnetic structure induced by changing the applied magnetic field. It also offers the possibility to sweep the frequency of the oscillating magnetic field that excites the precession of the magnetization at a constant external bias field. Here, we have reviewed uniform mode of magnetization dynamics using VNA-FMR technique in magnetic nanostructures such as trilayer thin films, nanosized stripes and one-dimensional nanowires. We discussed a theoretical background of magnetization dynamics in magnetic systems for better understanding. This involves the complete apprehension of LLG equations and the various damping mechanisms contributing to magnetic precession. We observed the magnetic excitation frequencies for Fe/Al/Fe trilayer films in the frequency domain with the help of VNA-FMR and BLS techniques. The acoustic and optic mode frequencies and their corresponding linewidths are determined by dispersion relation both at $k = 0$ (uniform precession mode) and $k \neq 0$ mode. We reported a very strong antiferromagnetic exchange coupling Al spacers between ferromagnetic Fe layers. The detailed discussion of ferromagnetic resonance in Permalloy nanostrips along the length and width of the strips is done. It is shown that for longer strips (1:10 and 1:333), when the field is applied along the width of the strip, a soft mode behaviour occurs. The frequency linewidths also change with the change in aspect ratio. The magnetization dynamics of one-dimensional Ni NWs of various lengths were done by a flip-chip based FMR technique. The resonance frequency increases linearly with the increase of magnetic field along the easy axis of NWs. The frequency linewidth is also larger for longer length samples. Various material parameters such as gyromagnetic ratio, Gilbert damping and magnetic anisotropies are determined from the same. These observed phenomena in various nanostructures imply that magnetic nanosystems could be the potential candidates for radiofrequency-assisted cancer treatment. In addition, these nanostructures can be helpful for microwave-assisted switching as well as magneto-electronic devices.

## REFERENCES

[1] Sarma SD: Spintronics: A new class of device based on electron spin, rather than on charge, may yield the next generation of microelectronics. *American Scientist*. 2001; 89(6):516–523.

[2] Awschalom DD, Flatté ME: Challenges for semiconductor spintronics. *Nature Physics*. 2007; 3(3):153.

[3] Zhao W, Prenat *G: Spintronics-based computing*. Berlin, Germany: Springer; 2015.

[4] Wolf SA, Awschalom DD, Buhrman RA, Daughton JM, Von Molnar S, Roukes ML, Chtchelkanova AY, Treger DM: Spintronics: A spin-based electronics vision for the future. *Science*. 2001; 294(5546):1488–1495.

[5] Liu JP, Fullerton E, Gutfleisch O, Sellmyer *DJ: Nanoscale magnetic materials and applications*. Springer Science and Business Media; 2009.

[6] Jiles DC: Recent advances and future directions in magnetic materials. *Acta Materialia*. 2003; 51(19):5907–39.

[7] Brataas A, Kent AD, Ohno H: Current-induced torques in magnetic materials. *Nature Materials*. 2012; 11(5):372.

[8] Žutić I, Fabian J, Sarma SD: Spintronics: Fundamentals and applications. *Reviews of Modern Physics*. 2004; 76(2):323.

[9] Chumak AV, Vasyuchka VI, Serga AA, Hillebrands B: Magnon spintronics. *Nature Physics*. 2015; 11(6):453.

[10] Baibich MN, Broto JM, Fert A, Van Dau FN, Petroff F, Etienne P, Creuzet G, Friederich A, Chazelas J: Giant magnetoresistance of (001) Fe/(001) Cr magnetic superlattices. *Physical Review Letters*. 1988; 61(21):2472.

[11] Berkowitz AE, Mitchell JR, Carey MJ, Young AP, Zhang S, Spada FE, Parker FT, Hutten A, Thomas G: Giant magnetoresistance in heterogeneous Cu-Co alloys. *Physical Review Letters.* 1992; 68(25):3745.

[12] Parkin SS, Li ZG, Smith DJ: Giant magnetoresistance in antiferromagnetic Co/Cu multilayers. *Applied Physics Letters.* 1991; 58(23):2710–2712.

[13] Van den Berg HA, Coehoorn R, Gijs MA, Grünberg P, Rasing T, Röll K: *Magnetic multilayers and giant magnetoresistance: Fundamentals and industrial applications.* Springer Science & Business Media; 2013.

[14] Fullerton EE, Kelly DM, Guimpel J, Schuller IK, Bruynseraede Y: Roughness and giant magnetoresistance in Fe/Cr superlattices. *Physical Review Letters.* 1992; 68(6):859.

[15] Ju HL, Kwon C, Li Q, Greene RL, Venkatesan T: Giant magnetoresistance in La1−xSrx-MnOz films near room temperature. *Applied Physics Letters.* 1994; 65(16):2108–2110.

[16] Gijs MA, Lenczowski SK, Giesbers JB: Perpendicular giant magnetoresistance of microstructured Fe/Cr magnetic multilayers from 4.2 to 300 K. *Physical Review Letters.* 1993; 70(21):3343.

[17] Tsymbal EY, Pettifor DG: Perspectives of giant magnetoresistance. In *Solid State Physics.* 2001; 56:113–237.

[18] White RL: Giant magnetoresistance: A primer. *IEEE Transactions on Magnetics.* 1992; 28(5):2482–2487.

[19] Kuanr BK, Maat S, Chandrashekariaih S, Veerakumar V, Camley RE, Celinski Z: Determination of exchange and rotational anisotropies in Ir Mn/Fe (t)/Ir Mn exchange coupled structures using dynamic and static techniques: Application to microwave devices. *Journal of Applied Physics.* 2008 Apr 1; 103(7):07C107.

[20] Kuanr AV, Kuanr BK: Influence of layering of NiO/NiFe on exchange bias. In 2005 *IEEE International Magnetics Conference (INTERMAG)* 2005 Apr 4 (pp. 1631–1632). IEEE.

[21] Kuanr BK, Buchmeier M, Buergler DE, Gruenberg P, Camley R, Celinski Z: Dynamic and static measurements on epitaxial Fe/Si/Fe. *Journal of Vacuum Science & Technology A: Vacuum, Surfaces, and Films.* 2003 Jul 30; 21(4):1157–1161.

[22] Buchmeier M, Kuanr BK, Gareev RR, Bürgler DE, Grünberg P: Spin waves in magnetic double layers with strong antiferromagnetic interlayer exchange coupling: Theory and experiment. *Physical Review B.* 2003 May 5; 67(18):184404.

[23] Kuanr BK, Buchmeier M, Gareev RR, Bürgler DE, Schreiber R, Grünberg P: Spin-wave modes and line broadening in strongly coupled epitaxial Fe/Al/Fe and Fe/Si/Fe trilayers observed by Brillouin light scattering. *Journal of Applied Physics.* 2003 Mar 15; 93(6):3427–3434.

[24] Grünberg P, Bürgler DE, Gareev R, Olligs D, Buchmeier M, Breidbach M, Kuanr B, Schreiber R: Experiments on the relation between GMR and interface roughness and on the interlayer exchange coupling across semiconductors. *Journal of Physics D: Applied Physics.* 2002 Sep 13; 35(19):2403.

[25] Kuanr BK, Gokhale S, Vedpathak M, Kuanr AV, Nimtz G: Temperature-dependent impedance anomaly, microwave GMR and exchange-coupling in thin Ni/Cu multilayered films. *Journal of Physics D: Applied Physics.* 2000 Jan 7; 33(1):34.

[26] Kuanr BK, Gruenberg P: Frequency variation of microwave GMR and exchange coupling of epitaxial Fe/Au/Py trilayer ultra-this films. In 1998 *IEEE International Magnetics Conference (INTERMAG)* 1998 Jan 6 (p. 125). IEEE.

[27] Kuanr BK, Kuanr AV, Grünberg P, Nimtz G: Swept-frequency FMR on Fe/Cr/Fe trilayer ultrathin films; microwave giant magnetoresistance. *Physics Letters A.* 1996 Sep 30; 221(3–4):245–252.

[28] Kuanr BK, Kuanr AV: FMR studies on Fe/Cr/Fe trilayer ultrathin films; magneto-crystalline anisotropy and electromagnetic impedance. *Journal of Magnetism and Magnetic Materials.* 1997 Jan 1; 165(1–3):275–279.

[29] Camley RE, Stamps RL: Magnetic multilayers: Spin configurations, excitations and giant magnetoresistance. *Journal of Physics: Condensed Matter.* 1993; 5(23):3727.

[30] Mosca DH, Petroff F, Fert A, Schroeder PA, Pratt Jr WP, Laloee R: Oscillatory interlayer coupling and giant magnetoresistance in Co/Cu multilayers. *Journal of Magnetism and Magnetic Materials.* 1991; 94(1–2):L1–L5.

[31] Miyazaki T, Tezuka N: Giant magnetic tunneling effect in Fe/Al2O3/Fe junction. *Journal of Magnetism and Magnetic Materials.* 1995; 139(3):L231–L234.

[32] Xiao JQ, Jiang JS, Chien CL: Giant magnetoresistance in nonmultilayer magnetic systems. *Physical Review Letters.* 1992; 68(25):3749.

[33] Baumgart P, Gurney BA, Wilhoit DR, Nguyen T, Dieny B, Speriosu VS: The role of spin-dependent impurity scattering in Fe/Cr giant magnetoresistance multilayers. *Journal of Applied Physics.* 1991; 69(8):4792–4794.

[34] Berkowitz AE, Mitchell JR, Carey MJ, Young AP, Zhang S, Spada FE, Parker FT, Hutten A, Thomas G: Giant magnetoresistance in heterogeneous Cu-Co alloys. *Physical Review Letters.* 1992; 68(25):3745.

[35] Dieny B, Speriosu VS, Parkin SS, Gurney BA, Wilhoit DR, Mauri D: Giant magneto-resistive in soft ferromagnetic multilayers. *Physical Review B.* 1991; 43(1):1297.

[36] Beaurepaire E, Merle JC, Daunois A, Bigot JY: Ultrafast spin dynamics in ferromagnetic nickel. *Physical Review Letters.* 1996; 76(22):4250.

[37] Hillebrands B, Ounadjela *K: Spin dynamics in confined magnetic structures I.* Springer Science & Business Media; 2003.

[38] Braun PF, Marie X, Lombez L, Urbaszek B, Amand T, Renucci P, Kalevich VK, Kavokin KV, Krebs O, Voisin P, Masumoto Y: Direct observation of the electron spin relaxation induced by nuclei in quantum dots. *Physical Review Letters.* 2005; 94(11):116601.

[39] Ando K, Takahashi S, Harii K, Sasage K, Ieda J, Maekawa S, Saitoh E: Electric manipulation of spin relaxation using the spin Hall effect. *Physical Review Letters.* 2008; 101(3):036601.

[40] Kuanr BK, Kuanr AV, Camley RE, Celinski Z: Magnetization relaxation in sputtered thin Fe films: An FMR study. *IEEE Transactions on Magnetics.* 2006; 42(10):2930–2932.

[41] Levitt *MH: Spin dynamics: Basics of nuclear magnetic resonance.* John Wiley & Sons; 2001 Nov 6.

[42] Hermenau J, Ternes M, Steinbrecher M, Wiesendanger R, Wiebe J: Long spin-relaxation times in a transition-metal atom in direct contact to a metal substrate. *Nano Letters.* 2018; 18(3):1978–1983.

[43] De Teresa JM, Ritter C, Ibarra MR, Algarabel PA, Garcia-Munoz JL, Blasco J, Garcia J, Marquina C: Charge localization, magnetic order, structural behavior, and spin dynamics of (La–Tb)2/3Ca1/3MnO3 manganese perovskites probed by neutron diffraction and muon spin relaxation. *Physical Review B.* 1997; 56(6):3317.

[44] Leggett AJ, Takagi S: Orientational dynamics of superfluid 3He: A "two-fluid" model. I. Spin dynamics with relaxation. *Annals of Physics.* 1977; 106(1):79–133.

[45] Schneider DJ, Freed JH: Spin relaxation and motional dynamics. *Lasers, Molecules and Methods.* 1989; 73:387–528.

[46] Nana AB, Marimuthu T, Kondiah PP, Choonara YE, Du Toit LC, Pillay V: Multifunctional magnetic nanowires: Design, fabrication, and future prospects as cancer therapeutics. *Cancers.* 2019 Dec;11(12):1956.

[47] Anastasova EI, Prilepskii AY, Fakhardo AF, Drozdov AS, Vinogradov VV: Magnetite nanocontainers: Toward injectable highly magnetic materials for targeted drug delivery. *ACS Applied Materials & Interfaces.* 2018 Aug 23; 10(36):30040–30044.

[48] Park JH, Yang SH, Lee J, Ko EH, Hong D, Choi IS: Nanocoating of single cells: From maintenance of cell viability to manipulation of cellular activities. *Advanced Materials.* 2014 Apr; 26(13):2001–2010.

[49] Zamani Kouhpanji MR, Stadler BJ: Magnetic nanowires for nanobarcoding and beyond. *Sensors*. 2021 Jan; 21(13):4573.

[50] Ye H, Wang Y, Xu D, Liu X, Liu S, Ma X: Design and fabrication of micro/nano-motors for environmental and sensing applications. *Applied Materials Today*. 2021 Jun 1; 23:101007.

[51] Ambhorkar P, Wang Z, Ko H, Lee S, Koo KI, Kim K, Cho DI: Nanowire-based biosensors: From growth to applications. *Micromachines*. 2018 Dec; 9(12):679.

[52] Mukhtar A, Wu K, Cao X: Magnetic nanowires in biomedical applications. *Nanotechnology*. 2020 Jul 29; 31(43):433001.

[53] Lin WS, Lin HM, Chen HH, Hwu YK, Chiou YJ: Shape effects of iron nanowires on hyperthermia treatment. *Journal of Nanomaterials*. 2013 Jan 1; 2013.

[54] Grünberg PE, Schreiber R, Pang Y, Brodsky MB, Sowers H: Layered magnetic structures: Evidence for antiferromagnetic coupling of Fe layers across Cr interlayers. *Physical Review Letters*. 1986; 57(19):2442.

[55] Xiong ZH, Wu D, Vardeny ZV, Shi J: Giant magnetoresistance in organic spin-valves. *Nature*. 2004; 427(6977):821.

[56] Park BG, Wunderlich J, Martí X, Holý V, Kurosaki Y, Yamada M, Yamamoto H, Nishide A, Hayakawa J, Takahashi H, Shick AB: A spin-valve-like magnetoresistance of an antiferromagnet-based tunnel junction. *Nature Materials*. 2011; 10(5):347.

[57] Zhong W, Liu W, Au CT, Du YW: Tunnelling magnetoresistance of double perovskite Sr2FeMoO6 enhanced by grain boundary adjustment. *Nanotechnology*. 2005; 17(1):250.

[58] Borisov K, Betto D, Lau YC, Fowley C, Titova A, Thiyagarajah N, Atcheson G, Lindner J, Deac AM, Coey JM, Stamenov P: Tunnelling magnetoresistance of the half-metallic compensated ferrimagnet Mn2RuxGa. *Applied Physics Letters*. 2016; 108(19):192407.

[59] Hanneken C, Otte F, Kubetzka A, Dupé B, Romming N, Von Bergmann K, Wiesendanger R, Heinze S: Electrical detection of magnetic skyrmions by tunnelling non-collinear magnetoresistance. *Nature Nanotechnology*. 2015; 10(12):1039.

[60] Nag A, Ray S: Magnetoresistance stories of double perovskites. *Pramana*. 2015; 84(6):967–975.

[61] Ren Y, Li ZZ, Xiao MW, Hu A: Oscillation effect and sign-change behaviour of the bias-dependent tunnelling magnetoresistance in ferromagnetic junctions. *Journal of Physics: Condensed Matter*. 2005; 17(26):4121.

[62] Tsymbal EY, Mryasov ON, LeClair PR: Spin-dependent tunnelling in magnetic tunnel junctions. *Journal of Physics: Condensed Matter*. 2003; 15(4):R109.

[63] Miyazaki T, Tezuka N: Giant magnetic tunneling effect in Fe/Al2O3/Fe junction. *Journal of Magnetism and Magnetic Materials*. 1995; 139(3):L231–L234.

[64] Deac AM, Fukushima A, Kubota H, Maehara H, Suzuki Y, Yuasa S, Nagamine Y, Tsunekawa K, Djayaprawira DD, Watanabe N: Bias-driven high-power microwave emission from MgO-based tunnel magnetoresistance devices. *Nature Physics*. 2008; 4(10):803.

[65] Parkin SS, Kaiser C, Panchula A, Rice PM, Hughes B, Samant M, Yang SH: Giant tunnelling magnetoresistance at room temperature with MgO (100) tunnel barriers. *Nature Materials*. 2004; 3(12):862.

[66] Bean JJ, Saito M, Fukami S, Sato H, Ikeda S, Ohno H, Ikuhara Y, McKenna KP: Atomic structure and electronic properties of MgO grain boundaries in tunnelling magnetoresistive devices. *Scientific Reports*. 2017; 7:45594.

[67] Graham DL, Ferreira HA, Freitas PP: Magnetoresistive-based biosensors and biochips. *TRENDS in Biotechnology*. 2004; 22(9):455–462.

[68] Zhang K, Li M, Chen YH, Wang HC, Zhao T: Inventors; Headway Technologies Inc, assignee. Novel way to reduce the ordering temperature for Co2MnSi-like Heusler alloys for CPP, TMR, MRAM, or other spintronics device applications. *United States patent application US 11/472, 126*, 2007.

[69] Tsang D: Applied Spintronics Tech Inc, assignee. Magnetic tunneling junction cell array with shared reference layer for MRAM applications. *United States patent US* 6,963,500, 2005.

[70] Zhang K, Li M, Zhou Y, Oikawa S, Yamada K, Koui K: Toshiba Corp, TDK Corp, assignee. Thin seeded Co/Ni multilayer film with perpendicular anisotropy for spin-tronic device applications. *United States patent US* 8,064,244, 2011.

[71] Khvalkovskiy AV, Apalkov D, Watts S, Chepulskii R, Beach RS, Ong A, Tang X, Driskill-Smith A, Butler WH, Visscher PB, Lottis D: Basic principles of STT-MRAM cell operation in memory arrays. *Journal of Physics D: Applied Physics*. 2013; 46(7):074001.

[72] Dieny B, Sousa RC, Herault J, Papusoi C, Prenat G, Ebels U, Houssameddine D, Rodmacq B, Auffret S, Buda-Prejbeanu LD, Cyrille MC: Spin-transfer effect and its use in spintronic components. *International Journal of Nanotechnology*. 2010; 7(4–8):591–614.

[73] Jabeur K, Di Pendina G, Bernard-Granger F, Prenat G: Spin orbit torque non-volatile flip-flop for high speed and low energy applications. *IEEE Electron Device Letters*. 2014; 35(3):408–410.

[74] Wang KL, Alzate JG, Amiri PK: Low-power non-volatile spintronic memory: STT-RAM and beyond. *Journal of Physics D: Applied Physics*. 2013; 46(7):074003.

[75] Kittel C: On the theory of ferromagnetic resonance absorption. *Physical Review*. 1948; 73(2):155.

[76] Nozaki Y, Ohta M, Taharazako S, Tateishi K, Yoshimura S, Matsuyama K: Magnetic force microscopy study of microwave-assisted magnetization reversal in submicron-scale ferromagnetic particles. *Applied Physics Letters*. 2007; 91(8):082510.

[77] Yoshioka T, Nozaki T, Seki T, Shiraishi M, Shinjo T, Suzuki Y, Uehara Y: Microwave-assisted magnetization reversal in a perpendicularly magnetized film. *Applied Physics Express*. 2009; 3(1):013002.

[78] Nistor C, Sun K, Wang Z, Wu M, Mathieu C, Hadley M: Observation of microwave-assisted magnetization reversal in Fe65Co35 thin films through ferromagnetic resonance measurements. *Applied Physics Letters*. 2009; 95(1):012504.

[79] Wang Z, Sun K, Tong W, Wu M, Liu M, Sun NX: Competition between pumping and damping in microwave-assisted magnetization reversal in magnetic films. *Physical Review B*. 2010; 81(6):064402.

[80] Taniguchi T, Kubota H: Spin torque oscillator for microwave assisted magnetization reversal. *Japanese Journal of Applied Physics*. 2018; 57(5):053001.

[81] Bigot JY, Vomir M: Ultrafast magnetization dynamics of nanostructures. *Annalen der Physik*. 2013; 525(1–2):2–30.

[82] Camley RE: Magnetization dynamics in thin films and multilayers. *Journal of Magnetism and Magnetic Materials*. 1999; 200(1–3):583–597.

[83] Camara IS, Tacchi S, Garnier LC, Eddrief M, Fortuna F, Carlotti G, Marangolo M: Magnetization dynamics of weak stripe domains in Fe-N thin films: A multi-technique complementary approach. *Journal of Physics: Condensed Matter*. 2017; 29(46):465803.

[84] Cinchetti M, Albaneda MS, Hoffmann D, Roth T, Wüstenberg JP, Krauß M, Andreyev O, Schneider HC, Bauer M, Aeschlimann M: Spin-flip processes and ultrafast magnetization dynamics in Co: Unifying the microscopic and macroscopic view of femtosecond magnetism. *Physical Review Letters*. 2006; 97(17):177201.

[85] Adeyeye AO, Haldar A, Shimon G: Magnetization dynamics of reconfigurable 2D magnonic crystals. *InSpin Wave Confinement*. 2017; 99–137.

[86] Barman A, Haldar A: Time-domain study of magnetization dynamics in magnetic thin films and micro-and nanostructures. *InSolid State Physics* 2014; 65:1–108. Academic Press.

[87] Tserkovnyak Y, Brataas A, Bauer GE, Halperin BI: Nonlocal magnetization dynamics in ferromagnetic heterostructures. *Reviews of Modern Physics*. 2005; 77(4):1375.

[88] Marko D: Magnetostatics and dynamics of ion irradiated NiFe/Ta multilayer films studied by vector network analyser ferromagnetic resonance. *Dissertation*, Dresden University, 2010.

[89] Landau LA, Lifshitz E. 1992. On the theory of the dispersion of magnetic permeability in ferromagnetic bodies. *Perspectives in Theoretical Physics*. Pergamon: 51–65.

[90] Gilbert TL: A Lagrangian formulation of the gyromagnetic equation of the magnetization field. *Physical Review*. 1955; 100:1243.

[91] Sharma M, Pathak S, Sharma M: FMR measurements of magnetic nanostructures. In *Ferromagnetic Resonance-Theory and Applications* 2013. InTech.

[92] Inaba N, Uesaka Y, Nakamura A, Futamoto M, Sugita Y, Narishige S: Damping constants of Co-Cr-Ta and Co-Cr-Pt thin films. *IEEE Transactions on Magnetics*. 1997; 33(5):2989–2991.

[93] Luo C, Feng Z, Fu Y, Zhang W, Wong PK, Kou ZX, Zhai Y, Ding HF, Farle M, Du J, Zhai HR: Enhancement of magnetization damping coefficient of permalloy thin films with dilute Nd dopants. *Physical Review B*. 2014; 89(18):184412.

[94] Heinrich B, Cochran JF, Hasegawa R: FMR linebroadening in metals due to two-magnon scattering. *Journal of Applied Physics*. 1985; 57(8):3690–3692.

[95] Nibarger JP, Lopusnik R, Silva TJ: Damping as a function of pulsed field amplitude and bias field in thin film Permalloy. *Applied Physics Letters*. 2003; 82(13):2112–2114.

[96] Skadsem HJ, Tserkovnyak Y, Brataas A, Bauer GE: Magnetization damping in a local-density approximation. *Physical Review B*. 2007; 75(9):094416.

[97] Martın JI, Nogues J, Liu K, Vicent JL, Schuller IK: Ordered magnetic nanostructures: Fabrication and properties. *Journal of Magnetism and Magnetic Materials*. 2003; 256(1–3):449–501.

[98] Wang SX, Taratorin *AM: Magnetic information storage technology: A volume in the electromagnetism series*. Elsevier; 1999 May 24.

[99] Huang XJ, Choi YK: Chemical sensors based on nanostructured materials. *Sensors and Actuators B: Chemical*. 2007; 122(2):659–671.

[100] Song MM, Song WJ, Bi H, Wang J, Wu WL, Sun J, Yu M: Cytotoxicity and cellular uptake of iron nanowires. *Biomaterials*. 2010; 31(7):1509–1517.

[101] Huynen I, Goglio G, Vanhoenacker D, Vander Vorst A: A novel nanostructured microstrip device for tunable stopband filtering applications at microwaves. *IEEE Microwave and Guided Wave Letters*. 1999; 9(10):401–403.

[102] Still T, Retsch M, Sainidou R, Jonas U, Fytas G: High frequency acoustics in nanostructures by spontaneous Brillouin light scattering. *The Journal of the Acoustical Society of America*. 2008; 123(5):3555.

[103] Alonso-Redondo E, Schmitt M, Urbach Z, Hui CM, Sainidou R, Rembert P, Matyjaszewski K, Bockstaller MR, Fytas G: A new class of tunable hypersonic phononic crystals based on polymer-tethered colloids. *Nature Communications*. 2015; 6:8309.

[104] Counil G, Crozat P, Devolder T, Chappert C, Zoll S, Fournel R: Lateral and longitudinal finite size effects in NA-FMR measurements. *IEEE Transactions on Magnetics*. 2006; 42(10):3321–3322.

[105] Bilzer C, Devolder T, Crozat P, Chappert C, Cardoso S, Freitas PP: Vector network analyzer ferromagnetic resonance of thin films on coplanar waveguides: Comparison of different evaluation methods. *Journal of Applied Physics*. 2007; 101(7):074505.

[106] Kalarickal SS, Krivosik P, Wu M, Patton CE, Schneider ML, Kabos P, Silva TJ, Nibarger JP: Ferromagnetic resonance linewidth in metallic thin films: Comparison of measurement methods. *Journal of Applied Physics*. 2006; 99(9):093909.

[107] Slonczewski JC: Origin of biquadratic exchange in magnetic multilayers. *Journal of Applied Physics*. 1993; 73(10):5957–5962.

[108] Choi, J., Gani, A. W., Bechstein, D. J., Lee, J. R., Utz, P. J., & Wang, S. X. (2016). Portable, one-step, and rapid GMR biosensor platform with smartphone interface. *Biosensors and Bioelectronics*, 85, 1–7.

[109] Su D, Wu K, Saha R, Peng C, Wang JP: Advances in magnetoresistive biosensors. *Micromachines*. 2020 Jan; 11(1):34.

[110] Arias R, Mills DL: Theory of spin excitations and the microwave response of cylindrical ferromagnetic nanowires. *Physical Review B*. 2001; 63(13):134439.

[111] Malkinski LM, Yu M, Vovk AY, Scherer DJ, Spinu L, Zhou W, Whittenburg S, Davis Z, Jung JS: Microwave absorption of patterned arrays of nanosized magnetic stripes with different aspect ratios. *Journal of Applied Physics*. 2007; 101(9):09J110.

[112] Kuanr B, Camley RE, Celinski Z: Narrowing of the frequency-linewidth in structured magnetic strips: Experiment and theory. *Applied Physics Letters*. 2005; 87(1):012502.

[113] Sharma M, Kuanr BK, Sharma M, Basu A: Relation between static and dynamic magnetization effects and resonance behavior in Ni nanowire arrays. IEEE Transactions on Magnetics. 2014; 50(4):1–0.

[114] Darques M, Spiegel J, De la Torre Medina J, Huynen I, Piraux L: Ferromagnetic nanowire-loaded membranes for microwave electronics. *Journal of Magnetism and Magnetic Materials*. 2009; 321(14):2055–2065.

[115] Shore D, Ghemes A, Dragos-Pinzaru O, Gao Z, Shao Q, Sharma A, Um J, Tabakovic I, Bischof JC, Stadler BJ: Nanowarming using Au-tipped Co 35 Fe 65 ferromagnetic nanowires. *Nanoscale*. 2019; 11(31):14607–14615.

[116] Sung H, Ferlay J, Siegel RL, Laversanne M, Soerjomataram I, Jemal A, Bray F: Global cancer statistics 2020: GLOBOCAN estimates of incidence and mortality worldwide for 36 cancers in 185 countries. *CA: A Cancer Journal for Clinicians*. 2021 May; 71(3):209–249.

[117] Hildebrandt B, Wust P, Ahlers O, Dieing A, Sreenivasa G, Kerner T, Felix R, Riess H: The cellular and molecular basis of hyperthermia. *Critical Reviews in Oncology/Hematology*. 2002 Jul 1; 43(1):33–56.

[118] Luderer AA, Borrelli NF, Panzarino JN, Mansfield GR, Hess DM, Brown JL, Barnett EH, Hahn EW: Glass-ceramic-mediated, magnetic-field-induced localized hyperthermia: Response of a murine mammary carcinoma. *Radiation Research*. 1983 Apr 1; 94(1):190–198.

[119] Wu H, Song L, Chen L, Huang Y, Wu Y, Zang F, An Y, Lyu H, Ma M, Chen J, Gu N: Injectable thermosensitive magnetic nanoemulsion hydrogel for multimodal-imaging-guided accurate thermoablative cancer therapy. *Nanoscale*. 2017; 9(42): 16175–16182.

[120] Shetake NG, Balla MM, Kumar A, Pandey BN: Magnetic hyperthermia therapy: An emerging modality of cancer treatment in combination with radiotherapy. *Journal of Radiation and Cancer Research*. 2016 Jan 1; 7(1):13.

[121] Xie J, Yan C, Yan Y, Chen L, Song L, Zang F, An Y, Teng G, Gu N, Zhang Y: Multimodal Mn-Zn ferrite nanocrystals for magnetically-induced cancer targeted hyperthermia: A comparison of passive and active targeting effects. *Nanoscale*. 2016; 8(38): 16902–16915.

[122] Hervault A, Dunn AE, Lim M, Boyer C, Mott D, Maenosono S, Thanh NT: Doxorubicin loaded dual pH-and thermo-responsive magnetic nanocarrier for combined magnetic hyperthermia and targeted controlled drug delivery applications. *Nanoscale*. 2016; 8(24):12152–12161.

[123] Barick KC, Singh S, Jadhav NV, Bahadur D, Pandey BN, Hassan PA: pH-responsive peptide mimic shell cross-linked magnetic nanocarriers for combination therapy. *Advanced Functional Materials*. 2012 Dec 5; 22(23):4975–4984.

[124] Venkatesha N, Poojar P, Qurishi Y, Geethanath S, Srivastava C: Graphene oxide-Fe3O4 nanoparticle composite with high transverse proton relaxivity value for magnetic resonance imaging. *Journal of Applied Physics.* 2015 Apr 21; 117(15):154702.

[125] Basel MT, Balivada S, Wang H, Shrestha TB, Seo GM, Pyle M, Abayaweera G, Dani R, Koper OB, Tamura M, Chikan V: Cell-delivered magnetic nanoparticles caused hyperthermia-mediated increased survival in a murine pancreatic cancer model. *International Journal of Nanomedicine.* 2012; 7:297.

[126] Balivada S, Rachakatla RS, Wang H, Samarakoon TN, Dani RK, Pyle M, Kroh FO, Walker B, Leaym X, Koper OB, Tamura M: A/C magnetic hyperthermia of melanoma mediated by iron (0)/iron oxide core/shell magnetic nanoparticles: A mouse study. *BMC Cancer.* 2010 Dec; 10(1):1–9.

[127] Chalkidou A, Simeonidis K, Angelakeris M, Samaras T, Martinez-Boubeta C, Balcells L, Papazisis K, Dendrinou-Samara C, Kalogirou O: In vitro application of Fe/MgO nanoparticles as magnetically mediated hyperthermia agents for cancer treatment. *Journal of Magnetism and Magnetic Materials.* 2011 Mar 1; 323(6):775–780.

[128] Laurent S, Dutz S, Häfeli UO, Mahmoudi M: Magnetic fluid hyperthermia: Focus on superparamagnetic iron oxide nanoparticles. *Advances in colloid and interface science.* 2011 Aug 10; 166(1–2):8–23.

[129] Kumar R, Chauhan A, Jha SK, Kuanr BK: Localized cancer treatment by radio-frequency hyperthermia using magnetic nanoparticles immobilized on graphene oxide: From novel synthesis to in vitro studies. *Journal of Materials Chemistry B.* 2018; 6(33): 5385–5399.

# 4 Nanoparticles in CT-Guided Interventional Devices and Processes

*Deepak Gupta, Priya Jadoun, Milan Singh,*
*Aakash Mathur, and Asheesh Kumar Gupta*

## 4.1 INTRODUCTION

Computed tomography (CT), invented by Sir Godfrey Hounsfield in 1972, is a diagnostic imaging examination with superior spatial and density resolution. It uses ionizing radiation called X-rays and takes a series of X-ray images. The images are processed via computational methods to generate high-quality and detailed images of the patient. It can generate 3D images of the spine, vertebrae, and internal organs; it also provides quality images of blood, blood vessels, and soft tissues. It is recommended for the diagnosis of appendicitis, cancer, trauma, heart disease, muscular-skeletal disorder, infectious disease, tumors, lung disease, chest problems, and internal organ injuries.

CT scanning is non-invasive, painless technique and takes 10–15 minutes to perform. It uses a thin beam to take cross-sectional images of a patient's body. The beam rotates 360 degrees around the body. The transmitted beam falls on the detectors located opposite the source. The detector transmits the data to the computer where it is processed to create contrasted individual, 2D, and 3D images. It helps doctors to spot and monitor the abnormality occurring in different areas of the body.

Types of CT scan based on procedures include contrast-enhanced CT scan (CECT scan), non-contrast CT scan (NCCT scan), high-resolution CT scan (HRCT scan), and coronary CT angiography (CCTA scan). In CECT, a contrast agent is introduced into the patient's vein to add contrast to the bloodstream for better imaging. In NCCT, the body is scanned without introducing contrast agents into the body. It is generally used for imaging spine, trauma/stroke locations, and the like. HRCT requires advanced techniques to produce images with enhanced resolution of organs like lungs. CCTA provides clear images of coronary arteries present in the lungs.

The kind of CT scan referred by the doctor depends on the area to be diagnosed as CT angiography, brain CT scan, CT scan neck, CT scan lungs, cardiac CT scan, CT scan spine, CT scan bone, CT scan abdomen, pelvis CT scan, and renal CT scan. Whenever a physician requires to take a better look inside the body of the patient, the physician orders CT scans (a) for quick diagnosis of the abnormality in life-threatening situations like car accidents, allergic reaction, and trauma, (b) for the identification of the precise location of an abnormality like bone fracture, tumor, cancer, blood clot,

DOI: 10.1201/9781003316398-4

infection, internal bleeding, and excess fluid, (c) to assess the response of the body to a specific treatment like chemotherapy and radiotherapy, and (d) CT scans play a pivotal role in the formulation of treatment plans which includes pre-operative planning, and hence contribute to safer and more successful surgical outcomes.

Since it is ionizing radiation, the radiologists always try to keep the CT dosage as low as possible to avoid the harmful effects of radiation on the patient's body. The scans are quick, and radiologists want to capture a detailed image in that instance. CT scan is used for both soft and hard tissues. However, imaging of soft tissues requires contrast materials also known as dyes to assist in the imaging process by CT. These materials make the blood vessels, organs, or tissues distinguishable by blocking the X-rays and making them stand out in CT images. This helps the physician to plan the treatment of specific blood vessels, organs, or tissues. The contrast materials are introduced into the patient's body orally, rectally, or intravenously (veins or arteries). At present iodine and bromine-sulphate-based compounds are clinically approved contrast agents. Iodine is generally injected intravenously, into the spinal spaces, and into other body cavities. Barium sulphate is usually administered orally in the form of powder or tablet and rectally in the form of liquid or paste. The selection of contrast material and the method of administration is based on the medical conditions and the patient's health.

Iodine-based intravenous (i.v.) contrast materials are given by injection for the imaging of vascular systems, soft tissues, heart, lungs, and solid abdominal and pelvic organs. Iodine-based intrathecal contrast materials are injected into fluid between the thin layers of tissue that cover the brain and spinal cord for the imaging of spinal or cisternal abnormality and cerebrospinal fluid leaks. Typically, 2 ml of contrast dye is administered per kg of body mass. For the imaging of the pharynx, esophagus, stomach, small intestine, and large intestine, barium-sulfate-based contrast materials are swallowed, whereas it is administered by enema for the imaging of the rectum and colon. The amount of X-ray attenuation by tissues before and after the intake of contrast materials is measured in Hounsfield units (HU) or CT numbers. The values of HU range between $-1000$ (for air) to $+1000$ (for bones).

To provide sufficient contrast to the CT images, it is inevitable to use large doses of iodinated compounds, which leads to adverse effects in patients. These compounds are rapidly removed by the kidneys within a few minutes. Iodine-based contrast agents like iohexol, omnipaque, lipiodol, and saline are routinely used as CT contrast agents. They are environmentally dangerous, and certain patients have an intolerance to iodine because of thyroid conditions. It is to be noted that iodinated molecules have non-localized distribution and their swift pharmacokinetics restrict their microvascular and targeting abilities [1]. After i.v. administration of iohexol, the sample reaches the kidney in 5 minutes. The blood vessels, heart, and liver display negligible enhancement of CT signal even after 2 hours postinjection, when entire iohexol reaches the bladder. The CT signal of the kidney and postcava is enhanced immediately after iohexol is injected, but it is excreted via a renal passage in 2 hours due to its short blood circulation time. Moreover, the synthesis and purification of iohexol involve complex chemistry and multiple steps [2].

This chapter presents a discussion of the application of NPs in CT imaging of different organs, tissues, and vascular structures. It discusses the development of NPs towards their theranostic effects in CT-based interventional devices and processes.

It presents the evolution in the properties of NPs that make them a suitable candidate for their application as superior contrast materials for CT-guided interventional processes. It enables readers to understand that contrast-based CT technology is developing with the innovations in contrast materials. It introduces novel contrast materials that possess anticancer properties and can eliminate tumors upon administration. However, their practical implications are still in their nascent stage which requires strenuous preclinical research, animal testing, and FDA approval before it serves humanity.

## 4.2    PROPERTIES OF NPS AS POTENTIAL CANDIDATEs FOR CT-GUIDED INTERVENTIONAL DEVICES

In comparison to the iodinated compounds, NPs display long circulation time, higher contrast densities, and a readily functional surface; thus, they are well-suited for *in vivo* tomography and target-specific processes. NPs as contrast materials in CT scans can provide a distinctive 3D image of the region under investigation. Even at modest X-ray doses, the targeted contrast agents consisting of inorganic NPs improve CT imaging contrast by multiple orders of magnitude. This reduces the duration of exposure and risks related to highly ionized X-rays. The benefits of inorganic NPs include a simpler production method and the capability to quickly tailor their characteristics and morphology by modifying their surfaces [3].

Metallic NPs have high density and high K-absorption edge (refer to Table 4.1) resulting in strong X-ray attenuation. However, they are toxic for *in vivo* purposes. Their surfaces can be modified with biocompatible organic molecules to limit their toxicity to human cells and tissues. Additionally decorating their surfaces with functional groups improves their circulation time, structural stability, morphology, biocompatibility, and targeting properties towards specific organs or tissue.

Using light-absorbing NPs, photothermal treatment (PTT) turns absorbed light energy into thermal energy. This thermal energy is produced in sufficient quantities and could ablate targeted cancer cells. The PTT is highly effective in tumor ablation procedures and causes minimal injury to normal tissues as compared to chemotherapy and radiotherapy. However, without imaging contrast agents, the poor accuracy of PTT could result in suboptimal treatment effects. As a result, the imaging-guided PTT procedure is a promising therapeutic strategy for precision therapy. It will assist in visualizing the precise size and position of tumors before and after treatment. This treatment uses nanomaterials possessing high near-infrared (NIR) absorption qualities, which allow them to create heat and cause tumor ablation when exposed to laser light. This procedure is non-invasive and has a high success rate. PTT can be used alone or in combination with other therapies, such as radiotherapy, because it is marginally invasive and has an enhanced therapeutic efficacy. To perform PTT, an accurate position of tumor diagnosis is required. Because of its great efficiency, low cost, and enhanced spatial and higher resolution, CT imaging is the most practical imaging tool for diagnosing tumors. As a result, developing a nanoplatform as a contrast agent for CT imaging is in high demand, not only for giving more comprehensive information for diagnosis at the molecular level but also for demonstrating promising clinical applications.

**TABLE 4.1**
**List of NPs, Atomic Number, and Their K-Absorption Edge Values**

| Sr. No. | Nanoparticle | Atomic Number | K-absorption Edge |
|---|---|---|---|
| 1 | Silica | 14 | 1.8 keV |
| 2 | Iron | 26 | 7.1 keV |
| 3 | Copper | 29 | 8.9 keV |
| 4 | Selenium | 34 | 12.6 keV |
| 5 | Molybdenum | 42 | 19.9 keV |
| 6 | Silver | 47 | 25.5 keV |
| 7 | Iodine | 53 | 33.2 keV |
| 8 | Gadolinium | 64 | 50.2 keV |
| 9 | Holonium | 67 | 55.6 keV |
| 10 | Tungsten | 74 | 69.5 keV |
| 11 | Rhenium | 75 | 71.6 keV |
| 12 | Platinum | 78 | 78.3 keV |
| 12 | Gold | 79 | 80.7 keV |
| 13 | Lead | 82 | 88.0 keV |
| 14 | Bismuth | 83 | 90.5 keV |

## 4.3 CHRONOLOGY ON THE USE OF NPS

### 4.3.1 IODINATED NPS

Quan Zou et al. (2018) [4] produced Iodinated Polypyrrole (IPPy) NPs in a single-step process where PPy NPs are produced by the polymerization of pyrrole monomers in the presence of $HIO_3$. The reaction substitutes the hydrogens with $I_2$ to produce IPPy NPs. The obtained I-PPy NPs exhibited uniform size, monodispersity, and good aqueous solubility. The *in vitro* cytotoxicity of I-PPy NPs was tested for the 4T1 cell line, Hela cell line, and 3T3 cell line using an MTT assay. Cell viability of 90% showed the high biocompatibility of I-PPy NPs. The CT signal increased linearly with the concentration of I-PPy. The CT scan of 4T1 cells incubated with I-PPy NPs was brighter than saline (control). *In vivo*, the BALB/c mice were injected with synthesized NPs. The contrast of the heart, kidney vein, and postcava were observed with enhanced contrast for 2 hours, and 12 hours for liver postinjection. When IPPy NPs were injected in the presence of an NIR laser, the temperature at the tumor region rose to 52 °C after 1 minute. The tumor of this group was suppressed in 14 days postinjection. The mice with IPPy (L+) remained alive for 60 days post-treatment. They exhibited no body weight loss or histopathological damage to major organs.

Xiaoju Men et al. (2020) [5] produced novel conjugated polymer dots (Pdots) functionalized doxorubicin (DOX)-based thermosensitive hydrogel nanoplatforms. Pdots are nontoxic, biodegradable, small-size materials with superior photophysical properties. DOX is usually used to slow down the growth of cancerous cells. Iohexol was grafted to Pdots–DOX to increase the imaging time. The nanocomposite hydrogel exhibited good stability and a high photothermal (PT) effect. Under NIR illumination, the nanocomposites were able to achieve local heating and controlled drug delivery. The *in vivo* imaging enhancement capabilities of Pdots–DOX–iohexol@ hydrogel were inspected in BALB/c mice infected with 4T1 tumor. The CT signal at the tumor region remained for 3 hours postinjection for the nanocomposite in comparison to iohexol, for which the signal vanished after 1 hour postinjection. The drug distribution at the tumor site remained identifiable for 24 hours postinjection. The nanocomposite caused no harm to the primary organs of the mice even after 14 days of treatment.

Wen Zhou et al. (2020) [7] produced two variants of semiconducting polymer NPs (SPN) called SPN-P and SPN-I from semiconducting polymer (PCPDTBT). Neither SPN exhibited precipitation or change in average diameter for 30 days, indicating their good physiological stabilities. *In vitro*, both SPN-P and SPN-I exhibited cellular uptake by the Lewis lung carcinoma cell line owing to efficient cell internalization. Both SPNs showed good biocompatibility without light irradiation to Lewis cells. The xenografted tumor-bearing c57bl/6 male mice model was used for *in vivo* CT imaging. After i.v. administration to the model, SPN-I can be detected from CT images for 4 hours postinjection in the tumor area. After 1 day the CT signal became threefold more than preinjection in the tumor area due to the prolonged accumulation of SPN-I. The SPN-I NPs under NIR irradiation have established their applications in photodynamic therapy (PDT).

Iodinated polyaniline (LC@I-PANi) NPs were synthesized by Liwen Fu et al. (2021) [6]. An excellent spherical morphology and colloidal stability were achieved via L-cysteine (LC). The hemolysis assay and cytotoxicity assessments confirmed that LC@I-PANi is biocompatible. Owing to the presence of LC, these NPs exhibited superior anti-protein performance. They offered strong PT ability *in vitro* at the cellular level. Due to the presence of iodine, the NPs demonstrated excellent CT imaging potential of LC@I-PANi NPs as demonstrated by Figure 4.1.

### 4.3.2   GOLD NPS

Gold NPs are biologically safe, easy to synthesize, and readily surface modifiable and can efficiently generate Auger electrons upon irradiation. Gold NPs had prolonged blood circulation time, controlled nephrotoxicity, and superior X-ray absorption coefficient compared to iodine and its compounds. Chenjie Xu et al. (2008) [8] produced 2-mercaptosuccinic acid (MSA) coated gold NPs (GNPs). The surface plasmonic characteristics of GNPs make them popular in imaging and therapy. The plasmonic absorptions of these GNPs ranged from 510 to 540 nm, demonstrating their optical activity. They synthesized GNPs by the Fens method in the presence of MSA. It served as both a reducing and a capping agent. The particle diameter of the MSA-GNPs ranged from 4 to 60 nm. They were small, nontoxic, and stronger CT imaging agents than the bigger GNPs and omnipaque. This was attributed to the larger surface area as the diameter of GNP decreases. These characteristics of GNPs established their candidature for delivery and multimodal imaging applications. Chen Peng et al. (2012) [9] synthesized dendrimer-encapsulated Au NPs (Au DENPs) modified with PEG for CT imaging. PEG, or polyethylene glycol, is a biodegradable polymer employed for improving biocompatibility, stability, renal scavenging ability, preventing opsonization, and blood circulation duration. The

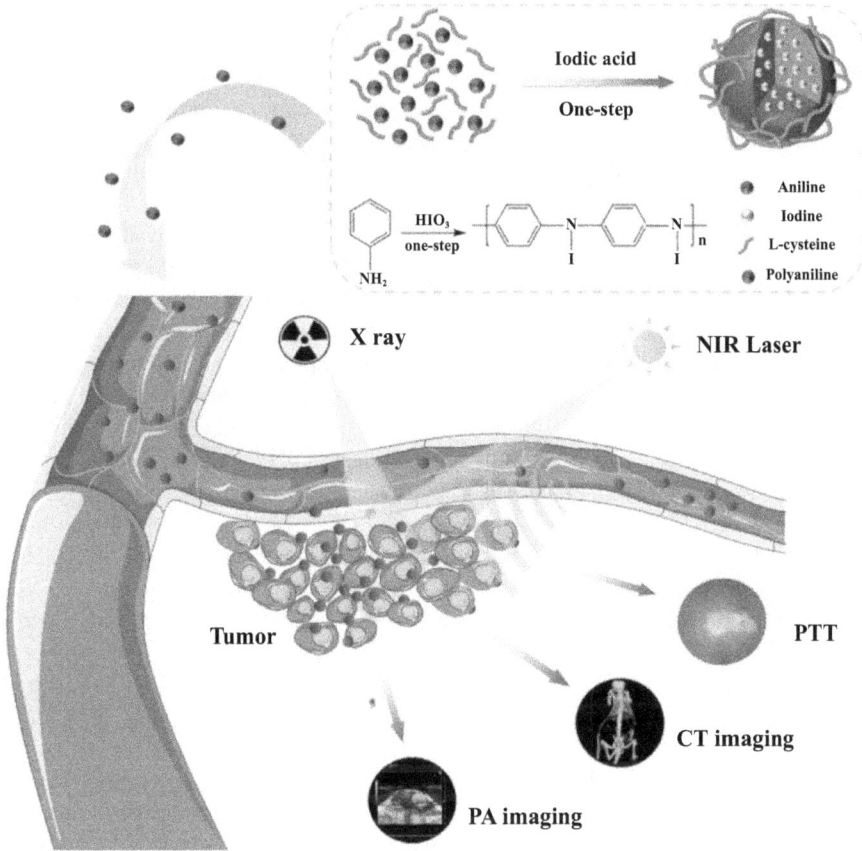

**FIGURE 4.1 (A)** Schematic illustration of the one-step synthesis of L-cysteine (LC).

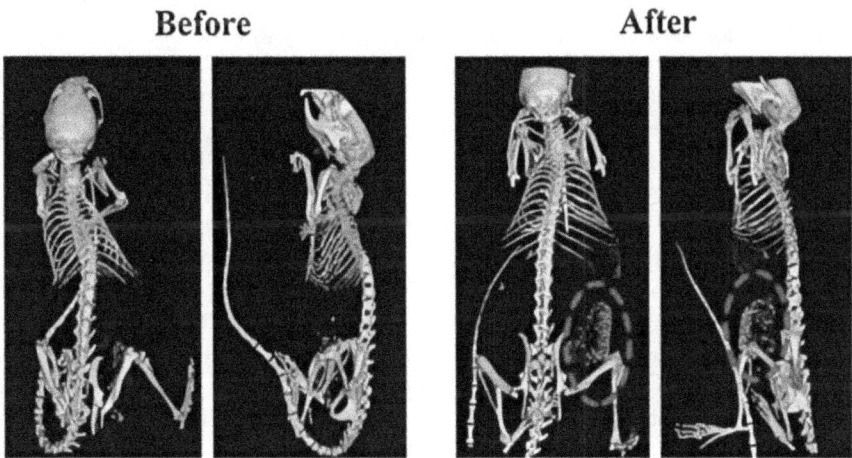

**FIGURE 4.1 (B)** *In vivo* CT images of tumors before and after intratumorally injection of LC.

decoration of dendrimer terminal amines by partial PEGylation enabled a large packing of Au within the dendrimer interior. The size of synthesized PEGylated Au DENPs ranged from 2 to 4 nm. These NPs were water-dispersible, robust in acidic environments, and stable in the 0–50 °C range, and exhibited non-cytotoxicity up to 100 mM concentration of PEG-Au-DENPs. Their X-ray absorption coefficient was significantly larger than omnipaque. The suitably extended half-life is ideal for imaging both blood pools and tumors. The i.v. administration of these NPs enabled blood pool imaging of both mice and rats. An enhanced CT image of the tumor-bearing xenotransplanted mice was obtained upon intraperitoneal and i.v. administration of PEG-Au-DENPs. These NPs were distributed uniformly in the tumor region and produced enhanced CT images for 6 hours postinjection due to the enhanced permeability and retention (EPR) effect. The renal uptake of Au indicated that the synthesized contrast material would be easily cleared via the reticuloendothelial route. Hui Liu et al. (2013) [10] produced acetylated folic acid (FA) decorated dendrimer stabilized Au NPs. They exhibited robustness at pH 5–8, at temperature range up to 50 °C, and in various aqueous media. The NPs showed desirable cytocompatibility with MTT assay in the 1000–3000 nm concentration range. KB cell lines treated with the synthesized NPs exhibited higher CT values due to the FA-mediated cellular uptake and target-oriented binding pathways. The FA-modified Au DSNPs produced targeted CT images of cancer cells *in vitro* and xenografted tumor models *in vivo*. The procedure is extendable to diverse biological systems. Wooram Park et al. (2016) [11] produced surface-modified branched Au NPs (BGNP) decorated with clostridium novyi-NT spores to achieve spore transmission into the solid tumor upon intratumoral administration. Branched GNPs had a larger surface area than AU NPs. The negatively charged BGNPs are immobilized on C. novyi-NT having positively charged spore walls via electrostatic interaction. The *in vitro* imaging of BGNP-coated spores on phantoms exhibited excellent contrast enhancement in comparison to water or spore (control). *In vivo* CT efficiency of BGNP-coated spores was studied on PC3 xenograft BALB/c nude mice by administering these NPs intratumorally to mice. It is to be noted that these spores required a hypoxic environment for germination. The administration was done critically to inject spores at the core of solid tumors for the effective germination of spores within hypoxic regions of tissues by needle catheter. The CT images exhibited spore deposition at the precise location of the tumor and enabled 3D rendered images of the tumor volume and other characteristic information like size and location. Their therapeutic efficiency was determined by histological and staining analyses. The germination and oncolytic capabilities of spores remained unaffected by the coating of BGNPs. Mild after-treatment toxicity signs were observed in the xenografted mice model but found to be manageable by antibiotics.

Prussian blue (PB) is a clinically approved drug and is used for PT tumor ablation. PBNPs and Au NPs were the two different types of intrinsic theranostic NPs, based on which Yan Dou et al. (2017) [12] developed PB@Au NPs, a core-satellite structure. They exhibited good colloidal stability. They had a diameter of 138.8 nm after PEGylation. The CT contrast for PB@Au was stronger than iohexol (control), as confirmed by the linear correlation between CT values and Au molar concentrations. To investigate the PT stability of PB@Au NPs were irradiated by NIR laser and the temperature increased up to 35 °C which was higher than pure water or Au

NPs solutions (control), while the temperature increased to 50 °C for CSNPs solutions and PBNP core which indicated good PT efficiency of CSNPs. The 90% of PB@Au NPs exited the body after 7 days of i.v. administration. No obvious toxicity within 14 days, revealed excellent biocompatibility of the NPs. *In vivo* CT imaging of mice carrying 4T1 xenografted tumors using PB@Au NPs revealed a high degree of contrast at the tumor site, and the CT value was observed to raise twofold, demonstrating the efficient accumulation through the EPR effect. Asahi Tomitaka et al. (2017) [13] produced magneto plasmonic NPs (MNP) with Au shells. The NPs were PEGylated to improve their stability and *in vivo* blood circulation time. The MNP@Au-PEG NPs were created with a motive to image the brain. For this, a blood–brain barrier (BBB) model was created for *in vitro* analysis. The modification of NPs did not affect the CT value and matches the CT values of Au NPs. The cell viability of MNP@Au-PEG was evaluated on human astrocytes by XTT assay. No significant decrease was observed in a cell at 100 µg/ml. The transmigration of MNP@Au-PEG across the BBB is only 33.8% without a magnetic field and increased by four times in its presence. Saji Uthaman (2018) et al. [14] stated that lymphatic networks are one of the primary pathways employing which cancer cells propagate during metastasis. It is crucial to eliminate lymph nodes near tumors to avoid metastasis and cancer recurrences. Mannan (M) was used to reduce and stabilize the GNPs to generate M-capped GNPs. These M-GNPs had a diameter of approximately 9.18 nm. The cell lines DC 2.4 and RAW 264.7 were studied with MTS assays to analyze the proportion of healthy cells in the presence of synthesized NPs. M-GNPs treatment resulted in cellular viability of 90% in both cell lines over a wide concentration range, indicating no cytotoxicity towards both cells. *In vivo*, CT phantom tests were performed on M-GNPs and the strongest signal of 303.2 HU is obtained at 20 mg mL$^{-1}$. *In vivo*, CT images with the high contrast of lymph nodes were observed after subcutaneously injecting M-GNPs into mice. The CT value in the popliteal lymph nodes (PLN) region increased sevenfold due to the efficient diffusion and higher accumulation of NPs. M-GNPs had remarkable cytocompatibility and lymph node targeting properties. Sen Liu et al. (2017) [3] produced Au@I coated Anti-RhoJ NPs (AIRA NPs) in a three-step process: First Au nanoclusters were synthesized, and these nanoclusters were anchored to the iodine forming Au@I. The complex is PEGylated to improve its aqueous stability. The proposed NPs were obtained after covalently binding Au@I with Anti-RhoJ. The AIRA NPs based radiosensitizer could suppress tumor angiogenesis. Tumor angiogenesis was triggered after the disability of pre-existed vessels and supports the development of peri-tumoral vessels which were responsible for cancer growth. They synthesized the Anti-RhoJ antibody to suppress the formation of both new and existing tumor vessels. The AIRA NPs were added to the human triple-negative breast cancer and human umbilical vein endothelial cell lines with MTT assay and incubated overnight at 37 °C to measure their cell viability. Even after 96 hours incubation, the NPs exhibited negligible cytotoxicity. The effectivity of AIRA NPs is measured via immunofluorescence of platelet-derived growth factor receptors and Vascular endothelial growth factor receptors. The CT values reduced remarkably after 30 days. PDX mice model when injected i.v. by AIRA NPs, the CT images captured enhanced signal in tumor site and different colors for new and pre-existed blood vessels 24 hours postinjection. This exhibited superior binding accuracy and tumor-targeting properties of AIRA NPs. The same group of mice was studied for *in vivo* anti-tumor effect of AIRA NPs during radiotherapy, the tumor volume decreased slightly after 6 days, and then rapidly with time. The solid tumor was ablated after 20 days postinjection. Tumor volume worsens for the bevacizumab (an FDA-approved drug) due to its rapid clearance by blood. Whereas tumor burden kept on increasing for the untreated (control) group. Jingyi Zhu et al. (2018) [15] produced dendrimer-entrapped Au NPs (Au DENPs) conjugated DOX. They covalently attach DOX to partly acetylated poly(amidoamine) dendrimers that had been prefunctionalized with FA to produce G5.NHAc-FA-DOX conjugates, which were then decorated with Au NPs to form DOX-Au DENPs. They exhibited a pH-responsive DOX releasing profile which increased under slightly acidic conditions. They exhibited cytotoxic effects on U87MG cells. DOX-Au DENPs were colloidally stable and had significantly large X-ray attenuation against omnipaque (control). Benqing Zhou (2018) [16] manufactured acetylated (Ac) polyethylenimine (PEI)-encapsulated Au NPs (Ac-PE-Au NPs) by utilizing branching PEI for *negative* CT contrast enhancement for orthotopic hepatic carcinoma (HCC) in comparison to normal liver. The Ac-PE-Au NPs (in absence of antifouling surface) demonstrated more absorption in normal liver than in HCC and made it readily distinguished in CT scans. The CT value of these NPs was greater than omnipaque for the identical Au or I concentrations. The Ac-PE-Au NPs exhibited a low hemolysis rate (5%) for 0–150 $\mu$M Au concentrations. The cytotoxicity of synthesized NPs displayed the cell viability of 75% for 80 $\mu$M and 65% for 100 $\mu$M Au concentrations, respectively with the MTT assay. Normal liver presented nearly twofold higher CT contrast in comparison to HCC, 3 hours postinjection. Yan Gao et al. (2020) [17] produced Au nanocages@PEG NPs for *in vivo* contrast-enhanced CT imaging. The Au nanocage structure had a higher surface area and better imaging capabilities than gold NPs. They were small-sized and had higher contrast, reduced toxicity, and negligible adverse effects. The CT imaging performance of AuNC@PEGs was superior due to a higher K-absorption edge and prolonged blood pool angiography time in comparison to traditional iodine agents. Hua-Qing Yin et al. (2020) [18] synthesized COP@Au@TCPP nanocomposites from Meso-tetrakis(4-carboxyl)-21H,23H-porphine (TCPP) and -cyclodextrin (-CD) functionalized Au NPs (CD-Au NPs), when integrated with COP-8 as shown in Figure 4.2. They employed $\beta$-CD as a linker to link TCPP, COP-8, and Au NPs to develop COP@Au@TCPP via hydrophobic processes. Through increased permeability and retention effect, they aggregated effectively at tumor locations. In tumor cells, CD-Au NPs acted as a catalyst to degrade $H_2O_2$ into $O_2$. The nanocomposites demonstrated 80% of cell viability and 9% higher CT values than iohexol (control) *in vitro* and *in vivo*. Figure 4.3 demonstrates the time-dependent CT images of healthy nude mice with and without COP@Au@TCPP as a contrast agent. They are mainly metabolized by the kidneys and bladder.

### 4.3.3 Copper-Based NPs

Copper-based NPs display enhanced fluorescence signals and optical imaging capabilities which are essential for image-guided applications. Copper chalcogenide NPs exhibit efficient light-to-heat transformation and can cause location-specific

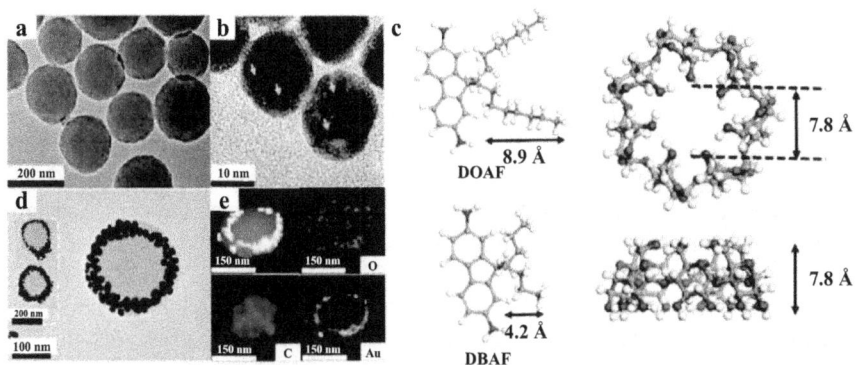

FIGURE 4.2 TEM image of (a) COP-8 and (b) CD-Au NPs. (c) The length of n-octyl group of DOAF and the n-butyl. The pore size (up) and height (down) of β-cyclodextrin. (d) TEM images of COP@Au@TCPP NPs. (e) TEM mapping of oxygen, carbon, and gold elements in COP@Au@TCPP. Reprinted with permission from [18]. Copyright (2020) @ ACS Publication.

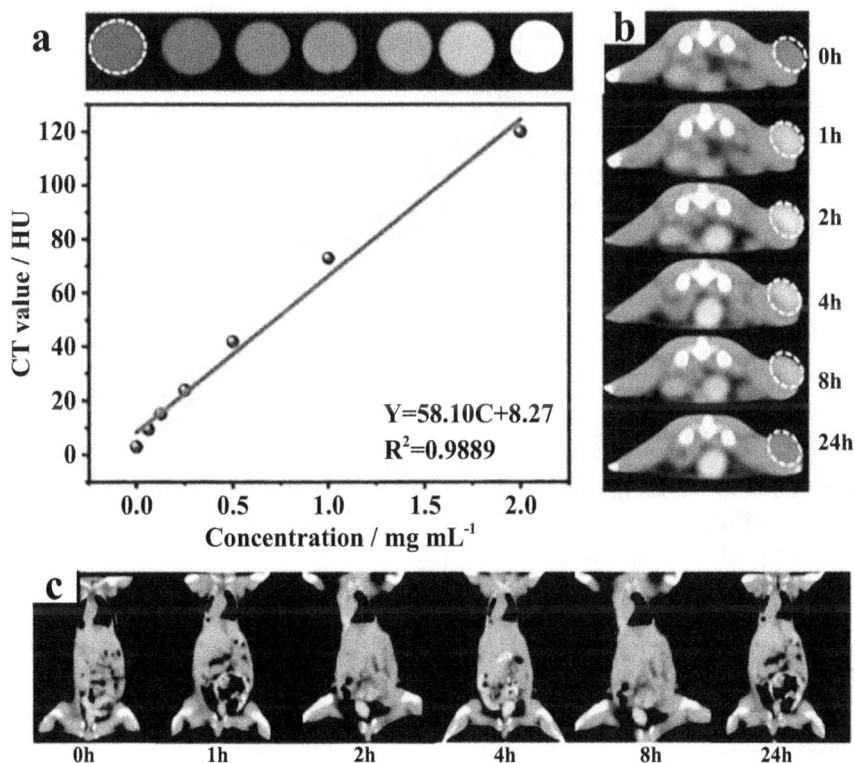

FIGURE 4.3 (a) CT images of COP@Au@TCPP under different concentration, and CT attenuation plot of COP@Au@TCPP (down). (b) Time-dependent CT imaging with COP@Au@TCPP as a positive control (tumor in the dashed circle). (c) Time-dependent CT images of healthy nude mice with COP@Au@TCPP as a content agent. Reprinted with permission from [18]. Copyright (2020) @ ACS Publications.

thermal ablation of the tumor. Shaohua Zhang et al. (2016) [19] produced ultrasmall PEGylated $Cu_{2-x}Se$ NPs. Both copper and selenium are biodegradable and essential for good health. Copper is required for the absorption of iron in the body, good blood vessels, and cardiovascular health, while selenium is necessary for the prostate, the liver, DNA synthesis, and so on. The ultrasmall NPs of 3.6 nm were obtained via a facile aqueous process. These NPs were decorated with HS-PEG-SH to enhance their colloidal stability, water solubility, and biocompatibility. At a copper concentration of 100 $gmL^{-1}$, they displayed minimal cytotoxicity towards 4T1 cells and 84% cell viability. The *in vitro* PTT effect of NPs was analyzed in 4T1 cells where laser irradiation killed 4T1 cells with high efficiency. They had a half-life of 8 hours, which prolonged their retention in the bloodstream. In order to conduct *in vivo* experiments, a solution of synthesized NPs was injected i.v. to the cancer-carrying BALB/c mice, and CT scans were obtained postinjection. The images clearly depicted the tumor with increased contrast caused by the EPR effect. *In vivo*, the HU value increased by 50 HU in 10 hours postinjection. The Cu and Se elements contributed to the improved imaging capabilities demonstrated by both *in vitro* and *in vivo* CT imaging studies. The primary accumulation of NPs in reticuloendothelial organs and extensive distribution in the kidney suggested that they could be easily eliminated via renal clearance. *In vivo* PTT investigations were conducted on mice bearing a 130 $mm^3$ tumor burden. Upon injection, the tumor temperature raised drastically to 57.6 °C under irradiation and demonstrated their superior PT efficacy in destroying cancer cells. To evaluate their therapeutic efficacy, hematoxylin and eosin (H&E) staining of tumor slices revealed significant karyopyknosis and necrosis 3 days following PT therapy. After 16 days of treatment, the tumor disappeared totally and did not return, but mice administered with PBS (control) or $Cu_{2-x}Se$ perished due to uncontrolled tumor growth. No visible inflammation or damage to major organs of mice treated was detected until 40 days after therapy, establishing that synthesized NPs are suitable CT contrast agents in a variety of interventional procedures.

### 4.3.4   IRON-BASED NPs

Shiyi Lu et al. (2018) [20] produced Au nanoflowers stabilized by dendrimer with very small iron oxide nanoparticles ($Fe_3O_4$/Au DSNFs). The $Fe_3O_4$ NPs were used for enhancing CT imaging and Au NFs for photoacoustic and PT conversion solutions. Besides these, NPs exhibited good colloidal stability and cellular uptake efficiency. Compared to omnipaque (control), $Fe_3O_4$/Au DSNFs displayed a much higher CT value. Both $Fe_3O_4$/Au DSNFs and omnipaque showed improvement in the brightness of CT signal for equal concentrations of Au and I. The cytocompatibility of $Fe_3O_4$/Au DSNFs was tested on 4T1 cells using CCK-8 assay, and the cell viability of 94% was observed at the highest Au concentration (2mM). It decreases to 34.7% upon laser irradiation highlighting Fe3O4/Au D SNFs efficiency as PT ablation material. *In vivo* CT imaging was performed for subcutaneous tumors. The tumor CT value increased 1.7 times in 60 minutes postinjection. Further, the intensities in the CT signal were greater than the surrounding tissues. Within 1 hour postinjection, NPs cleared via the reticuloendothelial system. NPs were completely removed from major organs 96 hours postinjection.

### 4.3.5 Silver NPs

Silver NPs have excellent antimicrobial properties. They have low density, which is required for renal clearance. Considering these properties, many studies are performed to study their properties towards CT contrast agent. Jessica C. Hsu et al. (2019) [21] produced glutathione-coated $Ag_2S$ NPs with customizable sizes (<5 nm) via the viscosity-mediated thermal decomposition approach. When normal liver and kidney cells were cultured for 4 hours, the $Ag_2S$ NPs exhibited superior water solubility and biocompatibility. They released little silver and had an insignificant effect on cell survival, irrespective of the particle sizes. Their clearance potential and distribution profile were studied by injecting them into female nude mice via tail vein. $Ag_2S$ NPs had prolonged bloodstream durability compared to iodine (control). *In vivo*, the CT contrast increased significantly in the bladder and slightly in the spleen and liver. The circulation half-life of $Ag_2S$-NP was 86.6 minutes. NPs were excreted quickly via urine. Eun Mi Lee et al. (2020) [22] developed a composite from Ag NPs and porous silicon microparticles (pSiMPs). These hybrid composites were biocompatible and the CT signal intensity of the AgNPs-pSiMPs hybrid composite was suitably high (>1000 HU) when dispersed in glycerol + DI $H_2O$. Glycerol in the mixture improved the composite's ability to localize in the lungs. The hybrid composite did not require surface modification before injection. In comparison to AgNPs (control) and lipiodol (control), they exhibited a strong CT signal at the precise location of administration in rabbits' lungs almost for an hour. AgNPs-pSiMPs had low toxicity and did not affect the immune system of mice.

### 4.3.6 Platinum-Based NPs

Platinum NPs share many properties of the gold NPs that make them a suitable candidate as a contrast agent in CT radiology. Zigui Wang et al. (2019) [23] developed a dual prodrug polymer NP (DPP NP) system for cancer-oriented light-responsive chemotherapy guided through Pt (II) drug-mediated CT imaging, in order to eliminate the use of an extra light-responsive molecule. The DPP NP promoted the light-activated splitting of the Pt(IV) based prodrug and then released bioactive Pt(II) drug near abnormal cells. The cleaved segment was acid hydrolyzed to produce demethyl cantharidin (DMC), which exhibited antitumor properties *in vitro* and *in vivo*. The DMC molecule was responsible for synergizing Pt drug-based chemotherapy known as Pt DMCT. Comparing DPP NP to cisplatin, distinct CT scans were obtained for long hours in mice. Light-activated synergistic effectiveness of DPP NP *in vivo* was tested in male KM mice bearing U14 xenograft tumors. Here they exhibited cancer inhibition effect upon irradiation. After Pt DMCT-guided laser irradiation, several mice with dual prodrug polymer were completely cured. The body weight and levels of physiological parameters remained the same for DPP NP-treated mice. When compared to cisplatin (control), the DPP NP exhibited improved synergistic antitumor activity and reduced toxicity.

### 4.3.7 Holonium-Based NPs

Dalong Ni et al. (2016) [24] produced novel $NaHoF_4$ NPs via the thermal decomposition method. $NaHoF_4$ were hydrophobic NPs. The NPs exhibit high monodispersity and uniform spherical morphology. The size of the NPs lies in the 3–7 nm range. Studies were performed with 3 nm $NaHoF_4$ NPs owing to the ease of renal clearance. *In vitro*, $NaHoF_4$ NPs after incubation for one day at 37 °C exhibited negligible hemolysis towards the human glioblastoma cell line (U87MG) in the MTT assay. To perform *in vivo* studies, its biocompatibility is enhanced by modifying its surface DSPEPEG with a molecular weight of 5000 g/mol ($DSPEPEG_{5K}$) resulting in the formation of $PEG-NaHoF_4$ NPs. CT signal of $PEG-NaHoF_4$ NPs increased with the $Ho^{3+}$ concentration. *In vivo* hemolysis of $PEG-NaHoF_4$ NPs was negligible. They had a non-detrimental effect on tissues. They were injected into the BALB/c mice carrying 4T1 tumors and produced nearly six times bright CT images, especially in tumor regions. No concerning changes in the histological organs of the mice were observed even after 30 days postinjection, demonstrating their biosafety.

### 4.3.8 TUNGSTEN-BASED NPS

Tungsten is a nontoxic heavy metal with high X-ray absorption. Anshuman Jakhmola et al. (2013) [2] synthesized tungsten oxide (WO3) NPs decorated with polycaprolactone (PCL) to develop a novel CT contrast material. They were decorated with biodegradable polycaprolactone (PCL) and PEGylated surfactants to promote aqueous stability. The PCL layer prevented particle aggregation and blood vessel blockage. The size of the NPs was around 100 nm. Even at concentrations of a minimum of 0.1 M, these NPs provided a good contrast for CT imaging. *In vivo*, they exhibited nontoxicity, high CT number, and were eliminated postingestion in some hours. PCL-WO3 NPs had a prolonged blood circulation time of up to 2 hours (compared to roughly 10 minutes of iodinated contrast agents) and they were quickly eliminated. They did not collect in the body, minimizing their long-term adverse effects. These properties make them promising candidates for CT angiography.

### 4.3.9 BISMUTH-BASED NPS

Like tungsten, bismuth is also a nontoxic heavy metal. It is the heaviest stable nonradioactive element with high X-ray absorption. Oded Rabin (2006) et al. [1] produced polymer-coated $Bi_2S_3$ NPs (BPNPs) by coating $Bi_2S_3$ nanocrystals with polyvinylpyrrolidone (PVP) for i.v. CT imaging. Polymer coating promoted the biological utility by limiting nanocrystal aggregation. The cardiac ventricles, venous and arterial structures, were well delineated in BPNP-guided CT procedures. *In vivo*, X-ray absorption was five times enhanced, and blood circulation duration was longer by 2 hours than iodine (control). Bismuth in the form of BPNPs was less hazardous to cells by at least one order of magnitude. At 24 hours, BPNPs moved to organs (like the liver and lymph nodes) with phagocytic cells. BPNPs had a better profile for hepatocytes than did the iodinated agent and hence could be used in the diagnosis of hepatic metastases. Pengpeng Lei et al. (2016) [25] produced upconversion NPs (UCNPs) from $NaGdF_4$: $Yb^{3+}$, $Tm^{3+}$, x% $Bi^{3+}$ (GYT-x% Bi) via solvothermal process. The purpose of adding $Bi^{3+}$ ions was an enhancement in the CT signal. x = 25 was the optimized percentage of Bi which remained constant during the studies. The obtained NPs were monodisperse. To evaluate the cytotoxicity they were incubated with HeLa cells using an MTT assay for one day. For concentrations up to 1000 µg/mL, 90% of the cells remained healthy, establishing their biocompatibility *in vitro*.

The elements Gd, Yb, and Bi ensured high CT signals. The citrate was coated over the NPs to increase their functionality and to obtain a higher HU value than iobitridol (control) *in vitro*. *In vivo* studies were performed on Kunming mice, and the CT signals were observed in the liver, spleen, and heart. After 5 minutes, the heart HU value rose by 24 HU. After 2 hours, good 3D rendering of CT images for liver and spleen were observed due to uptake by macrophages. Kunming mice were injected with H22 cells for *in vivo* tumor CT imaging. After injecting NPs intratumorally, the CT signal kept increasing in the tumor site for 2 hours after administration, possibly by the EPR effect. The CT signal sustained for 1 day postinjection exhibited citrus coated NPs superiority as a contrast material as shown in Figure 4.4. Besides CT number, citrate coating promoted excellent MR and NIR-to-NIR UCL imaging.

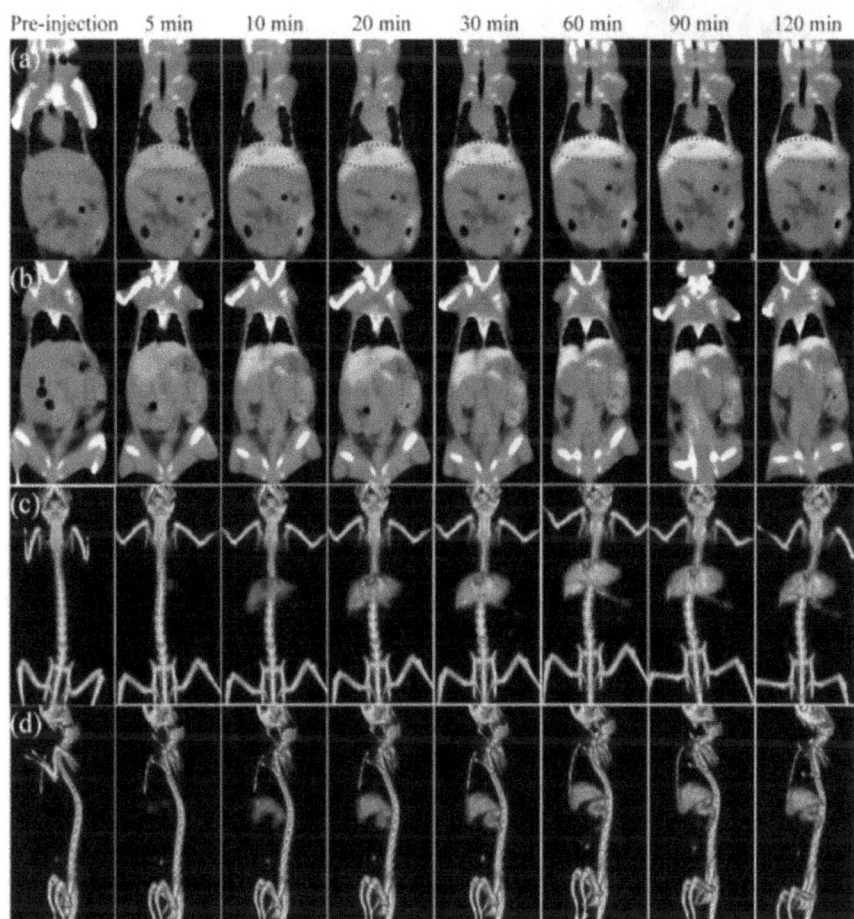

**FIGURE 4.4** *In vivo* CT coronal view images of a mouse after i.v. injection of the citrate-coated GYT-25% Bi UCNPs at timed intervals. (a) Heart (circles) and liver (ellipses). (b) Spleen (rectangles) and kidney. (c, d) The corresponding 3D renderings of *in vivo* CT images. Reprinted with permission from [25]. Copyright (2016) @ ACS Publications.

Chunyu Yang et al. (2018) [26] produced 1,2-dilauroyl-sn-glycero-3-phosphocholine (DLPC) decorated with Bi NPs producing Bi@DLPC NPs. They exhibited good aqueous dispersibility and relatively spherical shapes, excellent cellular uptake, and PT conversion efficiency. The cytotoxicity of 500 μg/mL Bi@DLPC NPs towards MDA-MB-231 and MCF-10A cell lines for 24 hours using MTT assay exhibited 87% of cell viability. When 100 μL of Bi@DLPC NPs were administered intramuscularly to MDA-MB-231 tumor-bearing mice, strong CT signals were observed for 12 hours postinjection; all results are shown in Figure 4.5. Their therapeutic efficiency was determined by histological and staining analyses, and the results are shown in Figure 4.6. Huiting Bi et al. (2018) [27] produced PEG-modified Bi NPs. The Bi-PEG

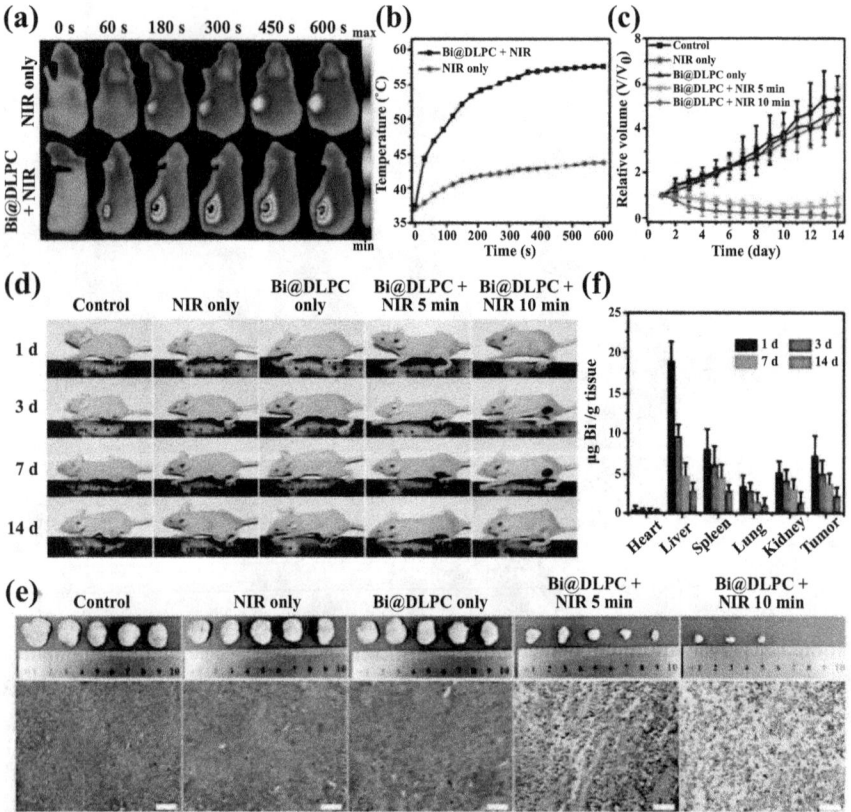

**FIGURE 4.5** *In vivo* antitumor efficacy of Bi@DLPC NPs. (a) Infrared thermal images and (b) temperature curves against time in tumor site of MDA-MB-231 tumor-bearing mice intravenously injected with Bi@DLPC NPs (1 mg/mL, 100 μL) at 6 hours under NIR laser irradiation (1 W/cm$^2$). (c) Normalized tumor volumes during the therapeutic period after different treatments. (d) Representative photos of mice from each experimental group during the treatments. (e) Photographs and their corresponding hematoxylin and eosin (H&E) staining of tumor tissues on the 14th day after different treatments (scale bar, 50 μm). (f) Biodistribution of Bi@DLPC NPs in major organs.

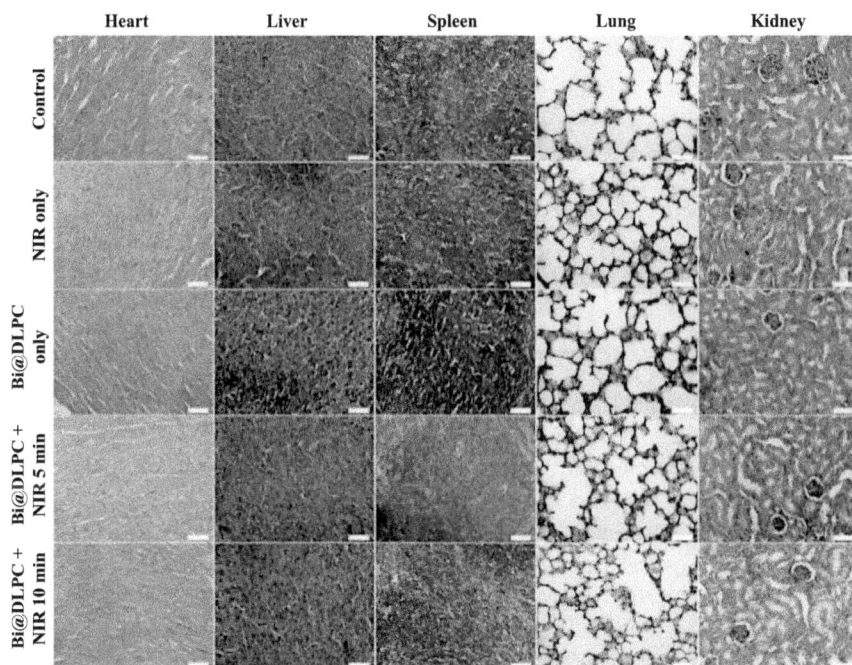

**FIGURE 4.6** Histology analysis of mice major organs (heart, liver, spleen, lung, kidney) at 14th day after different treatments (scale bar, 50 μm). Reprinted with permission from [26]. Copyright (2018) @ ACS Publications.

exhibited monodispersity. The CT signal showed a linear increase with Bi-PEG concentration. Upon incubating L929 cells for 1 to 2 days, the cell viability is over 92%. No obvious hemolysis was observed for the Bi-PEG NPs. When injected into mice, the main organs become visible only after 30 minutes and displayed good distribution efficiency of Bi-PEG; however, after 1 hour more organs started appearing in CT imaging and most importantly the entire liver became visible at this stage. After 3 hours, the sample was completely removed from the liver. This presents the long circulation time of Bi-PEG essential for contrast-based CT imaging. No pathological changes had been observed in the mice after 14 days postinjection, which exhibited the biosafety of the NPs. Hongying Guo et al. (2019) [28] produced $Bi_2S_3$-based NPs for their role as CT contrast agents *in vivo*. The outer surface of these NPs was functionalized by noncovalent amphiphilic Gd chelates to prepare $Bi_2S_3$-Gd NPs. These NPs were suitable for CT imaging-guided PTT. The size of $Bi_2S_3$-Gd NPs was obtained at ~170 nm. The high stability of these NPs was demonstrated by the 42.3 mV value of zeta potential. Under the NIR irradiation of 500s, their temperature increased to 23.2 °C at 300 μg/mL concentration, while the temperature of ultrapure water (control) increased only up to 6.6 °C; this showed that the $Bi_2S_3$-Gd NPs had a high PT efficacy. As $Bi_2S_3$ had a strong absorbance in the NIR region, there was no significant change in the absorbance of $Bi_2S_3$-Gd after NIR irradiation, suggesting good photostability. The cytotoxicity was analyzed in Hela and MCF-7 cell lines via

MTT assay; more than 80% of cell viabilities indicated no significant cytotoxicity. The PTT cytotoxicity in Hela cells under 2W/cm$^2$ 808 nm laser for 1 minute of irradiation exhibited a concentration-dependent rise in temperature. This suggested their high efficacy as a PTT agent for CT-guided cancer therapy. Moreover, CT imaging performed at different concentrations of Bi (0, 6, 12.5, 25, and 50 mM) demonstrated a linear behavior between HU value with Bi concentration—the obtained slope was larger than iodine-based contrast agents (control). This showed that Bi$_2$S$_3$-Gd NPs could be used for enhanced CT imaging as well as to suppress tumor development when exposed to NIR laser light.

### 4.3.10   GADOLINIUM-BASED NPs

Lanthanides (e.g. gadolinium) are ideal contrast agents for *in vivo* micro-CT as their K-edge coincides with the peaks of the X-rays used to perform HD *in vivo* micro-CT scans. Tsungrong Kuo et al. (2013) [29] synthesized gadolinium oxide (Gd$_2$O$_3$:Eu) NPs doped by europium, and their surfaces were decorated with an AS1411 aptamer to create conjugated A-GdO: Eu NPs. They had a size of ~10 nm. They were selective in targeting lung cancer cells (CL1–5) and exhibited strong CT signals. The X-ray attenuation of these NPs was roughly twofold more than iohexol (control) at equal concentrations. The high K-absorption edge of Gd in these NPs was the prime reason for the strong CT signal. In BALB/c mice, the red color of these NPs was seen in the tumor with naked eyes via CT modality. The MTT assay of the conjugated NPs showed no significant cytotoxic response for 100 µg/mL. They exhibited superior fluorescence, magnetic susceptibility, and good biocompatibility. Yu Cheng et al. (2019) [30] produced ultrasmall gadolinium oxide (Gd$_2$O$_3$). Their surfaces were decorated with bovine serum albumin (BSA) protein. The modified NPs were linked with MoS$_2$ and HA via disulphide bonds to produce Gd$_2$O$_3$/BSA@MoS$_2$-HA NPs. They exhibited substantial tumor uptake and glutathione-mediated removal capabilities. The degraded fragments were removed completely by the hepatobiliary system observed *in vitro* and *in vivo* surveys. *In vivo* experiments showed that the NPs possessed the outstanding capability of inhibiting tumor growth.

### 4.3.11   RHENIUM-BASED NPs

Rhenium sulfide is a potent drug that is utilized in preclinical studies for tumor radiotherapy. Xiaoyi Wang et al. (2019) [31] produced ReS$_2$ NPs by a facile method at room temperature. The NPs presented a uniform size of 3 nm with spherical morphology. ReS$_2$ NPs possess semiconducting properties with a bandgap in the NIR range and displayed PT efficiency with a temperature rose by 45 °C for 0.5 mg/mL ReS$_2$ solution. The NPs when incubated with 4T1 cells for 1 day using MTT assay exhibited 80% cell viability. And upon irradiation by laser, less than 15% of cells survived, highlighting its superior PT capabilities. The CT results of ReS$_2$ were better than iohexol (control). Kunming mice were given ReS$_2$ orally, and the CT images of the proximal small intestine and stomach were highlighted within 5 minutes of administration. After 3 hours, NPs moved to the distal small intestine and exited the body in 72 hours. The value of HU improved from 30 to 110 HU when 100 µL of

ReS$_2$ aqueous solution was injected into the tumor-bearing mice model. Under NIR illumination for 10 minutes, the BALB/c mice bearing 4T1 tumors exhibited a rise in tumor temperature by 31 °C as compared to 10 °C in PBS (control). Kunming mice monitored upon oral administration exhibited no weight loss and no histopathological damage to major organs, establishing that NPs were nontoxic *in vivo*.

## 4.3.12 LEAD-BASED NPS

Lead sulphide has peculiar electrical properties. Yibiao Zou et al. (2018) [32] used the atom transfer radical polymerization technique to produce lead sulphide (GCGLS) NPs by grafting gadolinium (III) complexes on its surface, as shown in Figure 4.7. The B16 cell line was incubated for 1 and 2 days with GCGLS NPs. They exhibited cell viability up to 80% *in vitro* even at high concentrations. Upon MTT assay execution of the aforementioned cell in the presence of NIR laser, 80% of the cell died and exhibited GCGLS NPs excellent PTT efficiency. The GCGLS NPs demonstrated higher X-ray attenuation than iodixanol (control), because of Pb content. The CT signal showed a linear increase with the concentration of NPs. GCGLS NPs accumulated in the tumor area. Under NIR laser illumination, the temperature of the tumor rose to 52 °C from 28.8 °C within 2 minutes and suppressed the tumor growth. GCGLS NPs started accumulating in the kidney 2 days postinjection and displayed good renal clearance. Besides CT imaging, the presence of Gd made GCGLS NPs a suitable candidate for MRI.

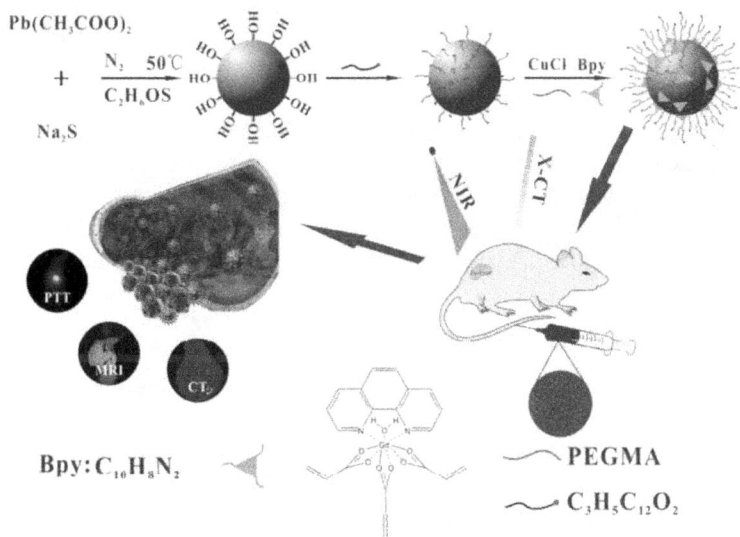

**FIGURE 4.7** Illustration of the preparation of GCGLS nanoparticles and used as dual-modality CT/MR imaging guided photothermal ablation. Copyright © (2018) ACS Publications.

### 4.3.13 SILICA-BASED NPs

Using $MoS_2$ quantum dots doped disulfide-based $SiO_2$ NPs ($MoS_2$@ss-$SiO_2$), Peishan Li et al. (2019) [33] created amino-modified biodegradable nanomaterials. Disulphide bonds induced good reactivity and biological activity. These bonds ruptured in the presence of glutathione and assisted in the release of payload. $MoS_2$ was added to scale their PTT efficiency and to make these NPs a suitable CT contrast material because Mo has high Z. The ultrasmall size of NPs assisted in easy renal clearance. In addition, to enhance tumor-targeting properties, hyaluronic acid (HA) and chlorin e6 (Ce6) were adsorbed on their surfaces to create a photodynamic solution of $MoS_2$@ss-$SiO_2$-Ce6/HA. *In vitro*, the temperature of $MoS_2$@ss-$SiO_2$-Ce6/HA NPs (1000 µg/mL) could rise from 25 °C to 56.8 °C upon laser irradiation. When these NPs were incubated with 4T1 cells for 1 day, the cell viabilities were high, showing the material's high biocompatibility. For *in vivo* CT imaging, they were injected in mice and found to accumulate more in the tumor region than in other organs. The CT signal reached its maximum value 12 hours postinjection. Upon irradiation with laser, the NPs showed tumor inhibition effect in 4T1 cancer-bearing mice for 21 days of treatment.

### 4.3.14 JANUS NPs

Janus NPs (JNPs) have two or more distinct physical properties on their surfaces, allowing two types of chemistry to occur on the same particle. JNPs have intrinsic imaging and therapeutic characteristics, making them theranostically desirable NP. Yanmin Ju (2017) et al. [34] synthesized monodisperse Au-$Fe_2$C JNPs of 20 nm diameter. They had a large absorption band in NIR. They had a PT transduction efficiency of 30.2% *in vitro*, which showed strong PTT effects. The conjugation of affibody with Au-$Fe_2$C JNPs resulted in deeper penetration and larger tumor accumulation efficacy at tumor locations in comparison to the non-target-oriented JNPs. The synthesized JNPs exhibited an incremental rise in the temperature from 4.99 °C to 49.95 °C after irradiation by laser for 5 minutes, while there was no increase in temperature for water (control). This showed the excellent PT efficiency of these JNPs. The cytotoxicity and PTT effectiveness were tested *in vitro* in $NIH_3T_3$ cells of mice and human breast cancer MDAMB-231 cells. No significant cytotoxicity was observed in either cell line at concentrations up to 400 g/mL, even after 24 hours of incubation.

## 4.4 SUMMARY AND FUTURE SCOPE

This chapter presents that metallic NPs are the most favorable substitute for iodinated compounds (Z = 53), which are used exceedingly in present-day CT imaging procedures. However, the obvious drawbacks of these compounds compelled researchers to discover a substitute for iodine-based contrast agents. Metallic NPs with high Z are the preferred materials; however, their toxicity to the human body remains a barrier. The barrier can be circumvented by decorating metallic NPs with organic compounds. The coating of one or more materials helps gain multimodal functionality with biocompatibility. The metallic particles that are the first preference of researchers include silver, gold, bismuth, gadolinium, and platinum. The fundamental desirable properties of the modified/functionalized NPs involve monodispersity, non-toxic to healthy cells, provide large CT imaging window, biodegradable or excretable, provide localized contrast to abnormality, scalable and cost-efficient production, target-specific, and multimodal imaging. The applications of NPs in CT-guided processes extend from a CT-enhanced contrast agent to nanomedicine, target-oriented dye, in CT-guided tumor vessel radiotherapy, PT cancer therapy, bacteriolytic tumor therapy, and multimodal imaging (viz. magnetic resonance imaging, photoacoustic imaging, fluorescence imaging). So far, the research has identified NPs CT-guided intervention devices, such as a needle for percutaneous biopsy, abscess draining, catheter-based angiography, and strategic thoracoscope placement in video-assisted thoracoscopic surgery. There is a tremendous opportunity for more advancements in this area. Issues such as low-radiation CT imaging and delineated tumor imaging are largely unresolved. In light of recent progress in imaging equipment and contrast agents, the usage of NP-assisted CT imaging for theranostic applications is anticipated to expand further.

## REFERENCES

[1] O. Rabin, J. Manuel Perez, J. Grimm, G. Wojtkiewicz, R. Weissleder, An X-ray computed tomography imaging agent based on long-circulating bismuth sulphide nanoparticles, *Nat. Mater.* 5 (2006) 118–122. https://doi.org/10.1038/nmat1571.

[2] A. Jakhmola, N. Anton, H. Anton, N. Messaddeq, F. Hallouard, A. Klymchenko, Y. Mely, T.F. Vandamme, Poly-ε-caprolactone tungsten oxide nanoparticles as a contrast agent for X-ray computed tomography, *Biomaterials.* 35 (2014) 2981–2986. https://doi.org/10.1016/j.biomaterials.2013.12.032.

[3] S. Liu, H. Li, L. Xia, P. Xu, Y. Ding, D. Huo, Y. Hu, Anti-RhoJ antibody functionalized Au@I nanoparticles as CT-guided tumor vessel-targeting radiosensitizers in patient-derived tumor xenograft model, *Biomaterials.* 141 (2017) 1–12. https://doi.org/10.1016/j.biomaterials.2017.06.036.

[4] Q. Zou, J. Huang, X. Zhang, One-step synthesis of iodinated polypyrrole nanoparticles for CT imaging guided photothermal therapy of tumors, *Small.* 14 (2018) 1803101. https://doi.org/10.1002/smll.201803101.

[5] X. Men, H. Chen, C. Sun, Y. Liu, R. Wang, X. Zhang, C. Wu, Z. Yuan, Thermosensitive polymer dot nanocomposites for trimodal computed tomography/photoacoustic/fluorescence imaging-guided synergistic chemo-photothermal therapy, *ACS Appl. Mater. Interfaces.* 12 (2020) 51174–51184. https://doi.org/10.1021/acsami.0c13252.

[6] L. Fu, S. Yang, S. Jiang, X. Zhou, Z. Sha, C. He, One-step synthesis of multifunctional nanoparticles for CT/PA imaging guided breast cancer photothermal therapy, *Colloids Surf. B Biointerfaces.* 201 (2021) 111630. https://doi.org/10.1016/j.colsurfb.2021.111630.

[7] W. Zhou, Y. Chen, Y. Zhang, X. Xin, R. Li, C. Xie, Q. Fan, Iodine-rich semiconducting polymer nanoparticles for CT/fluorescence dual-modal imaging-guided enhanced photodynamic therapy, *Small.* 16 (2020) 1905641. https://doi.org/10.1002/smll.201905641.

[8] C. Xu, G.A. Tung, S. Sun, Size and concentration effect of gold nanoparticles on X-ray attenuation as measured on computed tomography, *Chem. Mater.* 20 (2008) 4167–4169. https://doi.org/10.1021/cm8008418.

[9] C. Peng, L. Zheng, Q. Chen, M. Shen, R. Guo, H. Wang, X. Cao, G. Zhang, X. Shi, PEGylated dendrimer-entrapped gold nanoparticles for in vivo blood pool and tumor imaging by computed tomography, *Biomaterials.* 33 (2012) 1107–1119. https://doi.org/10.1016/j.biomaterials.2011.10.052.

[10] H. Liu, M. Shen, J. Zhao, J. Zhu, T. Xiao, X. Cao, G. Zhang, X. Shi, Facile formation of folic acid-modified dendrimer-stabilized gold–silver alloy nanoparticles for potential cellular computed tomography imaging applications, *The Analyst*. 138 (2013) 1979. https://doi.org/10.1039/c3an36649a.

[11] W. Park, S. Cho, X. Huang, A.C. Larson, D.-H. Kim, Branched gold nanoparticle coating of *Clostridium novyi* -NT spores for CT-guided intratumoral injection, *Small*. 13 (2017) 1602722. https://doi.org/10.1002/smll.201602722.

[12] Y. Dou, X. Li, W. Yang, Y. Guo, M. Wu, Y. Liu, X. Li, X. Zhang, J. Chang, PB@Au Core-satellite multifunctional nanotheranostics for magnetic resonance and computed tomography imaging in vivo and synergetic photothermal and radiosensitive therapy, *ACS Appl. Mater. Interfaces*. 9 (2017) 1263–1272. https://doi.org/10.1021/acsami.6b13493.

[13] A. Tomitaka, H. Arami, A. Raymond, A. Yndart, A. Kaushik, R.D. Jayant, Y. Takemura, Y. Cai, M. Toborek, M. Nair, Development of magneto-plasmonic nanoparticles for multimodal image-guided therapy to the brain, *Nanoscale*. 9 (2017) 764–773. https://doi.org/10.1039/C6NR07520G.

[14] S. Uthaman, H.S. Kim, V. Revuri, J.-J. Min, Y. Lee, K.M. Huh, I.-K. Park, Green synthesis of bioactive polysaccharide-capped gold nanoparticles for lymph node CT imaging, *Carbohydr. Polym*. 181 (2018) 27–33. https://doi.org/10.1016/j.carbpol.2017.10.042.

[15] J. Zhu, G. Wang, C.S. Alves, H. Tomás, Z. Xiong, M. Shen, J. Rodrigues, X. Shi, Multifunctional dendrimer-entrapped gold nanoparticles conjugated with doxorubicin for pH-responsive drug delivery and targeted computed tomography imaging, *Langmuir*. 34 (2018) 12428–12435. https://doi.org/10.1021/acs.langmuir.8b02901.

[16] B. Zhou, Z. Xiong, P. Wang, C. Peng, M. Shen, X. Shi, Acetylated polyethylenimine-entrapped gold nanoparticles enable negative computed tomography imaging of orthotopic hepatic carcinoma, *Langmuir*. 34 (2018) 8701–8707. https://doi.org/10.1021/acs.langmuir.8b01669.

[17] Y. Gao, J. Kang, Z. Lei, Y. Li, X. Mei, G. Wang, Use of the highly biocompatible Au Nanocages@PEG nanoparticles as a new contrast agent for in vivo computed tomography scan imaging, *Nanoscale Res. Lett*. 15 (2020) 53. https://doi.org/10.1186/s11671-020-3286-2.

[18] H.-Q. Yin, P.-P. Cao, X.-Y. Wang, Y.-H. Li, X.-B. Yin, Computed tomography imaging-guided tandem catalysis-enhanced photodynamic therapy with gold nanoparticle functional covalent organic polymers, *ACS Appl. Bio Mater*. 3 (2020) 2534–2542. https://doi.org/10.1021/acsabm.0c00244.

[19] S. Zhang, C. Sun, J. Zeng, Q. Sun, G. Wang, Y. Wang, Y. Wu, S. Dou, M. Gao, Z. Li, Ambient aqueous synthesis of ultrasmall PEGylated $Cu_{2-x}Se$ nanoparticles as a multifunctional theranostic agent for multimodal imaging guided photothermal therapy of cancer, *Adv. Mater*. 28 (2016) 8927–8936. https://doi.org/10.1002/adma.201602193.

[20] S. Lu, X. Li, J. Zhang, C. Peng, M. Shen, X. Shi, Dendrimer-stabilized gold nanoflowers embedded with ultrasmall iron oxide nanoparticles for multimode imaging-guided combination therapy of tumors, *Adv. Sci*. 5 (2018) 1801612. https://doi.org/10.1002/advs.201801612.

[21] J.C. Hsu, E.D. Cruz, K.C. Lau, M. Bouché, J. Kim, A.D.A. Maidment, D.P. Cormode, Renally excretable and size-tunable silver sulfide nanoparticles for dual-energy mammography or computed tomography, *Chem. Mater*. 31 (2019) 7845–7854. https://doi.org/10.1021/acs.chemmater.9b01750.

[22] E.M. Lee, J. Lee, Y. Kim, K.S. Yi, J. Cho, J. Kim, J.M. An, D. Lee, S.J. Kim, E. An, Y.J. Hong, H. Jo, S.H. Lee, Y. Jung, C.-H. Choi, J.S. Kang, J. Hur, D. Kim, Hybrid composite of silver nanoparticle–porous silicon microparticles as an image-guided localization agent for computed tomography scan of the lungs, *ACS Biomater. Sci. Eng*. 6 (2020) 4390–4396. https://doi.org/10.1021/acsbiomaterials.0c00611.

[23] Z. Wang, G. Kuang, Z. Yu, A. Li, D. Zhou, Y. Huang, Light-activatable dual prodrug polymer nanoparticle for precise synergistic chemotherapy guided by drug-mediated computed tomography imaging, *Acta Biomater.* 94 (2019) 459–468. https://doi.org/10.1016/j.actbio.2019.05.047.

[24] D. Ni, J. Zhang, W. Bu, C. Zhang, Z. Yao, H. Xing, J. Wang, F. Duan, Y. Liu, W. Fan, X. Feng, J. Shi, PEGylated NaHoF4 nanoparticles as contrast agents for both X-ray computed tomography and ultra-high field magnetic resonance imaging, *Biomaterials.* 76 (2016) 218–225. https://doi.org/10.1016/j.biomaterials.2015.10.063.

[25] P. Lei, P. Zhang, S. Yao, S. Song, L. Dong, X. Xu, X. Liu, K. Du, J. Feng, H. Zhang, Optimization of Bi $^{3+}$ in upconversion nanoparticles induced simultaneous enhancement of near-infrared optical and X-ray computed tomography imaging capability, *ACS Appl. Mater. Interfaces.* 8 (2016) 27490–27497. https://doi.org/10.1021/acsami.6b08335.

[26] C. Yang, C. Guo, W. Guo, X. Zhao, S. Liu, X. Han, Multifunctional bismuth nanoparticles as theranostic agent for PA/CT imaging and NIR laser-driven photothermal therapy, *ACS Appl. Nano Mater.* 1 (2018) 820–830. https://doi.org/10.1021/acsanm.7b00255.

[27] H. Bi, F. He, Y. Dong, Y. Dai, L. Xu, R. Lv, S. Gai, P. Yang, J. Lin, Bismuth nanoparticles with "light" property served as a multifunctional probe for X-ray computed tomography and fluorescence imaging, *Chem. Mater.* 30 (2018) 3301–3307. https://doi.org/10.1021/acs.chemmater.8b00565.

[28] H. Guo, X. Zhao, H. Sun, H. Zhu, H. Sun, Synthesis of gadolinium-based Bi $_2$ S $_3$ nanoparticles as cancer theranostics for dual-modality computed tomography/magnetic resonance imaging-guided photothermal therapy, *Nanotechnology.* 30 (2019) 075101. https://doi.org/10.1088/1361-6528/aaf442.

[29] T. Kuo, W. Lai, C. Li, Y. Wun, H. Chang, J. Chen, P. Yang, C. Chen, AS1411 aptamer-conjugated Gd2O3:Eu nanoparticles for target-specific computed tomography/magnetic resonance/fluorescence molecular imaging, *Nano Res.* 7 (2014) 658–669. https://doi.org/10.1007/s12274-014-0420-4.

[30] Y. Cheng, T. Lu, Y. Wang, Y. Song, S. Wang, Q. Lu, L. Yang, F. Tan, J. Li, N. Li, Glutathione-mediated clearable nanoparticles based on ultrasmall Gd $_2$ O $_3$ for MSOT/CT/MR imaging guided photothermal/radio combination cancer therapy, *Mol. Pharm.* 16 (2019) 3489–3501. https://doi.org/10.1021/acs.molpharmaceut.9b00332.

[31] X. Wang, J. Wang, J. Pan, F. Zhao, D. Kan, R. Cheng, X. Zhang, S.-K. Sun, Rhenium sulfide nanoparticles as a biosafe spectral CT contrast agent for gastrointestinal tract imaging and tumor theranostics in vivo, *ACS Appl. Mater. Interfaces.* 11 (2019) 33650–33658. https://doi.org/10.1021/acsami.9b10479.

[32] Y. Zou, H. Jin, F. Sun, X. Dai, Z. Xu, S. Yang, G. Liao, Design and synthesis of a lead sulfide based nanotheranostic agent for computer tomography/magnetic resonance dual-mode-bioimaging-guided photothermal therapy, *ACS Appl. Nano Mater.* 1 (2018) 2294–2305. https://doi.org/10.1021/acsanm.8b00359.

[33] P. Li, L. Liu, Q. Lu, S. Yang, L. Yang, Y. Cheng, Y. Wang, S. Wang, Y. Song, F. Tan, N. Li, Ultrasmall MoS2 nanodots doped biodegradable SiO2 nanoparticles for clearable FL/CT/MSOT imaging guided PTT/PDT combination tumor therapy, *ACS Appl. Mater. Interfaces.* 11 (2019). https://doi.org/10.1021/acsami.8b18924.

[34] Y. Ju, H. Zhang, J. Yu, S. Tong, N. Tian, Z. Wang, X. Wang, X. Su, X. Chu, J. Lin, Y. Ding, G. Li, F. Sheng, Y. Hou, Monodisperse Au–Fe $_2$ C Janus nanoparticles: An attractive multifunctional material for triple-modal imaging-guided tumor photothermal therapy, *ACS Nano.* 11 (2017) 9239–9248. https://doi.org/10.1021/acsnano.7b04461.

# 5 Applications of Lipid Nanocarriers in Medical Imaging

*S. Mohana Lakshmi, C. K. Ashok Kumar, and Shvetank Bhatt*

## 5.1 INTRODUCTION

Nanotechnology has achieved great progress, notably in the fields of material science, photonics, and treatment processes (Puri et al., 2021) as well as in the pharmaceutical and food industries. Nanomedicine (applying nanotechnology in medicine) has fuelled the upsurge in the research of various drug-carrying nanoparticles (Germain et al., 2020). Pharmacological release; absorption, distribution, metabolism and excretion (ADME); and other physical, chemical, and biological mechanisms that impact the bioavailability of bio-components have been highlighted in studies to help enhance the bioavailability of drug compounds. Nanocarriers, which are employed as alternative drug delivery systems, have received a lot of interest recently because of their improved drug penetration and distribution into specific parts of the skin. These nanocarriers might lead to longer half-lives, more precise short- and long-term release kinetics, and site-specific medicinal component delivery (Ragelle et al., 2017).

Nanocarriers are colloids with diameters ranging from 1 to 1000 nm. Medical nanocarriers are generally 1–100 nm in size. The determination of the type of nanocarrier is based on the interaction between the polar charged body composed of phospholipids and the solvent, or the nonpolar uncharged lipid groups in the solvent. The presence of specific physicochemical characters, such as nanoparticles with integrated lipid core matrix and higher compatibility with the biological environment, pose lipid-based nanocarriers (LNs) as drug carriers (Jahangir et al., 2020).

### 5.1.1 LIPID-BASED NANOCARRIERS

These LNs (Figure 5.2), which are composed of homogeneous bilayer lipids in solid cores, can entrap a wide range of drugs. The hydrophilic drug can be enclosed in the aqueous region, when the lipophilic medication is confined to the lipid leaflets. Furthermore, LNs securely deliver drugs to the tumour site, gradually release them, and then break them down (Wacker, 2013 ; Devaraj GN et al., 2002). Liposomes, niosomes, self-nanoemulsifying and self-microemulsifying drug delivery systems (SNEDDS and SMEDDS), micelles,

DOI: 10.1201/9781003316398-5

nanoemulsions, polymeric nanoparticles, and albumin-based formulations are all examples of lipid-based nanomaterials that have lately burst in popularity (Bordat et al., 2019; McClements DJ., 2012).

## 5.1.2 Diagnostic Tools

Novel drugs that treat various diseases and their delivery systems are continually being investigated, enhancing the efficacy and safer environment of therapies while also providing additional advantages, such as the administration of existing drugs via more preferred routes, such as oral administration. Simultaneously, new diagnostic methods are being thoroughly researched to allow early diagnosis of problematic situations and prompt treatment. *In vivo* imaging and real-time monitoring and proper and efficient diagnosis also allow the notion of tailored therapy, assuring the appropriate dose and dosage, customized for the individual patient, and delivering 100% efficacy with minimal hazards. However, to achieve optimal efficacy, the two disciplines must be merged and utilized synergistically, especially in the case of complicated disorders (Haider et al., 2020). In recent decades, disease detection and therapy have come a long way. Some of the diagnostic methods available include ultrasound scans (US), computer tomography scans (CT scan), magnetic resonance imaging scans (MRI scans), and proton emission tomography scans (PET scans).

Ultrasound sonography is a low-effective, real-time imaging method that relies on the backscattering of ultrasonic waves partially (frequencies ranging from 2 to 12 MHz) by various organs due to impedance mismatches (Massoud and Gambhir, 2003). Ultrasound contrast chemicals are frequently required to get a better image and discriminate between sick and healthy tissues due to the limited variation in echogenicity between various soft tissues.

MRI is a non-surgical imaging technology that can scan tissues at any depth with excellent clarity and contrast. Due to its limited sensitivity in comparison to other types of imaging techniques and to reduce scan run times, MRI has a thriving research community focused on developing MRI contrast agents. Because different tissues have distinct relaxation characteristics, MRI may be used to recreate pictures of structures like organs and lesions, as well as analyse perfusion and flow-related disorders. Contrast agents that boost the T1-signal or reduce the T2-signal, resulting in an enhancement of bright-positive or dark-negative contrasts, are used to improve MRI (Villaraza et al., 2010).

PET is also a non-surgical nuclear imaging technology that uses a radio labelling of a substance to see deeper tissues with the greatest sensitivity and generate a 3-D picture of patients. Quantitative information may be retrieved from pictures using statistical reconstructive methodologies and correlation factors, and radioisotope concentration can be quantified in the exact region of interest (Andreozzi et al., 2011).

## 5.1.3 Contrasting Agents and Dyes

By acquiring multimodal pictures, medical imaging overcomes intrinsic limits in terms of specificity, sensitivity, scanning run time, and resolution. This has sparked the creation of innovative scanners that can use various imaging modalities in tandem (e.g., PET and MRI scans) as well as the emergence of a scientific sector devoted to the production of multimodal imaging probes (contrast agents) (Sanchez et al., 2017). The list of contrasting agents and their medical applications are mentioned in Table 5.1.

Inorganic metallic agents such as iron oxides are remarkable among the contrast agents used in MRI (Figure 5.1) because they liberate electrons that freely excite and may be exploited as imaging criteria. Various nanoparticulate formulations of minute sizes like quantum dots, based on semiconductors, are also widely used in diagnostics *in vivo* (Ding et al., 2014; Park SM et al., 2017).

### 5.1.3.1 Fluorine

[19]Fluorine has been treated as a preferential MRI contrasting agent due to its exceptional magnetic properties, including 100% abundance in nature with $1^{1/2}$ spin, sensitivity, and protonic equivalence of gyromagnetic ratio. However, it has a drawback in that a high dose (10–50 mM) is required for appropriate intensity of the signal compared to various contrasting agents (Hequet et al., 2020; Sánchez A et al., 2018).

Perfluorocarbons (PFCs) have been used as contrast agents in ultrasound sonography and MRI. Because liquid PFCs (long perfluorinated carbon chains) are more resistant to pressure variations and mechanical stresses than gaseous PFCs (short perfluorinated carbon chains), they have been utilized instead of gaseous PFCs. PFC droplets encapsulated by nanoparticulate devices, like nanodroplets that are coated with phospholipids and cholesterol, are intended to provide liquid PFC via the parenteral route (Diou et al., 2012).

### 5.1.3.2 Gadolinium

Though Gd is used extensively for labelling and contrasting in diagnosing diseases, the primary disadvantage of $Gd^{3+}$ is that it is chemically similar to endogenous metals (such as Ca and Zn), which may result in transmission halts in the neuromuscular junctions (Bellin et al., 2006; De Smet M et al., 2013).

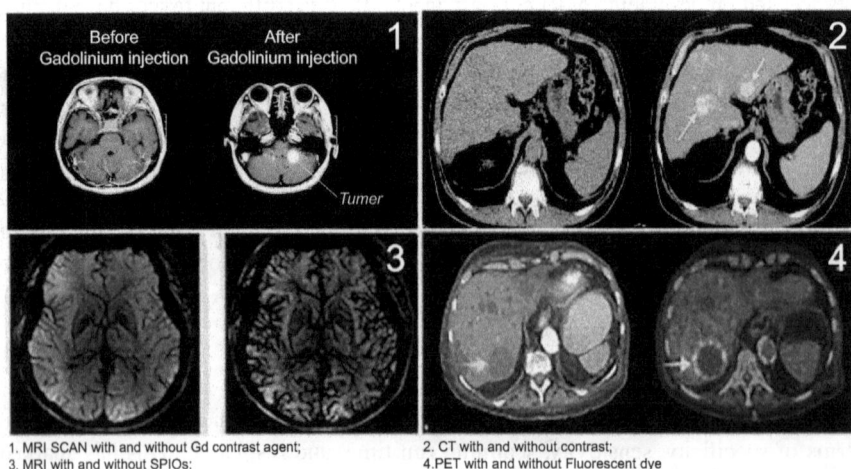

1. MRI SCAN with and without Gd contrast agent;
2. CT with and without contrast;
3. MRI with and without SPIOs;
4. PET with and without Fluorescent dye

**FIGURE 5.1**   Use of contrasting agents in tandem with diagnostic tools.

### 5.1.3.3 SPIOs

Superparamagnetic iron oxide nanoparticles (SPIOs) are investigated for their *in vitro* and *in vivo* applications in imaging, as a kind of T2 MRI contrast agent. They have a strong magnetic moment, which can boost the relaxivities of protons by up to ten times (Villaraza et al., 2010).

Detection of the metastasis from the primary cancers to the sentinel invasion to the lymph nodes and proper visualization of the physiological process like cell apoptosis, expression of genes has been achieved through the SPIOs encapsulated inside the magnetic nanoparticles with a broad spectrum of applications and enhanced efficacy (Reddy et al., 2012).

### 5.1.3.4 Fluorophores

The medical use of two fluorophores (indocyanine green (ICG) and fluorescein) has been authorized by the FDA for use with PET scanning. ICG's use is restricted by a number of drawbacks, including concentration-based aggregation, low *in vitro* stability in aqueous solvents, poor quantum yield, and strong nonspecific plasma protein binding, which leads to fast removal from the body. ICG has been successfully encapsulated using lipid micelles, like glycocholic acid and phosphatidylcholine, to circumvent these issues (Kirchherr et al., 2009).

## 5.2 NOVEL APPROACHES IN THERANOSTICS (THERAPEUTICS AND DIAGNOSTICS)

Nanotechnology in therapeutics along with imaging in diagnosis in the same arena is used for image-guided drug administration to follow the drug distribution in biological spaces, and pharmacokinetic profiles of the administered therapeutic drug in

**TABLE 5.1**

**List of Contrasting Agents and Their Medical Applications**

| Contrasting Agent | Contrast Coupling | Application |
| --- | --- | --- |
| SPIOs | MRI-T2 contrasting agent | The detection of tumours and metastasis |
| Fluorine, $^{19}$F | MRI-contrasting Ultrasound-contrasting agent | Cell labelling and lymph node detection |
| Perfluorocarbons | MRI-contrasting Ultrasound-contrasting agent | Detection and imaging of physiological and pathological changes |
| Gadolinium, $Gd^{3+}$ | MRI-contrasting | Evaluation of a broad spectrum of musculoskeletal (MSK) disease processes |
| Fluorophores (indocyanine green (ICG) and fluorescein) | PET and CT scans | Detection of primary hepatocellular carcinoma |

real time is one of the most recent growing concepts in nanoparticulate drug delivery systems. As a result, the *in vivo* theranostic technique can revolutionize medication delivery systems and diagnostic imaging.

However, in addition to therapeutic uses, nanomedicine formulations are increasingly being employed for imaging and, more recently, theranostic methods, which are systems and techniques that integrate illness detection and therapy. Formulations using theranostic nanomedicine could be employed for everything from non-invasive assessment of low-molecular-weight drug biodistribution and target site accumulation, to visualization of distribution and release of drugs at the target site, to optimization strategies for triggered drug release, to the accurate prediction and monitoring of therapeutic responses in real time. Nanotheranostic formulations are thus regarded as the most suited systems for preclinical medicine because of the understanding that they may aid in better analysis of various aspects of the process of drug delivery and the development of better systems for drug delivery. They may also aid in realizing the efficacy of "customized medicine" and the development of highly effective and low-toxic regimens of treatment for patients (Ahmad et al., 2016).

### 5.2.1   THERANOSTIC APPLICATIONS OF LIPID NANOPARTICLES

Lipid nanoparticles (LNPs) are the most common type of nanocarrier used to transport therapeutic and diagnostic compounds. Biocompatibility, biodegradability, safety, and co-delivery of molecules that are both hydrophobic and hydrophilic in nature are among their main advantages over other classes. The concept of co-delivery enabled a theranostic approach that addresses the issues of interaction between therapy and diagnosis in real time. The nanolipid systems are bio-inspired, as they resemble natural physiological characteristics and show significant potential as vehicles of diagnostic drug delivery (Neubi et al., 2018).

Advanced nanolipid systems may also have functional capabilities that enable them to react to external and internal physiological stimuli, resulting in stimulus-responsive or stimuli-sensitive nanosystems. Heat, pH changes, magnetic fields, ultrasounds, and light are some of the stimuli that cause structural rearrangements after signal reception from the nanosystem. These results in changes in its features, such as size and membrane integrity, allowing the integrated or encapsulated contents to be released spatiotemporally (Naziris et al., 2016).

As a result, it makes sense to alter diagnostic instruments by including or conjugating other functional lipid nanocarrier components to make them acceptable for concurrent diagnosis. Through surface functionalization with targeting moieties, LNPs can enhance the Pka parameters, effectiveness, and toxicity of both diagnostic and therapeutic drugs, while bypassing biophysiological hurdles and targeting accurately particular tissues and cells (Riaz et al., 2018). This novel paradigm, which combines better imaging (using contrast drugs), specific targeting (using biomarker agent), and therapeutic drug delivery in a single lipid nanocarrier system, appears to hold a lot of promise for overcoming numerous obstacles in successful disease treatment.

Nanocarriers have been used recently as imaging modalities, with promising results in both preclinical and clinical settings. For the past decade, diagnostic nanoparticles have been in the market. Various imaging contrasting agents (e.g.,

SPIOs, paramagnetic metal ions (PMIs), and near infrared (NIR) probing drugs) can be encapsulated in nanocarrier systems to improve the noise-to-signal ratio in the targeted cells compared with healthy surrounding cells. The improved resolution of the images reveal tiny lesions that were previously invisible using existing approaches.

Since SLNs are lipid nanovesicles, the imaging agent combined with the therapeutic agent in lipid vesicles must be compatible with standard procedures of diagnosis such as CT, X-rays, and MRI (Kim et al., 2017). Lipid nanocarrier systems are shown to be more effective than polymer based and inorganic nanoparticles in terms of better safety, higher biocompatibility, and efficient biodegradability, as well as other theranostic concerns (Harrington et al., 2001).

They are classified into many categories, as follows:

1. Depending on the mechanism of action of the drugs and the functional nature that is linked directly to the properties of the biomaterials that are used in the formulations that determine the final output of nanostructure of the self-assembly.
2. Depending on the encapsulated or incorporated therapeutic drug or diagnostic agent that is associated with the desired therapeutic application or the intended diagnostic significance.

Because nanoparticles have dimensions that are comparable to those of biomolecules including RNAs, DNAs, and polyproteins, they help with medication administration and imaging in surgery. In today's theranostics, polymer-based nanoparticles, liposomes, dendrimers, and nanoparticles made of silica are employed. Drug delivery technologies in tandem with image guidance also include pH-triggered nanoparticles, as well as magnetic and photo-responsive theranosomes that are most beneficial in cancer therapies. SLNs containing prostacyclin can also be employed for atherosclerosis therapy using image guidance by suppressing the aggregation of platelets.

A lipid nanocarrier formulation was chosen as a theranostic approach for its biocompatibility and safety. Longer circulation half-life inside the body, strong encapsulation capabilities for drugs and diagnostic agents, large segregation at tumour locations, and improved sensitivity to numerous functions are all key benefits of the lipidic nanocarrier system (Ovais et al., 2018). The list of lipid-based nanoparticles used in drug delivery and medical imaging is found in Table 5.2.

## 5.2.2 LIPOSOMES

Liposomes are phospholipid vesicle systems that were developed in the late twentieth century and consist of two layers of overlapping phospholipids similar to those present in the plasma membrane covering human cells. Liposomes are biocompatible as a result and can help medicine diffuse across plasma membranes. Liposomes are self-assembled vesicles with an aqueous core and one or more concentric lipid bilayers.

Liposomes have a standard shape with a nonpolar lipid bilayer and a polar central core, ranging in size from 20 nm to not more than 100 nm. Liposomes can encapsulate lipophilic drugs in a lipid bilayer while holding and stabilizing hydrophilic pharmaceuticals in the hydrophilic core. As a result, liposomes may transport both

**TABLE 5.2**

**List of Lipid-Based Nanoparticles Used in Drug Delivery and Medical Imaging**

| Nanoparticles | Size | Structure | Advantages | Disadvantages |
|---|---|---|---|---|
| Liposomes | 20–100 nm | Nonpolar lipid bilayers with hydrophilic central core | • High biocompatibility<br>• High drug encapsulation<br>• Versatile for various drugs | High cost to produce Smaller size leads to low efficiency load |
| Solid lipid nanoparticles (SLNs) | 50–1000 nm | Solid lipids coated with phospholipids and surfactants | Biocompatible Controlled drug release Stability | Drug rejection Limited drug encapsulation |
| Nanostructured lipid carriers (NLCs) | 10–1000 nm | Solid and liquid lipid cores surrounded by phospholipids and surfactants | Best encapsulation Better stability Low toxicity | Drug expulsion Surface instability |
| Micro/ nanoemulsion | <100 nm | Interdispersed phases of oil and water | Better penetration Better absorption | Toxicity due to surfactants |
| Nanoassemblies | 100–1000 nm | Phospholipid bilayers assembled with DNA/peptide chains/drugs due to interactions | Easily dispersible Low cost to make | Uncontrolled particle size Lack of stability |
| Micelles | <100 nm | Phospholipid and surfactant in solvents | Targeted drug delivery | Low drug load |

nonpolar and polar medicines via the water-soluble lumen and outer lipid bilayers, increasing their versatility. The limited bilayer area of liposomes, however, makes significant drug loading of hydrophobic medications difficult. The projected greater drug-loading capacity must be balanced against the liposome size distribution of the particles and their relative stability.

Furthermore, because the surface of the liposomes may be altered with appropriate moieties to directly contact a targeted molecule of a certain cell, system, or organ, the liposomes have the advantage of easy customization and targeting. Other compounds, such as cholesterol, can be added to their formulations in addition to phospholipids, enhancing the permeability and penetration of hydrophobic medications across the two layers of the membrane and increasing the storage and stability of this type of nanoparticles in human blood (Yingchoncharoen et al., 2016)

Phosphatidylcholine (PC), ethanolamine (PE), and inositol (PI) are the three main phospholipid (PL) components. For liposomal synthesis, natural herbal phospholipids such as soybean and sunflower PLs are extensively used (PI). Herbal PLs, on the other hand, are more easily oxidized and have less physical and storage stability because of their high degree of unsaturation (Wu et al., 2020). Liposomes made from animal-based phospholipids, such as those obtained from milk and eggs, as well as those derived from marine sources, are becoming increasingly popular. The presence of more Polyunsaturated Fatty Acids (PUFA) in marine sources has been demonstrated, making them a possible functional component (Burri et al., 2012).

### 5.2.2.1    Liposomes for Theranostic Approach

Because of their unique features, liposomes are by far the most well known and multifunctional. They are made up of unilamellar bilayer lipids that surround an aqueous central core and have many benefits, including biodegradability, long-term therapeutic release, very low toxicity, and the capability to incorporate both hydrophobic and hydrophilic drug molecules. Their surfaces can also be altered to aid in disease treatment (Silva et al., 2019).

Liposomes have been studied in animals and humans on various dimensions as imaging and treatment tools due to their diverse features. Immediate and precise diagnosis, monitoring of spread of disease and treatment with drugs and *in vivo* pharmacokinetics in real time, and thorough information on diseases are all possibilities with molecular imaging employing multimodal probes. Multimodal imaging, which may be used to extract accurate and exact assessments of hallmarks of numerous illnesses, is enabled by the versatility of liposomal surface functionalization to attach diverse molecular probes. The goal of this study was to concentrate on liposomes' clinical applications and their prospective usage as imaging agents and in image-guided medication delivery systems.

In spite of their capability to load various large-small molecules, they were explored to act as carriers for a variety of diagnostic agents, like $^{64}$Cu (Petersen et al., 2012). Copper incorporated into the liposomal formulations carrying therapeutic drugs for various kinds of diseases is extremely beneficial from a theranostic perspective. Moreover, cancer theranostics with copper liposomes is a significant advantage in both treating the cancer and identifying the growth of the tissue using a common single formulation.

A theranostic nanoparticulate system with incorporated siRNA-ABC (PEGylated siRNA-nanoparticulate system) was composed, which was assembled by cationic PEGylated liposomes that are prepared by using DOPE-Rhodamine and Gd.DOTA.DSA. Later, this system was used to entrap Alexa fluor 488-labeled anti-survivin siRNA. This produced system was found to be capable of mediating functional siRNA delivery to tumours, resulting in significant phenotypic (pharmacodynamic) tumour size reductions compared to controls, while biodistribution of nanoparticles and behaviour of siRNA was observed simultaneously using real-time imaging (Kenny et al., 2011).

Moreover, the concept of theranostics has evolved more recently and was used in one study to entrap the anticancer drug, doxorubicin and [Gd(HPDO3A)(H2O)] that is a contrasting agent for MRI. This was formulated in the form of drug nanoparticles that were triggered with the application of temperature (De smet et al., 2010). The contrasting agent efficiently improved the quality of imaging using MRI alongside reducing the noise and other disturbances that hinder the reports, leading to improper diagnosis or estimation of the stage of cancer.

Gadonano-F LNPs were made using the lipid components listed earlier. For MRI-positive contrast imaging, gadolinium metallo-chelating lipid (Gd.DOTA.DSA) is used, while rhodamine conjugated lipids (DOPE-Rhoda) is used for fluorescence imaging. The association of folate with the surface of Gadonano-FLNP is achieved by the use of conjugated lipid with polyethylene glycol (folate-PEG2000-DSPE), guaranteeing that LNPs are capable of tumour cell targeting that is specific to FR *in vivo* (Kamaly et al., 2010).

Wong et al. conducted research to compare the distribution of [64]Cu-conjugated liposomes with the [18]F-FDG and to evaluate the clinical efficacy of the prepared liposomes. The evaluation was performed in terms of the therapeutic efficacy and diagnostic ability to detect the volume of tumours by acting as a contrasting agent. They suggested that the results were comparable to both formulations and there was heterogeneity of the tracers in imaging the tumours (Wong et al., 2013).

In the aqueous phase of multilamellar liposomes, iodine-125 diatrizoate or 125-I-iotrol was used. The leakage of 125I activity from phosphatidylcholine/cholesterol/stearylamine (P/C/S) liposomes was two to three times greater than that from SM vesicles. These were injected into the experimental rats, and the results suggested that the drug clearance was enhanced to 0.5 times the activity from the spleen and liver and from the whole body it was ~ 24 hr (Zalutsky et al., 1987).

The 111In-labeled liposomal carrier and co-encapsulated doxorubicin and contrasting agent [Gd(HPDO3A)] were measured in the blood and organs of tumour-bearing rats. The biodistribution of radiolabelled liposomes was studied using SPECT/CT scans, with the spleen and liver showing the greatest uptake of 111In-labeled TSLs. After 48 h, the MR-HIFU-treated tumours had four times more liposomal absorption than the control. The concentration of the drug doxorubicin has been raised seven times (Se Smet et al., 2013).

Liposomes encapsulating gadobenate dimeglumine and doxorubicin were developed (Gd-Dox-STL). The release profile of DOX was similar in both formulations. The difference in the T1-relaxation time was higher than the difference for Gd-DOX-TSL, according to MRI studies. Although both liposome relaxities at 42 °C were identical, Gd-DOX-relaxivity STL's at 37 °C was two times lower than Gd-DOX-LTSL's (Kim et al., 2016). Nordling-David et al. used liposomes that encapsulated temozolomide (TMZ) along with Gd-DTPA liposomes to treat glioblastoma. Treatment of rats that are bearing glioblastoma with either a free drug or the liposomes of the drug resulted in much stronger tumour suppression (Nordling-David MM et al., 2017).

Shao et al. used 2-[1-hexyloxyethanol]-2-devinyl pyropheophor-bide-a (HPPH) to make manganese-conjugated liposomes. N-HPPH-lipid, which was also utilized to produce liposomes that conjugated with manganese, was obtained by amine modification of this derivative. They found a 150% increase in T1 relaxivity (mM • s)−1 with the amino alteration (Shao et al., 2017)

Han et al. created the EC1-GLuc recombinant protein by combining the EC1-peptide, an ErbB2 ligand, and Gaussia luciferase (GLuc). To make the EC1-GLuc-liposome, pure EC1-GLuc was combined with a liposome that chelates nickel. For bioluminescence imaging, *in vitro* tests suggested that the produced liposomes preferentially targeted and internalized ErbB2-overexpressing SKOv3 cells. The SKOv3 cells were efficiently administered with a fluorescent dye that is impermeable in the cells (HPTS) contained in the EC-GLuc-liposome. Furthermore, HPTS was successfully transported through EC1-GLuc-liposome into metastatic SKOv3 tumours *in vivo* and targeted them for bioluminescence imaging (Han et al., 2014).

### 5.2.3 SOLID LIPID NANOPARTICLES (SLN)

The SLNs which were developed in 1990 are researched further to combine the benefits of polymers and nanocarriers, such as increased drug loading, controlled delivery, biocompatibility of the lipid emulsion, and better bioavailability of the drug (Yuan et al., 2007). At room temperature, SLNs are a form of colloidal drug delivery system made up of lipids that are solid. Biocompatible compounds such as fatty acids, triglycerides, biowaxes, and steroids are often used in SLN systems. This whole matriculate system is stabilized by a mixture of polymers and surfactants. The size of these particles ranges from 50 to 1000 nanometres. Due to their nanosize and huge surface area, SLN is good for coating with ligand molecules, antibodies, and other groups (Rostami et al., 2014).

SLN has many advantages, including simplicity of synthesis, pharmaceutical stability, higher drug percentage, effective drug release and higher stability, sustained release of both hydrophobic and hydrophilic pharmaceuticals, prevention of degradation of labile drugs, cost affectivity, and nontoxicity. The SLN method may also successfully encapsulate anticancer agents and other chemicals with limited solubility due to their high lipid content (Wong et al., 2007). SLNs, on the other hand, have a restricted drug-loading capacity as well as drug ejection due to crystallization during storage.

#### 5.2.3.1 SLNs for Theranostic Approach

Due to biodegradability and biocompatibility, the simultaneous approach for the use as *in vivo* diagnostic agents and for therapeutic administration is a well-accepted method for safer medical uses, especially with the novel possibilities of multi-use attributed to the nanomedicine. It's possible that SLNs may provide an enhanced administration of theranostic drugs based on safety and cost effectiveness, with manufacturing that can be readily hiked up and adapted to accurate tasks like directed drug targeting (Albuquerque et al., 2015).

By co-formulating methotrexate and SPIOs in the form of nanoparticles (SPIONs) inside cetyl palmitates and stearic acid lipids, specific effort to accomplish a combination treatment and diagnosing of rheumatoid arthritis was presented. Anti-CD64, which selectively holds affinity to the receptor on the cellular surface and is excessively expressed in macrophages infected with RA, was used to further categorize these solid lipid nanoparticles (Albuquerque et al., 2015).

*In vitro* investigation revealed relaxometric qualities comparable to Endorem®. A superparamagnetic iron oxide SLN of minute size was synthesized, and *in vitro* examination displayed relaxometric capabilities similar to Endorem®. MRI scans of the CNS using both SLN and Endorem® revealed that superparamagnetic formulations had a slower down in clearance from the blood than Endorem®. The CNS absorption of SLN lasted until the end of the trial, according to MRI data (135 min). These findings support SLN's capacity to cross the BBB; it may be employed as an MRI contrasting agent in the CNS (Peira et al., 2003).

The SLN contains BAT (6-[p-(bromoacetamido)benzyl]-1,4,8,11-tetraazacyclo-tetradecane-N, N′, N″, N‴-tetraacetic acid), a copper-specific chelator coupled with a synthetic lipid. Following purification, the SLNs (diameter 150 nm) was effectively labelled with $^{64}$Cu (66.5% radiolabelled yield) and showed more than 96% labelled formulations. PET imaging was performed at 0.5, 3, 20, and 48 h after the $^{64}$Cu-SLNs were injected into mice via IV route to perform investigations on the diagnostic and therapeutic properties of the drug (Andreozzi et al., 2011).

SLNs incorporating [Gd–DOTA(H$_2$O)]$^2$ and [Gd–DTPA(H$_2$O)]$^2$ were produced and studied relaxometrically by the authors. The assessment of the quantity of Gd-3 complex in the SLNs was performed, and it was discovered that the Gd(III) chelates did not exhibit any changes in their longitudinal relaxivity, at physiological pH, after their incorporation into SLN. There was a noticeable rise in the relaxivity of the para-magnetic SLN that contained highly mobile [Gd–DTPA(H$_2$O)]$^2$ (Morel et al., 1998).

To increase the IR-780 delivery to target the tumours, SLNs coupled with c(RGDyK) were created as the most efficient carrier systems. The varied func-tionality of SLNs with cRGD-IR-780 pairing had desired dispensability in single medium, stability, and targeting of the cell lines considerably expressed the v3 inte-grin. Additionally, cell viability and PTT treatment assays indicated that U87MG cells or tumour transplantation can be destroyed using cRGD-IR-780 SLNs when exposed to laser light (Kuang et al., 2017).

The surface of the SLNs was loaded with Gd(3) complex formulations to create highly effective MRI probes (pSLNs). An *in vitro* study with tailored pSLNs demon-strated their use as molecular imaging probes (Rolla et al., 2013).

**Lipid-based Nanoparticles**

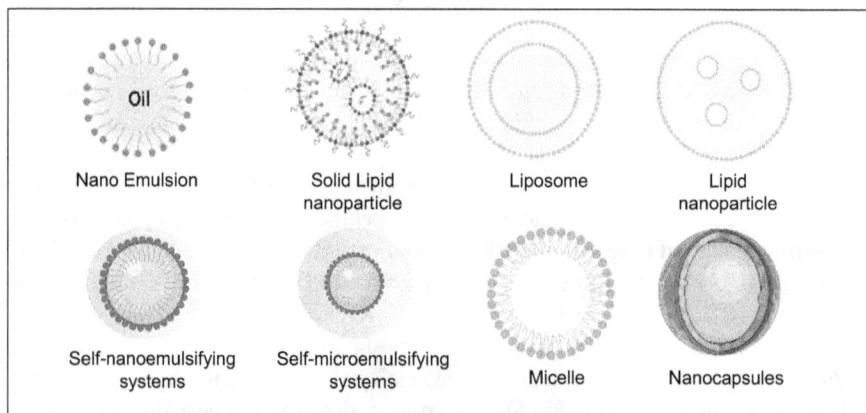

FIGURE 5.2    Various forms of lipid-based nanoparticles.

### 5.2.4 NANO-STRUCTURED LIPID CARRIERS (NLCS)

NLCs have been extensively studied as a medication delivery mechanism. They are considered more successful than prior nanocarriers with lipid bases due to their increased drug-loading capacity, efficient stability, and superior biocompatibility (Haider et al., 2020).

Researchers altered SLN by swapping a small amount of liquid for solid lipids, resulting in an NLC with enhanced biocompatibility and formulation properties. The average particle sizes, which range from 10 to 1000 nm, are equivalent to SLNs.

Many lipids are used in topical lipid nanoparticulate compositions, including fatty acids, steroids, waxes, and triglycerides such as ethyl oleate, glyceryl tricaprylate, isopropyl myristate, and glyceryl dioleate. Because pharmaceuticals are more soluble in oil than solid lipids, traditional NLC has a larger drug encapsulation capability than SLN (Yuan et al., 2007). High drug-loading ability was achieved due to the packing of liquid lipids in the systems, and they may assist in reducing drug ejection during storage by inhibiting lipid crystallization. They offer superb control of the release of the drug and lower *in vivo* toxicity, making them ideal for site-specific targeting. The disadvantage of NLC is that it is difficult to surface functionalize (Puri et al., 2009). However, there are many disadvantages, including the expulsion of drugs during the storage due to the polymorphic transition of the lipid from the nanocarrier matrix and low loading capacity. The list of lipid-based nanoparticles with theranostic application is found in Table 5.3.

#### 5.2.4.1 NLC for Theranostic Approach

NLCs that target CXCR4 containing AMD3100 in the outer shell are composed (AMDNLCs). AMDNLCs containing IR780 lowered cancer cell metastasis while also facilitating tumour targeting and photothermal therapeutic results. When subjected to repetitive irradiation to the laser, the incorporated IR780 generated more heat than the unencapsulated IR780. *In vivo*, CXCR4-targeted NLCs demonstrated promising photothermal anti-tumour and anti-metastatic activity (Li et al., 2017).

### 5.2.5 NANOCAPSULES

When taken orally, nanocapsules are used to boost the Bioavailability (BA) of various drugs (Roger et al., 2011). Lipid nanocapsules are nanovesicles with lipid and water phases separated by a polymeric shell and stabilized by surfactants. The nanocapsules were positively charged and had a diameter of approximately 100 nm. Many of nanocarriers' physicochemical properties, and hence their biological identity, are controlled by their surface properties. As a result, the nanocapsules' lipid polymer outer shell impacts their properties, such as surface charge and hydrophilicity, as well as their gastrointestinal behaviour, such as interactions with mucus and/or enzymes. The derivatives of chitosan, polyethylene glycol (PEG), caprolactone, PLA, and Eudragit-100, are just a few polymers used to make lipid nanocapsules.

### 5.2.5.1  Nanocapsules for Theranostic Approach

The presence of the NFL-TBS is shown by nanocapsules packed with a far-red fluorochrome (DiD) and paclitaxel. (Balzeau et al., 2013). As measured by Fluorescence-Activated Cell Sorting (FACS) analysis, the peptide boosts their absorption in the *in vitro* glioblastoma cells, reducing their growth. In other investigations, LNCs were utilized to transfer genes such as long-circulating DNA, also known as DNA-LNCs, or plasmid DNA using *in vivo* bio-fluorescence imaging (BFI) (David et al., 2013; Morille M et al., 2010).

### 5.2.6  NANOEMULSIONS

Nanoemulsions with droplet sizes of <100 nm are attaining popularity in the pharmaceutical industry due to the unique qualities of nanosized droplets, with their huge surface area (Ashaolu et al., 2021). Oil in water (O/W), water in oil (W/O), and multi-phase emulsions with inter-dispersed microdispersions between oil and aqueous phases (O/W/O and W/O/W) are the three kinds of nanoemulsions based on their components (Zhang et al., 2018).

Based on the charge on the surface of the nanosized droplets, nanoemulsions are classed as anionic, neutral, or cationic nanoemulsions. Vegetable oils, such as sesame oil, soybean oil, and others, are widely used for the dissolution of lipophiles (Constantinides et al., 2004). In comparison to other nanocarriers, nanoemulsions are simple to create and do not require the use of nonpolar/co-solvents, especially when utilizing low-toxicity vegetable oils. Nanoemulsions are thermodynamically unstable, leading to improper stabilization like flocculation and SMEDDS and SNEDDS.

Anhydrous nanoemulsions are otherwise called self-nanoemulsifying drug delivery systems. They consist of oil and surfactant mixed with optional cosurfactant or solubilizers. The fundamental feature of SNEDDS is that they generate nanoemulsions deliberately when poured into the polar or aqueous phase with gentle mixing, such as during GI motility (Guo et al., 2019). Similarly, microemulsions in aqueous media are used by self-microemulsifying drug delivery systems (Khan et al., 2012).

Because of their composition, they have various benefits over ordinary nanoemulsions. First, because they do not contain water, they have superior long-term stability profiles and may be administered in lower amounts. They can also be included in dose carriers like Hydroxypropyl Methylcellulose (HPMC) and gelatin capsules, which increases their acceptability and, as a result, the compliance of the patient (Date et al., 2010). Furthermore, traditional SMEDDS and SNEDDS have significant drawbacks, like long-term component capsule shell incompatibility and notable precipitation of the drug under storage at lower temperatures.

### 5.2.6.1  Nanoemulsions for Theranostic Approach

For MRI, Near-Infrared Fluorescence (NIRF), and therapeutic usage, Gianella et al. designed a theranostic nanomodule made of an O/W nanoemulsion filled with crystals of iron oxide, Cy7 dye, and prednisolone valerate acetate. The efficacy of this nanotheranostic formulation, which coupled the sensitivity of optical fluorescence imaging and the spatial resolution of MRI, was examined on a colon cancer model (Gianella et al.,

2011). Nanoparticle perfluorocarbon emulsions with a core consisting of at least 50% of liquid perfluorocarbons along with a combination of soybean oil and triacetin and paclitaxel encapsulated were studied by Dayton and colleagues (Dayton et al., 2006). "Lipidots™," a revolutionary O/W nanoemulsion technique in which soybean oil and wax are coated with lecithin and PEG to generate exceptionally brilliant fluorescent nanoprobes with minimal cytotoxicity and an excellent pharmacokinetic profile *in vivo*, has also been researched to encapsulate near infrared dyes (Navarro et al., 2012).

## 5.2.7 MICELLES

Micelles are colloidal dispersions that develop spontaneously at specific concentrations and temperatures from amphiphilic or surfactant substances. They are distinguished by two separate parts that have opposing solvent affinities. In the presence of suitable surfactants, fatty acids of long chains or phospholipids are used to create lipid micelles. Amphiphilic molecules exist independently at low concentrations in an aqueous medium, whereas aggregation occurs at concentrations above the critical micelle concentration.

However, because of the small size of their core, they are unable to carry large quantities of medications. Phospholipids are used to make lipid micelles. A decrease in the amount of free energy present in the system is what drives the development of micelles. This decrease in free energy is caused by the elimination of hydrophobic fragments from the aqueous environment and the establishment of a hydrogen bond network in water. Lipid-based micelle manufacturing is a straightforward procedure that produces colloids with extremely comparable diameters in aqueous medium (Torchilin et al., 2007).

### 5.2.7.1 Micelles for Theranostic Approach

The docetaxel (M-DOC) was created and described as a lipid-based nanomicelle-loaded docetaxel. The pharmacokinetics and anticancer effectiveness of docetaxel *in vivo* were assessed using optical imaging. The emulsion-diffusion process was used to make the M-DOC. In optical imaging, the fluorescent dye 1,1-dioctadecyl-3,3,3',3'-tetramethylindotricarbocyanine iodide resembled M-DOC (Ma et al., 2012).

## 5.2.8 SELF-ASSEMBLED NANOPARTICLES

Because of their ease of use, spontaneity, scalability, adaptability, and low cost, self-assembled nanomaterials are finding use in a high number of fields, including nanotechnology, imaging methods, biosensors, and the medicinal sciences, amongst others. Several different types of physical interactions are responsible for the process by which amphiphiles self-assemble into nanostructures such as micelles, vesicles, and hydrogels. The self-assembly process is responsible for the formation of the vast majority of the biological nanostructures, including the building of cell membranes through the assembly of phospholipid bilayers, the helical structure of DNA, the folding of polypeptide chains, and many more. It is also possible that self-assembly is responsible for the contact that occurs between a ligand and its receptor (Azevedo and da Silva, 2018; Decuzzi P and Ferrari M, 2008).

### 5.2.8.1 Self-Assembled Nanoparticles for Theranostic Approach

The targeting ability of nanoparticles loaded with docetaxel and labelled with an NIR fluorescent dye (cyanine) in MCF-7/ADR tumour mice revealed interaction between receptors of HA and CD44. Doxorubicin-loaded nanoparticles containing surfaces of PEG showed higher effectiveness, which was likely due to doxorubicin's extended half-life and decreased clearance, as well as accumulation in the tumour through drug targeting (Cho et al., 2012).

When the SPIOs/SQ gemcitabine NPs were delivered into tumour animals intravenously and followed by a magnetic field in the extracorporeal region, the anticancer medication showed excellent anticancer effect at extremely low dosages. Furthermore, the MRI responsiveness of implanted SPIOs, with their T2-imaging features, makes them a good choice for theranostic approach, as cytotoxicity may be seen readily with MRI (Arias et al., 2011).

## 5.2.9 QUANTUM DOTS

Fluorescent QDs, which are tiny inorganic semiconductor nanocrystals (1–10 nm), have various unique optical features that make them ideal for *in vivo* imaging. The emission colour of QDs may be accurately controlled by the amount of radiation from the UV to near IR due to quantum confinement phenomena. QDs are very bright and photosensitive. They also have a large absorption band and a limited emission band, making them suitable for multiplexing. Finally, because of their huge surface area, QDs may be used to combine multiple contrast agents to create multimodality imaging probes. Biocompatible QD conjugates have been utilized to successfully map sentinel lymph nodes, target tumours, image tumour angiogenesis, and monitor metastatic cells (Bentolila et al., 2009).

The fields of molecular diagnostics and nanotherapeutics have benefited from the use of these semiconductor nanocrystals. The majority of the early research that was done in biology and medicine utilizing (Yi-Ping KW and Leong, 2010) QDs concentrated on the establishment of innovative biosensing formats in order to increase the detection sensitivity to its maximum level. In contrast, quantum dots have the potential to act as more than passive bio-probes or labels in the context of biological imaging and cellular research. The high surface-to-volume ratio of QDs makes it possible to create a "smart" multifunctional nanoplatform, in which the QDs serve as a nanoscaffold for therapeutic and diagnostic (theranostic) modalities as well as an imaging agent. This article provides a brief review that focuses on the potential and limitations of using functionalized QDs as fluorescent contrast agents for imaging or as nanoscale vehicles for the administration of therapeutics, with a particular emphasis on the promise and challenges posed by QD-based theranostics (Yi-Ping KW and Leong, 2010).

Using novel v3-targeted near infrared light-emitting solid lipid nanoparticles coupled to the v3 integrin-specific ligand cyclic Arg-Gly-Asp, we investigated how v3 targeting affects the biodistribution of nanoparticles (cRGD). For the purpose of live animal imaging, near-infrared quantum dots were produced in both non-targeted SLN as well as actively targeted (RGD-SLN) and blocked RGD-SLN (Shuhendler et al., 2012).

## TABLE 5.3
### Theranostic Application of Lipid-Based Nanoparticles

| Lipid Based Nanocarrier System | Labelling Agent | Coadministered Drug | Application | Reference |
| --- | --- | --- | --- | --- |
| Liposomes | $^{64}$Cu | 18F-fluorodeoxyglucose | Cancer | (Wong et al., 2013) |
| | DOPE-Rhodamine fluorescence | Antisurvivin siRNA | Cancer | (Kenny et al., 2011) |
| | Gd(HPDO3A)(H2O) | Doxorubicin | Cancer | (De smet et al., 2010). |
| | Gd.DOTA.DSA | Gadonano-F | Tumours | (Kamaly et al., 2010) |
| | Gd(HPDO3A)(H$_2$O) | Doxorubicin | Cancer | (Se Smet et al., 2013). |
| | Gd-BOPTA | Doxorubicin | Cancer | (Kim et al., 2016). |
| | GD-DTPA | Temozolomide | Glioblastoma | (Nordling et al., 2017). |
| | Nickel-chelating liposome | EC1-GLuc recombinant protein | SKOv3 tumours | (Han et al., 2014). |
| Solid Lipid Nanoparticles | SPIONs | Methotrexate | Rheumatoid Arthritis | (Albuquerque et al., 2015). |
| | IR780 | c(RGDyK) | U87MG tumours | (Kuang et al., 2017) |
| | Gd–DTPA(H$_2$O) | -- | -- | (Morel et al., 1998) |
| Nanostructured Lipid Carriers | IR780 | Plerixafor | Tumour | (Li et al., 2017). |
| Nanocapsules | Far-red fluorochrome | Paclitaxel | Glioblastoma | (Balzeau et al., 2013). |
| | Biofluorescence | Plasmid DNA | Immunity | (David et al., 2013) |
| Nanoemulsions | Iron oxide crystals | Prednisolone valerate acetate | Colon cancer | (Gianella et al., 2011). |
| | Perfluorocarbon | paclitaxel | Cancer | (Dayton et al., 2006) |
| Micelles | 1,1-dioctadecyl-3,3,3,3'-tetramethylindotricarbocyanine iodide | Docetaxel | Breast tumour | (Ma et al., 2012) |
| Self-assembled nano particles | NIR fluorescent dye (cyanine 5.5) | Docetaxel | MCF-7/ADR tumour | (Cho et al., 2012) |
| | SPIOs | Gemcitabine | Cancer | (Arias et al., 2011) |

In order to create LDL-mimetic SLNs, the natural components of low-density lipoprotein (LDL) are recompounded. These SLNs have a stable core/shell nanostructure that contains quantum dots and paclitaxel within the lipid shell. Anionic siRNA molecules are complexed electrostatically with the outer surface of the SLNs. According to the findings of an analysis, the SLN/siRNA complexes that were generated are able to successfully transport paclitaxel and Bcl-2 targeted siRNA into carcinoma cells in the lungs, where they then exhibit synergistic anticancer effects by activating caspase-mediated death (Bae et al., 2013).

The aqueous cores of 100 nm cationic (DOPC:Chol:DOTAP); sterically stabilized, fluid-phase (DOPC:Chol:DSPE-PEG2000); and sterically stabilized, gelphase liposomes were encapsulated with functionalized PEG-lipid coated QD (Al-Jamal et al., 2009). Due to the quick and extensive retention and accumulation within the tumour, f-QD-L hybrid vesicles have a lot of potential for tumour imaging applications.

Zhang and coworker developed cisplatin and quantum dots encapsulated liposomes (CdSe/ZnS QDLs) for theranostic application in brain and skin. The cellular uptake of CdSe/ZnS QDLs demonstrated effective internalization and significant fluorescence in melanoma cells (Zhang et al., 2012).

Fluorine-18 has been connected to QD micelles of phospholipids and is frequently utilized for therapeutic imaging. After injection of this probe into mice and established that the kinetics of their cellular absorption could be tracked using PET and in vivo fibered confocal fluorescence imaging. The QDs that are encapsulated in phospholipid micelles enable a very flexible surface chemistry for conjugating a variety of drugs and bio-moieties with a regulated QD:molecule valency (Duconge et al., 2008).

Studies were conducted to investigate the possibility of co-loaded NLC with quantum dots (CdTe/CdS/ZnS) and paclitaxel as a parenteral multifunctional delivery system. Oleic acid, glyceryl monostearate, and phosphatidylcholine served as the matrix of the lipid for the loaded NLC, which was made via low temperature-solidification and an emulsion-evaporation approach. The prepared NLC's capacity to precisely target and detect the H22 cancer cells was demonstrated in *in vivo* and *ex vivo* imaging investigations (Olerile et al., 2017). Theranostic approach of lipid-based nanoparticles are depicted in Figure 5.3.

## 5.3   LIMITATIONS

The possibility that lipidic nanoparticles will join together is a significant downside, particularly in the event that the size of the created nanoformulation is less than 100 nm. The fusion causes an increase in the amount of non-uniformity and dispersion in size, in addition to the release of encapsulated contents from the lipid vesicles, which may be seen as a consequence of the change. The stability of lipidic nanoparticles over extended periods of time is an additional challenge that has to be solved. It is expected that seemingly undamaged lipid nanoparticles would display a different biodistribution as a result of changes in physicochemical characteristics and surface coating features throughout extended periods of storage. This is because of the changes that occur during storage (Carregal-Romero et al., 2018).

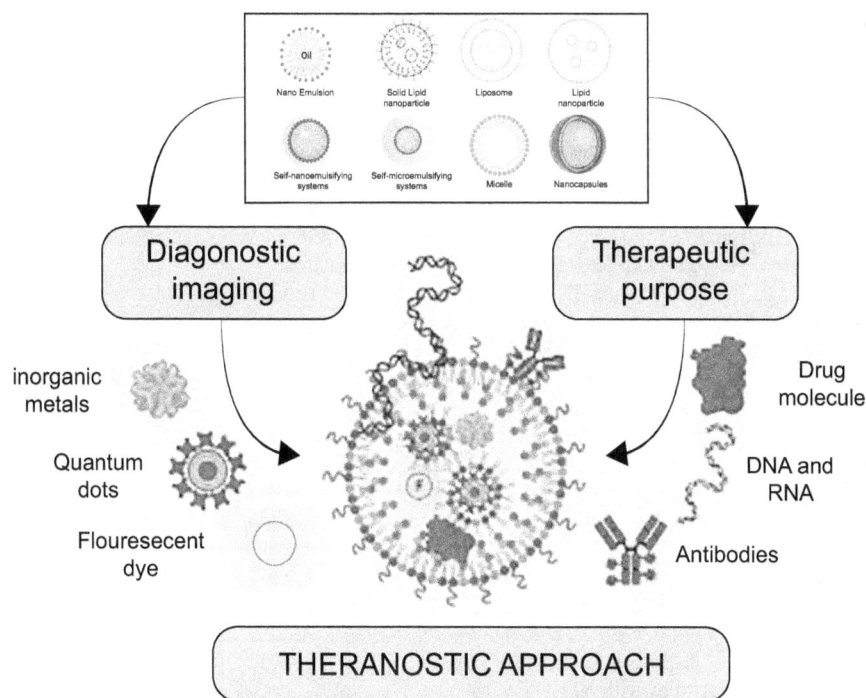

**FIGURE 5.3**   Theranostic approach of lipid-based nanoparticles.

## 5.4   CONCLUSION

The study into LNP technologies and applications has made remarkable strides recently, which has resulted in the creation of more sophisticated nanosystems that exhibit a variety of useful features. In addition, the present issue with various diseases like cancer has brought a significant amount of attention to these nanoparticles. Nevertheless, it seems that there is still some distance to go until successful development of nanotheranostic drug delivery systems. In regard to the issue of the toxicity and poor bioavailability, there is not a great deal of research in delivering the drugs to the site of action and preventing them from degradation. The class of nanoparticles based on lipids has received the greatest attention in terms of research and diagnostics. The lipid-based nanocarrier designs are usually bio-inspired, biocompatible, and adaptable, and they are the most reliable option for the development of new technologies and instruments for nanodiagnostics.

## 5.5   ACKNOWLEDGEMENTS

We thank Amity Institute of Pharmacy, Amity University Madhya Pradesh Gwalior for extending all the support to author this chapter.

# REFERENCES

Ahmad S, Tripathy DB, Mishra A. Sustainable nanomaterials. *Sustainable Inorganic Chemistry*, 2016, 205.

Albuquerque J, Moura CC, Sarmento B, Reis S. Solid lipid nanoparticles: A potential multifunctional approach towards rheumatoid arthritis theranostics. *Molecules* (Basel, Switzerland). 2015, 20 (6): 11103–11118. https://doi.org/10.3390/molecules200611103

Al-Jamal WT, Al-Jamal KT, Tian B, Cakebread A, Halket JM, Kostarelos K. Tumor targeting of functionalized quantum dot–liposome hybrids by intravenous administration. *Mol. Pharmaceut.* 2009, 6: 520–530. doi: 10.1021/mp800187d

Andreozzi E, Seo JW, Ferrara K, Louie A. Novel method to label solid lipid nanoparticles with 64cu for positron emission tomography imaging. *Bioconjug Chem.* 2011, 22 (4): 808–818. doi: 10.1021/bc100478k

Arias JL, Reddy LH, Othman M, Gillet B, Desmaele D, Zouhiri F, Dosio F, Gref R, Couvreur P. Squalene based nanocomposites: A new platform for the design of multifunctional pharmaceutical theragnostics. *ACS Nano.* 2011, 5 (2): 1513–1521.

Ashaolu TJ. Nanoemulsions for health, food, and cosmetics: A review. *Environ Chem Lett.* 2021, 19: 3381–3395. https://doi.org/10.1007/s10311-021-01216-9

Azevedo HS, da Silva RM. *Self-assembling biomaterials: Molecular design, characterization and application in biology and medicine.* Sawston: Woodhead Publishing, 2018.

Bae KH, Lee JY, Lee SH, Park TG, Nam YS. Optically traceable solid lipid nanoparticles loaded with siRNA and paclitaxel for synergistic chemotherapy with in situ imaging. *Adv. Healthc. Mater.* 2013, 2: 576–584. doi: 10.1002/adhm.201200338.

Balzeau J, Pinier M, Berges R, Saulnier P, Benoit J-P, Eyer J. The effect of functionalizing lipid nanocapsules with NFL-TBS.40–63 peptide on their uptake by glioblastoma cells. *Biomaterials.* 2013, 34: 3381–3389. doi: 10.1016/j.biomaterials.2013.01.068.

Bellin MF. MR contrast agents, the old and the new. *Eur J Radiol.* 2006, 60 (3): 314–323.

Bentolila LA, Ebenstein Y, Weiss S. Quantum dots for in vivo small-animal imaging. *J. Nucl. Med.* 2009, 50: 493–496. doi: 10.2967/jnumed.108.053561.

Bordat A, Boissenot T, Nicolas J, Tsapis N. Thermoresponsive polymer nanocarriers for biomedical applications. *Adv Drug Deliv Rev.* 2019, 138: 167–192. doi: 10.1016/j.addr.2018.10.005.

Burri L, Hoem N, Banni S, Berge K. Marine Omega-3 phospholipids: Metabolism and biological activities. *Int. J. Mol. Sci.* 2012, 13: 15401–15419.

Carregal-Romero S, Plaza-García S, Piñol R, Murillo JL, Ruiz-Cabello J, Padro D, Millán A, Ramos-Cabrer P. MRI study of the influence of surface coating aging on the in vivo biodistribution of iron oxide nanoparticles. *Biosensors.* 2018, 8: 127. doi: 10.3390/bios8040127.

Cho HJ, Yoon IS, Yoon HY, Koo H, Jin YJ, Ko SH, Shim JS, Kim K, Kwon IC, Kim DD. Polyethylene glycol-conjugated hyaluronic acid-ceramide self-assembled nanoparticles for targeted delivery of doxorubicin. *Biomaterials.* 2012, 33 (4): 1190–1200.

Constantinides PP, Tustian A, Kessler DR. Tocol emulsions for drug solubilization and parenteral delivery. *Adv Drug Deliv Rev.* 2004, 56: 1243–1255. doi: 10.1016/j.addr.2003.12.005.

Date AA, Desai N, Dixit R, Nagarsenker M. Self-nanoemulsifying drug delivery systems: Formulation insights, applications and advances. *Nanomedicine.* 2010, 5: 1595–616.

David S, Passirani C, Carmoy N, Morille M, Mevel M, Chatin B, et al. DNA nanocarriers for systemic administration: Characterization and *in vivo* bioimaging in healthy mice. *Mol. Ther. Nucleic Acids.* 2013, 2: e64. doi: 10.1038/mtna.2012.56.

Dayton PA, Zhao S, Bloch SH, Schumann P, Penrose K, Matsunaga TO, Zutshi R, Doinikov A, Ferrara KW. Application of ultrasound to selectively localize nanodroplets for targeted imaging and therapy. *Mol Imaging.* 2006, 5 (3): 160–174.

Decuzzi P, Ferrari M. The receptor-mediated endocytosis of nonspherical particles. *Biophys J Elsevier*. 2008, 94: 3790–3797.

De Smet M, Langereis S, van den Bosch S, Bitter K, Hijnen NM, Heijman E, Grull H. SPECT/CT imaging of temperature-sensitive liposomes for MR-image guided drug delivery with high intensity focused ultrasound. *J. Control. Release*. 2013, 169: 82–90.

De Smet S, Langereis S, van den Bosch, Grüll H. Temperature-sensitive liposomes for doxorubicin delivery under MRI guidance. *Journal of Controlled Release*. 2010, 143 (1): 120–127.

Devaraj GN, Parakh S, Devraj R, Apte S, Rao B, Rambhau D. Release studies on niosomes containing fatty alcohols as bilayer stabilizers instead of cholesterol. *J. Colloid Interface Sci*. 2002, 251: 360–365.

Ding K, Jing L, Liu C, Hou Y, Gao M. Magnetically engineered Cd-free quantum dots as dual-modality probes for fluorescence/magnetic resonance imaging of tumors. *Biomaterials*. 2014, 35: 1608–1617. doi: 10.1016/j.biomaterials.2013.10.078.

Diou O, Tsapis N, Fattal E. Targeted nanotheranostics for personalized cancer therapy. *Expert Opin Drug Deliv*. 2012, 9 (12): 1475–1487.

Duconge F, Pons T, Pestourie C, Herin L, Theze B, Gombert K, Mahler B, Hinnen F, Kuhnast B, Dolle F, et al. Fluorine-18-labeled phospholipid quantum dot micelles for in vivo multimodal imaging from whole body to cellular scales. *Bioconjug Chem*. 2008, 19 (9): 1921–1926.

Germain M, Caputo F, Metcalfe S, Tosi G, Spring K, Aslund AKO, et al. Delivering the power of nanomedicine to patients today. *J Control Release*. 2020, 326:164–171. doi: 10.1016/j.jconrel.2020.07.007.

Gianella A, Jarzyna PA, Mani V, Ramachandran S, Calcagno C, Tang J, Kann B, Dijk WJ, Thijssen VL, Griffioen AW, et al. Multifunctional nanoemulsion platform for imaging guided therapy evaluated in experimental cancer. *ACS Nano*. 2011, 5 (6): 4422–4433.

Guo Y, Mao X, Zhang J, Sun P, Wang H, Zhang Y, et al. *Oral delivery of lycopene-loaded microemulsion for brain-targeting: Preparation, characterization, pharmacokinetic evaluation and tissue distribution*. Drug Deliv Taylor & Francis. 2019, 26: 1191–1205.

Haider M, Abdin SM, Kamal L, Orive G. Nanostructured lipid carriers for delivery of chemotherapeutics: A review. *Pharmaceutics*. 2020, 12: 288. doi: 10.3390/pharmaceutics12030288

Haider N, Fatima S, Taha M, Rizwanullah, Firdous J, Ahmad R, Mazhar F, Khan MA. Nanomedicines in diagnosis and treatment of cancer: An update. *Curr. Pharm. Des*. 2020, 26: 1216–1231

Han XJ, Wei YF, Wan YY, Jiang LP, Zhang JF, Xin HB. Development of a novel liposomal nanodelivery system for bioluminescence imaging and targeted drug delivery in ErbB2-overexpressing metastatic ovarian carcinoma. *Int. J. Mol. Med*. 2014, 34: 1225–1232. doi: 10.3892/ijmm.2014.1922

Harrington KJ, Mohammadtaghi S, Uster PS, Glass D, Peters AM, Vile RG, Stewart JS. Effective targeting of solid tumors in patients with locally advanced cancers by radiolabeled pegylated liposomes effective targeting of solid tumors in patients with locally advanced cancers by radiolabeled pegylated liposomes. *Clin. Cancer Res*. 2001, 7: 243–254.

Hequet E, Henoumont C, Djouana Kenfack V, Lemaur V, Lazzaroni R, Boutry S, Vander Elst L, Muller RN, Laurent S. Design, characterization and molecular modeling of new fluorinated paramagnetic contrast agents for dual 1H/19F MRI. *Magnetochemistry*. 2020, 6: 8. doi: 10.3390/magnetochemistry6010008.

Ho YP, Leong KW. Quantum dot-based theranostics. *Nanoscale*. 2010 Jan, 2(1): 60–8. doi: 10.1039/b9nr00178f. Epub 2009 Oct 6. PMID: 20648364; PMCID: PMC2965651.

Jahangir MA, Taleuzzaman M, Kala C, Gilani SJ. Advancements in polymer and lipid-based nanotherapeutics for cancer drug targeting. *Curr Pharm Des*. 2020, 26: 5119–5127. doi: 10.2174/1381612826999200820173253.

Kamaly N, Kalber T, Kenny G, Bell J, Jorgensen M, Miller A. A novel bimodal lipidic contrast agent for cellular labelling and tumour MRI. *Org Biomol Chem.* 2010 Jan 7, 8(1): 201–11. doi: 10.1039/b910561a. Epub 2009 Nov 5. PMID: 20024151

Kamaly N, Miller AD. Paramagnetic liposome nanoparticles for cellular and tumour imaging. *Int J Mol Sci.* 2010, 11(4): 1759–1776.

Kenny N, Kamaly TL, Kalber et al. Novel multifunctional nanoparticle mediates siRNA tumour delivery, visualisation and therapeutic tumour reduction in vivo. *Journal of Controlled Release.* 2011, 149 (2): 111–116.

Khan AW, Kotta S, Ansari SH, Sharma RK, Ali J. Potentials and challenges in self-nanoemulsifying drug delivery systems. *Expert Opin Drug Deliv.* 2012, 9: 1305–1317.

Kim HR, You DG, Park SJ, Choi KS, Um W, Kim JH, Park JH, Kim YS. Mri monitoring of tumor-selective anticancer drug delivery with stable thermosensitive liposomes triggered by high-intensity focused ultrasound. *Mol. Pharm.* 2016, 13: 1528–1539.

Kim M, Jang J, Cha C. Carbon nanomaterials as versatile platforms for theranostic applications. *Drug Discov. Today.* 2017, 22: 1430–1437. doi: 10.1016/j.drudis.2017.05.004.

Kirchherr AK, Briel A, Mader K. Stabilization of indocyanine green by encapsulation within micellar systems. *Mol Pharm.* 2009, 6 (2): 480–491.

Kuang Y, Zhang K, Cao Y, Chen X, Wang K, Liu M, Pei R. Hydrophobic IR-780 dye encapsulated in cRGD-conjugated solid lipid nanoparticles for NIR imaging-guided photothermal therapy. *ACS Appl. Mater. Interfaces.* 2017, 9: 12217–12226. doi: 10.1021/acsami.6b16705.

Li H, Wang K, Yang X, Zhou Y, Ping Q, Oupicky D, et al. Dual-function nanostructured lipid carriers to deliver IR780 for breast cancer treatment: Anti-metastatic and photothermal anti-tumor therapy. *Acta Biomater.* 2017, 53: 399–413. doi: 10.1016/j.actbio.2017.01.070.

Ma M, Hao Y, Liu N, Yin Z, Wang L, Liang X, et al. A novel lipid-based nanomicelle of docetaxel: Evaluation of antitumor activity and biodistribution. *Int. J. Nanomed.* 2012, 7: 3389–3398. doi: 10.2147/IJN.S29827.

Massoud TF, Gambhir SS. Molecular imaging in living subjects: Seeing fundamental biological processes in a new light. *Genes Dev.* 2003, 17 (5): 545–580.

McClements DJ. Nanoemulsions versus microemulsions: Terminology, differences, and similarities. *Soft Matter.* 2012, 8: 1719–1729.

Morel S, Terreno E, Ugazio E, Aime S, Gasco MR. NMR relaxometric investigations of solid lipid nanoparticles (SLN) containing gadolinium (III) complexes. *Eur. J. Pharm. Biopharm.* 1998, 45: 157–163. doi: 10.1016/S0939-6411(97)00107-0.

Morille M, Montier T, Legras P, Carmoy N, Brodin P, Pitard B, et al. Long-circulating DNA lipid nanocapsules as new vector for passive tumor targeting. *Biomaterials.* 2010, 31: 321–329. doi: 10.1016/j.biomaterials.2009.09.044.

Navarro FP, Mittler F, Berger M, Josserand V, Gravier J, Vinet F. Texier I: Cell tolerability and biodistribution in mice of indocyanine green-loaded lipid nanoparticles. *J Biomed Nanotechnol.* 2012, 8 (4): 594–604.

Naziris N, Pippa N, Pispas S, Demetzos C. Stimuli-responsive drug delivery nanosystems: From bench to clinic. *Curr. Nanomed.* 2016, 6: 166–185.

Neubi GMN, Opoku-Damoah Y, Gu X, Han Y, Zhou J, Ding Y. Bio-inspired drug delivery systems: An emerging platform for targeted cancer therapy. *Biomater. Sci.* 2018, 6: 958–973.

Nordling-David MM, Yaffe R, Guez D, Meirow H, Last D, Grad E, Salomon S, Sharabi S, Levi-Kalisman Y, Golomb G, Mardor Y. Liposomal temozolomide drug delivery using convection enhanced delivery. *J. Control. Release.* 2017 Sep 10;261: 138–146. doi: 10.1016/j.jconrel.2017.06.028. Epub 2017 Jun 27. PMID: 28666727.

Olerile LD, Liu Y, Zhang B, Wang T, Mu S, Zhang J, et al. Near-infrared mediated quantum dots and paclitaxel co-loaded nanostructured lipid carriers for cancer theragnostic. *Colloids Surf. B Biointerfaces.* 2017, 150: 121–130. doi: 10.1016/j.colsurfb.2016.11.032.

Ovais M, Khalil AT, Ayaz M, Ahmad I, Nethi SK, Mukherjee S. Biosynthesis of metal nanopar-ticles via microbial enzymes: A mechanistic approach. *Int. J. Mol. Sci.* 2018, 19: 4100. doi: 10.3390/ijms19124100.

Park SM, Aalipour A, Vermesh O, Yu JH, Gambhir SS. Towards clinically translatable in vivo nanodiagnostics. *Nat. Rev. Mater.* 2017, 2: 17014. doi: 10.1038/natrevmats.2017.14.

Peira E, Marzola P, Podio V, Aime S, Sbarbati A, Gasco MR. In vitro and in vivo study of solid lipid nanoparticles loaded with superparamagnetic iron oxide. *J. Drug Target.* 2003, 11: 19–24.

Petersen AL, Hansen AE, Gabizon A, Andresen TL. Liposome imaging agents in per-sonalized medicine. *Adv. Drug Delivery Rev.* 2012, 64: 1417–1435. doi: 10.1016/j.addr.2012.09.003.

Puri A, Loomis K, Smith B, Lee J-H, Yavlovich A, Heldman E, et al. Lipid-based nanopar-ticles as pharmaceutical drug carriers: From concepts to clinic. *Critic Rev Therap Drug Carrier Syst.* 2009, 26: 6. doi: 10.1615/CritRevTherDrugCarrierSyst.v26.i6.10.

Puri N, Gupta A, Mishra A. Recent advances on nano-adsorbents and nanomembranes for the remediation of water. *Journal of Cleaner Production.* 2021, 322: 129051.

Ragelle H, Danhier F, Préat V, Langer R, Anderson DG. Nanoparticle-based drug delivery systems: A commercial and regulatory outlook as the field matures. *Expert Opin. Drug Deliv.* 2017, 14: 851–864.

Reddy LH, Arias JL, Nicolas J, Couvreur P. Magnetic nanoparticles: Design and characteriza-tion, toxicity and biocompatibility, pharmaceutical and biomedical applications. *Chem Rev.* 2012, 112 (11): 5818–5878.

Riaz MK, Riaz MA, Zhang X, Lin C, Wong KH, Chen X, Zhang G, Lu A, Yang Z. Surface functionalization and targeting strategies of liposomes in solid tumor therapy: A review. *Int. J. Mol. Sci.* 2018, 19: 195.

Roger E, Lagarce F, Benoit JP. Development and characterization of a novel lipid nanocapsule formulation of Sn38 for oral administration. *Eur J Pharm Biopharm.* Elsevier B.V. 2011, 79: 181–188.

Rolla, G, Botta, M, Tei, L, Cabella, C, Ghiani, S, Brioschi, C, Maiocchi, A. Paramagnetic solid lipid nanoparticles as a novel platform for the development of molecular MRI probes. *Chemistry* (Weinheim an der Bergstrasse, Germany). 19: 11189–11193. doi: 10.1002/chem.201301837.

Rostami E, Kashanian S, Azandaryani AH, Faramarzi H, Dolatabadi JEN, Omidfar K. Drug targeting using solid lipid nanoparticles. *Chem Phys Lipids.* 2014, 181: 56–61. doi: 10.1016/j.chemphyslip.2014.03.006.

B. L. Sanchez-Gaytan, F. Fay, S. Hak, A. Alaarg, Z. A. Fayad, C. Pérez-Medina, W. J. M. Mulder, Y. Zhao, Angew. *Chem. Int. Ed.* 2017, 56, 2923. https://doi.org/10.1002/anie.201611288

Sánchez A, Ovejero Paredes K, Ruiz-Cabello J, Martínez-Ruíz P, Pingarrón JM, Villalonga R, Filice M. Hybrid decorated Core@Shell Janus nanoparticles as a flexible platform for targeted multimodal molecular bioimaging of cancer. *ACS Appl. Mater. Interfaces.* 2018, 10: 31032–31043.

de Smet M, Langereis S, van den Bosch S, Bitter K, Hijnen NM, Heijman E, Grüll H. SPECT/CT imaging of temperature-sensitive liposomes for MR-image guided drug delivery with high intensity focused ultrasound. *J Control Release.* 2013 Jul 10, 169 (1–2): 82–90. doi: 10.1016/j.jconrel.2013.04.005. Epub 2013 Apr 15. PMID: 23598044.

Shao S, Do TN, Razi A, Chitgupi U, Geng J, Alsop RJ, Dzikovski BG, Rheinstadter MC, Ortega J, Karttunen M, et al. Design of hydrated porphyrin-phospholipid bilayers with enhanced magnetic resonance contrast. *Small.* 2017, 13.

Shuhendler AJ, Prasad P, Leung M, Rauth AM, DaCosta RS, Wu XY. A novel solid lipid nanoparticle formulation for active targeting to tumor $\alpha v\beta 3$ integrin receptors reveals cyclic RGD as a double-edged sword. *Adv. Healthc. Mater.* 2012, 1: 600–608. doi: 10.1002/adhm.201200006.

Silva CO, Pinho JO, Lopes JM, Almeida AJ, Gaspar MM, Reis C. Current trends in cancer nanotheranostics: Metallic, polymeric, and lipid-based systems. *Pharmaceutics*. 2019, 11: 22. doi: 10.3390/pharmaceutics11010022.

Torchilin VP. Micellar nanocarriers: Pharmaceutical perspectives. *Pharm Res*. 2007, 24 (1): 1–16.

Villaraza AJ, Bumb A, Brechbiel MW. Macromolecules, dendrimers, and nanomaterials in magnetic resonance imaging: The interplay between size, function, and pharmacokinetics. *Chem Rev*. 2010, 110 (5): 2921–2959.

Wacker M. Nanocarriers for intravenous injection--the long hard road to the market. *Int J Pharm*. 2013 Nov 30;457(1):50-62. doi: 10.1016/j.ijpharm.2013.08.079. Epub 2013 Sep 12. PMID: 24036012.

Wong AW, Ormsby E, Zhang H, Seo JW, Mahakian LM, Caskey CF, Ferrara KW. A comparison of image contrast with (64)Cu-labeled long circulating liposomes and (18)F-FDG in a murine model of mammary carcinoma. *Am. J. Nucl. Med. Mol. Imaging*. 2013, 3: 32–43.

Wong HL, Bendayan R, Rauth AM, Li Y, Wu XY. Chemotherapy with anticancer drugs encapsulated in solid lipid nanoparticles. *Adv Drug Deliv Rev*. 2007, 59: 491–504. doi: 10.1016/j.addr.2007.04.008.

Wu Y, Mou B, Song S, Tan CP, Lai O-M, Shen C, Cheong L-Z. Curcumin-loaded liposomes prepared from bovine milk and krill phospholipids: Effects of chemical composition on storage stability, in-vitro digestibility and anti-hyperglycemic properties. *Food Res. Int.* 2020, 136: 109301.

Yingchoncharoen P, Kalinowski DS, Richardson DR. Lipid-based drug delivery systems in cancer therapy: What is available and what is yet to come. *Pharmacol. Rev.* 2016, 68: 701–787. doi: 10.1124/pr.115.012070.

Yi-Ping KW. Leong, Quantum dot-based theranostics. *Nanoscale*, 2010, 2: 60–68.

Yuan H, Wang L-L, Du Y-Z, You J, Hu F-Q, Zeng S. Preparation and characteristics of nano-structured lipid carriers for control-releasing progesterone by melt-emulsification. *Colloids Surf B Biointerf*. 2007, 60: 174–179. doi: 10.1016/j.colsurfb.2007.06.011.

Zalutsky MR, Noska MA, Seltzer SE. Characterization of liposomes containing iodine-125-labeled radiographic contrast agents. *Investig. Radiol.* 1987, 22: 141–147.

Zhang LW, Wen CJ, Al-Suwayeh SA, et al. Cisplatin and quantum dots encapsulated in liposomes as multifunctional nanocarriers for theranostic use in brain and skin[J]. *J Nanopart Res*. 2012, 14 (7): 882. doi: 10.1007/s11051-012-0882-9.

Zhang MW, Corona PT, Ruocco N, Alvarez D, de Molina PM, Mitragotri S, et al. Controlling complex nanoemulsion morphology using asymmetric cosurfactants for the preparation of polymer nanocapsules. *Langmuir*. 2018, 34: 978–990. doi: 10.1021/acs.langmuir.7b02843.

# 6 Lipid-Based Nanoparticles as Drug Delivery Agents

*Bhavna Kumari and Anupam Prakash*

## 6.1 INTRODUCTION

Several nanotechnology platforms in the world of clinical biology, including both diagnostics and therapeutics, have gained a huge amount of interest in recent years. Norio Taniguchi developed the term "nanotechnology" in 1974 (Naik et al., 2009). Nanotechnology is defined as the process of applying materials at the atomic, molecular, and supramolecular sizes to create ingredients with unique properties. A more inclusive definition of nanotechnology was adopted by the National Nanotechnology Initiative, which described it as the manipulation of matter having at least one dimension of between 1 and 100 nanometres (nanoparticles) (Sharma et al., 2015). Nanomedicine is the name for the use of nanotechnology in medicine, which includes anything from nanodiagnostics to nanoscale therapies and innovative nanoparticle drug delivery systems. Lipid-based nanoparticles are one type of nanoparticle that are composed of lipids and have been widely employed in medication delivery and medical imaging. Compared to bigger particulates with the same chemical makeup, lipid-based nanoparticles have unique physicochemical properties such as biocompatibility and biodegradability (Buse et al., 2010), and active pharmaceutical loading in a lipid system can improve water solubility and hence the bioavailability of drugs. Lipid-based solutions also' help to prevent active pharmaceutical ingredients' oxidation, breakdown, and degradation, while also improving storage and delivery (Rawat et al., 2006). The main components of a biological membrane are lipids. They have the ability to self-assemble, which may be used in the laboratory to make self-assembling LNPs that are proving to be a superior alternative for delivering active pharmaceuticals to target cells *in vivo*. Biopharmaceutical agents (such as DNA, RNA interference effectors, and mRNA vaccine), small molecule medications (e.g. antitumour chemotherapy drugs), and maybe contrasting imaging agents (e.g. gadolinium (III) ($Gd^{3+}$), (18F), radiopharmaceuticals, and/or fluorescence probes) are examples of some active pharmaceutical ingredients (APIs) (Turánek et al., 2015). Nanoparticles have also proven to be effective in molecular imaging. It is vital to have a high-quality and resolution picture in order to diagnose, treat, and analyse diseases. In a number of conventional imaging techniques, contrast agents are utilized to create images of the areas. In these imaging procedures, contrast agents are required. Existing contrasting agents are unable to deliver precisely

DOI: 10.1201/9781003316398-6

to the intended tissue or cells and spread throughout the body, resulting in a toxic appearance. Nanotechnology has made it feasible to create contrasting agents that can overcome these issues and deliver the necessary image quality. Nanoparticles as contrasting agents have various advantages, including enhanced contrast, the ability to bear large payloads, and lengthy circulation duration (Cormode et al., 2009). Lipid-based nanoparticles are essential for researchers since they are one of the numerous forms of nanoparticles that have all those capabilities for ideal imaging. Theranostic nanoparticles are multifunctional nanosystems that integrate diagnostic and therapeutic abilities into a single biocompatible and biodegradable nanoparticle for more disease treatment (Chen et al., 2014). In this chapter, we have focused on various types of lipid-based nanoparticles and their use in medical imaging and medication delivery for the treatment of various neurodegenerative illnesses.

## 6.2   TYPES OF LIPID-BASED NANOPARTICLES

The number and diversity of potential LNPs for usage in biological environments is nearly endless, as are the number and variety of lipids possible. A wide range of lipid-based nanoparticles (LBNPs) are often utilized as drug transporters and include nanoemulsions, liposomes, solid lipid nanoparticles, and nanostructured lipid carriers (Haider et al., 2020).

### 6.2.1   LIPOSOMES

Liposomes are the oldest type of lipid nanoparticle (Tenchov et al., 2021). Alec Bangham proposed liposomes as a model phospholipid membrane system in 1965 (Bangham, 1998). The Greek words *lipos* and *soma* are the origin of the word "liposome" and respectively indicate fat and body (Rahimpour et al., 2012). They are essentially fat structures with the ability to encapsulate both therapeutic and diagnostic agents. Liposomes (shown in Figure 6.1) are small, spherical phospholipid bilayer vesicles that range in size from 20 nm to 1000 nm and have emerged as a viable carrier for delivering active biological compounds (Large et al., 2021). Because of their amphiphilic nature, liposomes are excellent drug carriers for substances of varied polarity (hydrophilic and hydrophobic). Liposomes provide a diverse pharmaceutical delivery approach due to their strong biocompatibility and biodegradability. Phosphatidylethanolamine, phosphatidylglycerol, phosphatidylcholine, phosphatidylserine, and phosphatidylinositol are examples of natural phospholipids that are utilized to make different kinds of liposomes (Ahmed et al., 2019; Tenchov et al., 2021; Kumar, 2019).

Liposomes protect encapsulated drugs from external surroundings, and they may be functionalized with multiple targeting ligands for tissue-specific delivery, prolonging the entrapped agents' time in systemic circulation. Substances like cholesterol can be introduced to the nanoparticle to reduce fluidity and improve permeability of hydrophobic pharmaceuticals via the phospholipid bilayer. Doxil is a lipid nanoparticle formulation of the anticancer chemical doxorubicin used in the treatment of ovarian cancer and was the first liposomal medicine to be authorized (García-Pinel et al., 2019).

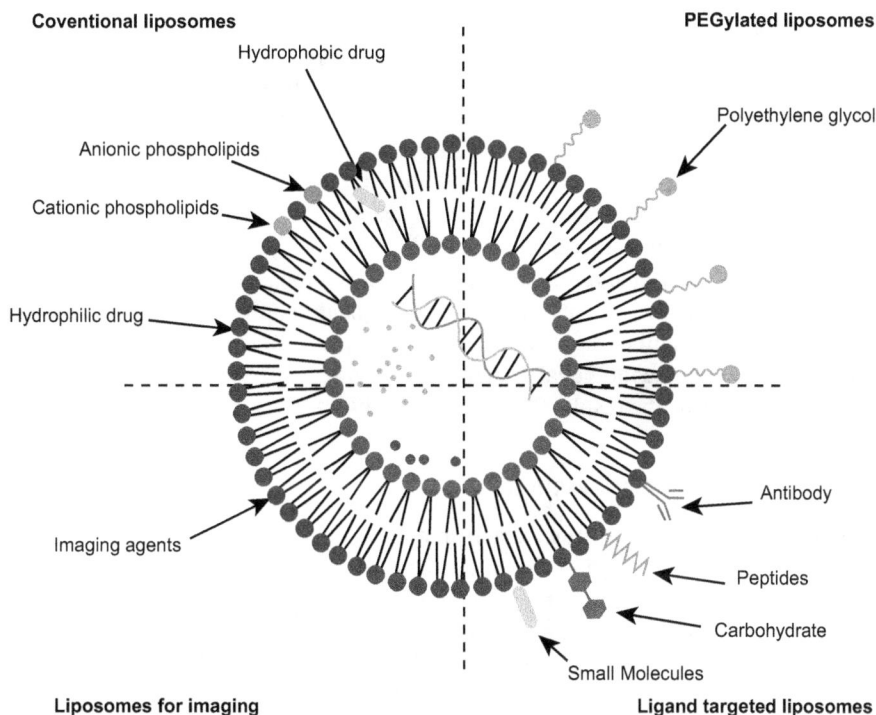

**Coventional liposomes**

Hydrophobic drug

Anionic phospholipids

Cationic phospholipids

Hydrophilic drug

Imaging agents

Liposomes for imaging

**PEGylated liposomes**

Polyethylene glycol

Antibody

Peptides

Carbohydrate

Small Molecules

Ligand targeted liposomes

**FIGURE 6.1** Liposome.

## TABLE 6.1
## Examples of Liposome Formulation

| Composition of Liposome | Drug Delivered | Uses | Reference |
| --- | --- | --- | --- |
| PEGylated disteroylphosphateidyl ethanolamine, hydrogenated soy phosphatidylcholine, and cholesterol | Doxorubicin (doxil) | Breast cancer metastases | (Barenholz, 2012) (Fernandes et al., 2016) |
| Hydrogenated soy phosphatidylcholine, cholesterol, distearoylphosphatidyl glycerol, and α tocopherol | Amphotericin B | Broad-spectrum polyene antibiotic | (Moore et al., 2008) |
| Egg phosphatidylcholine and cholesterol | Doxorubicin (myocet) | Antitumour activities | (Bulbake et al., 2017) |
| Dimyristoyl phosphatidylcholine and egg phosphatidylglycerol | Verteporfin | Photosensitizer | (Participants et al., 2005) |
| Cationic lipid, PEG-lipid, cholesterol, and distearoylphosphatidylcholine (DSPC) | mRNA (Pfizer and Moderna vaccine) | Covid-19 vaccine | (Cross, 2021) |

More sophisticated liposomes, such as OX26 immunoliposomes, have been developed in recent years to actively target the brain (Huwyler et al., 1996). Examples are represented in Table 6.1. Based on the number of lipid bilayers present and the different sizes they have, liposomes can be classified as either unilamellar (small unilamellar vesicles (SUV) with diameters of range 20–100 nm, large unilamellar vesicles (LUV) with diameters of range 100–1000 nm, or giant unilamellar vesicles (GUV) with diameters >1000 nm) (Kumar, 2019). Liposomes are grouped into five based on their delivery system: immunoliposomes, pH-sensitive liposomes, long-circulating liposomes, positively charged liposomes, and conventional liposomes (Sriraman et al., 2014). PEGylation is a method that uses covalent bonding to link polyethylene glycol to the hydrophilic terminus of individual lipids. By limiting their continued uptake by the reticuloendothelial system, the latter were successful in improving the half-life and therapeutic results (Torchilin et al., 1994).

### 6.2.2 SOLID LIPID NANOPARTICLES

Liposomes are effective as drug carriers, but they need complex manufacturing procedures using organic solvents, have low drug entrapment efficiency, and are hard to undertake on a large scale (represented in Figure 6.2). Solid lipid nanoparticles (SLNs) were created to address some of these issues (Muller et al., 2011).

In comparison to the oily phase, drug mobility diminishes in the solid lipid state, leading in controlled drug release and good stability. Higher proportion of drug loading, compatibility with hydrophilic or hydrophobic pharmaceuticals, processability and cytocompatibility as they do not demand the use of organic solvents in their production, simplicity and adaptability for industrial applications, and cost effectiveness are just a few of the benefits of SLNs. Each of these features is beneficial to drug administration (Üner, 2006). Solid lipid nanoparticles ranging from 1 to 1000 nm in

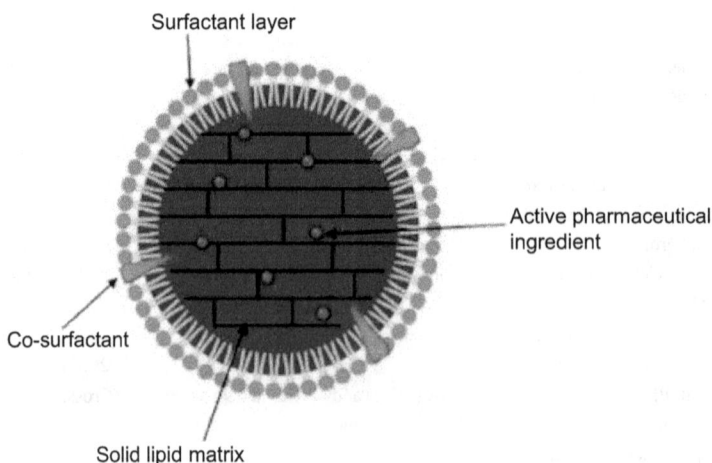

**FIGURE 6.2** Solid lipid nanoparticles.

size are constituted of lipids that are solid at both normal temperature and human body temperature, such as mono-, di-, and triglycerides, tricaprin, tripalmitin, complex glyceride mixes, and waxes, but are stabilized with non-toxic surfactants and polymers (e.g. poloxamers, soya lecithin, Tween 80 and sodium dodecyl sulphate) (Saupe et al., 2006). Because the vast majority of solid lipids are naturally occurring lipids, their cytotoxic activity is lower than that of artificial polymers. Solid lipid nanoparticles shield the biologically active drug from the surroundings and preserve it from deterioration, while also enhancing its bioavailability. They can be taken orally or intravenously. When medications are paired with the capacity of SLNs to pass the blood–brain barrier, the therapeutic effects are dramatically improved. When integrated with SLNs, camptothecin, an antitumour medication, had a longer mean residence duration in the brain than did a control solution (Yang et al., 1999). Because SLN can be targeted to specific organs, tissues, or cells with the introduction of targeting ligands, they can assist in reducing the systemic toxicity of many drugs. Another benefit is the ability to produce SLN in dry powder form, which may then be placed into pellets, capsules, or pills for improved drug administration. SLNs can be used topically as well. Vitamin E–containing SLNs, for example, produce a sticky coating on the skin's surface. The release of vitamin E is sustained for a long time with this formulation. Occlusion, which is produced by the creation of a thin film on the skin's surface, also helps vitamin E penetrate deeper into the skin (Jores et al., 2004). Because SLNs are also able to cross the blood–brain barrier, they are ideal candidates for use as a CNS MRI contrasting material. Examples are summarized in Table 6.2.

### 6.2.3  Nanostructured Lipid Carrier

SLN formulations may be limited because of unwanted particle development caused by clumping or coagulation, resulting in sudden "burst release" of the medication.

**TABLE 6.2**

**Examples of SNL Formulation**

| Composition of Solid Lipid Nanoparticles | Drug Delivered | Uses | References |
|---|---|---|---|
| Insulin | PEG'Glycolgrafted chitosan | High insulin loading value | Zhang et al. (2008) |
| Methotrexate | Glyceryl behenate, Campritol 888ATO, lecithin | *In vitro* drug release, topical treatment of psoriasis | Lu et al. (2006) |
| Rizatriptan | Tristearin, Phospholipon80 | | Nair et al. (2011) |
| Indirubin | Cetyl palmitate, polysorbate 80 | Used against human glioblastoma cells | Rahiminejad et al. (2019) |
| Niclosamide | 4, 6-diamidino-2-phenylindol (DAPI), Niclo, stearylamine, pluronic F-68 | Anticancer efficacy against triple negative breast cancer cells | Pindiprolu et al. (2019) |

Expulsion of the loaded drug solution occurs often during the storage of SLN, which is a major drawback of the solid lipid nanoparticles. To get around these limitations, changes to the SLN structure were needed. Further research led to the introduction of nanostructured lipid carriers. NLCs are a kind of lipid-based nanocarrier that represent its second generation (Buse et al., 2010). Nanostructured lipid carriers are similar to SLNs in size (10–1000 nm) and include both solid and liquid lipids at room temperature (represented in Figure 6.3), whereas SLNs are made up of only solid lipids (Kumar, 2019). NLCs are made by combining spatially incompatible LL and SL mixtures in which the oil molecules, for example oleic acid, olive oil, and caprylic/capric triglyceride (liquid lipid), do not contribute to the crystalline matrix of SL, and the SL crystals that may be stearic acid, cetearyl alcohol, tripalmitin, hydrine, glyceryl, or monostearate do not dissolve in the LL. For producing a stable drug-loaded NLC, proper lipid selection is necessary (Haider et al., 2020). The lip-ids used in NLCs are chosen according to the drug's nature, whether hydrophobic or hydrophilic. NLCs are divided into three categories based on their compositions and ratios of both lipid combinations: imperfect, amorphous, and multiple structure (Haider et al., 2020).

Low levels of liquid-phase lipid are present in the imperfect variant of NLC (oil). It has a disordered crystalline structure and drug storage areas because it is composed of lipids with different fatty acid chain lengths, including saturated and unsaturated lipids. A substantial portion of the active drug(s) is easily integrated inside such areas (Gaba et al., 2015) (Üner 2006).

When the amount of liquid lipid component(s) in a formulation is increased, the outcome is an amorphous kind of NLC that lacks crystalline structure and has oil-soluble gaps inside the solid lipid structure. For medications with a high solubility in liquid lipids, NLC formulations offer a better drug-loading capacity, and the absence of a crystalline structure minimizes undesirable drug ejection during the cooling process (Gaba et al., 2015) (Üner 2006).

Oil-in-fat-in-water carriers called multiple NLCs are made of a solid lipid matrix that encloses several liquid oil nano-compartments. The existence of these oil

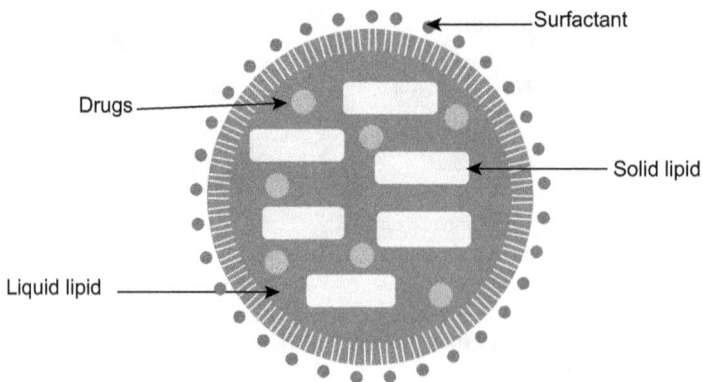

**FIGURE 6.3**  Nanostructured lipid carrier.

nano-compartments boosts the drug-loading capacity because highly lipophilic compounds are more soluble in LL than in SL. Additionally, the solid matrix that surrounds the nano-compartments functions as a barrier to stop drug leakage and enable controlled drug release (Gaba et al., 2015) (Üner 2006).

Due to the liquid lipids in the formulation, NLCs have a greater drug-loading capacity. They also restrict drug leakage during storage by avoiding lipid crystallization. The NLC matrix is formed up of low-cost lipids with a negligible toxicological risk. This is regarded as one of the most significant advantages. Aside from that, they are site-specific, are physically stable, and have the potential to improve bioavailability. They are simple to make and may be employed in industrial applications. The intranasal administration of NLC formulations bypassed the blood–brain barrier and successfully reached the site of action, according to brain distribution research. An *in vivo* hematotoxicity experiment demonstrated that the developed olazopine-loaded nanostructural lipid carrier (OLZ-NLCs) formulation was safe and effective for delivering OLZ in the brain through the intranasal route for the treatment of schizophrenia without causing agranulocytic angina or leukopenia (Gadhave et al., 2019). As a result, OLZ-NLC has a promising future for the safe and cost-effective administration of antipsychotic medications for the treatment of CNS diseases. A few examples are summarized in Table 6.3. Using lipid nanocarriers, appropriate surfactants, and extending the nasal residence duration, studies demonstrate that the rate of nasal absorption may be increased. As a result, the combination of NLC and the intranasal route of drug administration may be a preferable option for delivering duloxetine for depression therapy (Haider et al., 2020).

## TABLE 6.3
## Examples of NLC Formulation

| Composition of Nanostructured Lipid Particles | Drug Delivered | Uses | References |
|---|---|---|---|
| Dimethyl sulfoxide, trypsin-EDTA, antibiotics (penicillin and streptomycin) | Thymoquinone | To avoid the low bioavailability | Haron et al. (2018) |
| Glycerol monostearate (GM), soybean phosphatidylcholine (SPC), oleic acid | Etoposide (VP16) | For the treatment of gastric tumours | Wang et al. (2014) |
| Isopropyl myristate, isopropyl palmitate, and isopropyl stearate | 5-FU-stearic acid prodrug | Shows greater anticancer activity *in vitro* and *in vivo* than all free medicines or uncoated NLC | Ying Qu et al. (2015) |
| Compritol® 888 ATO, PEG-SA, soybean phosphatidylcholine (S100), oleic acid, glycerin monostearate, Lapa, DOX | Lapachone and DOX | To overcome multiresistance drugs | Li et al. (2018) |

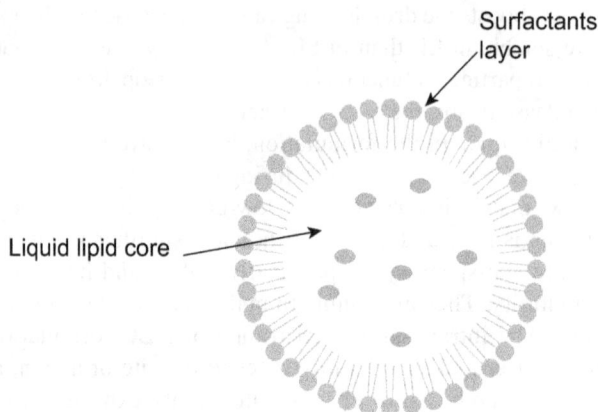

**FIGURE 6.4**    Nanoemulsion.

### 6.2.4  NANOEMULSIONS

Miniemulsion is another name for nanoemulsion, which is composed of two immiscible liquid phases, one or more surfactants, and possibly one or more co-surfactants
(see Figure 6.4). Oil and water are usually the two immiscible liquids. They have
diameters ranging from 20 to 500 nm. Scientists are interested in nanoemulsion
because of its advantages, which include optical clarity, simplicity of manufacture,
thermodynamic stability, large surface area, and improved bioavailability of bioactive chemicals. Lecithin (phosphatidylcholine produced from egg yolk or soybean),
Cremophorl EL (polyoxyl-35 castor oil), grape-seed oil, sodium deoxycholate (bile
salt), pine-nut oil, casein, gums, starch derivatives, and block copolymers are all
typical emulsifiers used in pharmaceutical NE formulations (Karami et al., 2019).
Several *in vivo* investigations have shown that NEs are one of the most promising
delivery mechanisms for CNS-active medicines employed as part of effective treatments against difficult-to-treat CNS illnesses such as brain tumours.

## 6.3  METHODS OF PREPARATION OF
## LIPID-BASED NANOPARTICLES

Lipid-based nanoparticles can be made using a variety of ways. The manufacturing
procedures should guarantee that the nanoparticle formulations are homogeneous in
composition, size, and structure, while limiting toxicity and optimizing therapeutic
effects. Methods for making lipid-based nanoparticles may be divided into two categories: top-down and bottom-up techniques. A top-down strategy to nanoparticle
creation involves breaking up a micron or bigger sized piece of material, or removing
substances from the particulate's surface (Real et al., 2022). This method involves
high-pressure homogenization (Souto & Müller 2006), media milling, lithography,
and so on, while bottom-up manufacturing of lipid-based medicinal nanoparticles is
a typical approach. It is a self-assembly method that assembles simple components

into bigger structures using chemical or physical forces at the nanoscale. Due to a wider range of equipment and greater scale-up options, high-pressure homogenization is the most preferred (AZoNano.com, 2022).

### 6.3.1   HIGH-PRESSURE HOMOGENIZATION

High-pressure homogenization with piston-gap homogenizers are generally used to create solid lipid nanoparticles. With high pressure ranging from 100 to 2000 bar, liquid travels through a tiny gap (homogenizers) of a few micrometres in size, resulting in the dispersion of small and uniformly sized particles. The target chemicals are introduced into lipid by distribution and dissolution in a melted lipid mixture. High-pressure homogenization is widely used for industrial purposes. The key benefit is effective particle size reduction, low contamination, and industrial practicality, such as low-cost and large-scale manufacturing (Jenning et al., 2002).

### 6.3.2   HOT HIGH-PRESSURE HOMOGENIZATION

HPH that is performed above the lipid melting point is referred to as hot HPH. The fat and medicine are melted at the same temperature and combined with an aqueous surfactant. The high-shear device produces a heated pre-emulsion. The pre-emulsion is typically homogenized using a piston gap homogenizer to create a hot colloidal emulsion (Kumar 2019). The fluid droplets are homogenized repeatedly until the desired size of nanoparticles has been formed from the fluid droplets. To create SLNs/NLCs, a heated colloidal emulsion is cooled to room temperature before the droplets are solidified. At pressures between 500 and 1500 bar, three to five rotations are adequate. High-pressure homogenization raises the sample's temperature (about 108 °C for 500 bar). For instance, at high temperatures, the medicine cryptotanshinone—which is not well soluble in water—is incorporated into SLNs. This is insufficient for hydrophilic medications (Souto & Müller, 2006). Additionally, medicines that are temperature sensitive may be ruined by high temperatures (Kasongo et al., 2012).

### 6.3.3   COLD HIGH-PRESSURE HOMOGENIZATION

The shortcomings of hot high-pressure homogenization are addressed by this method. The initial step in HPH is dissolving the API in the lipid, which is the same for both methods. However, the substance is instantly frozen utilizing dry ice or liquid nitrogen after it has been dissolved in lipid, creating a homogenous dispersion of the molecule in the lipid matrix. However, in both conditions, the temperature should be maintained above the lipid's melting point throughout the first phase (Kumar et al., 2019). The solid is then ground by milling it into microscopic particles. Following that, the nanoparticle is combined with a cold aqueous surfactant solution. The dispersion is homogenized under high pressure to produce SLNs. In contrast to heat homogenization, cold homogenization entails homogenizing solid lipids rather than a lipid melt. Pharmaceuticals that are water-soluble or thermolabile respond optimally to this cold HPH methodology (Kasongo et al., 2012).

## 6.4   ROUTES OF ADMINISTRATION

Pharmaceuticals carried in lipid nanoparticles can be delivered by a variety of methods, including topical, oral, intravenous, ocular, respiratory, and brain drug administration depending upon the type of drug and its target site. A few examples are summarized in Table 6.4. Many medications, such as peptides and proteins, cannot be taken orally due to enzymatic breakdown in the gastrointestinal tract. Repeated intravenous administration can be used to administer such medications. Because of their nano size, SLNs may easily circulate in the vascular system and provide excellent therapeutic activity when given intravenously. Stealth SLNs (sterically stabilized SLNs) were used to deliver doxorubicin across the BBB for the treatment of leukaemia and solid tumours in humans. Stealth SLNs outperformed non-stealth SLNs in terms of drug transport via the BBB because they had a lower uptake by murine macrophages (Zara et al., 2002). Oral administration of SLN-based capsules, pills, and pellets has been effectively prepared (Takeuchi et al., 1996). A hydrophilic coating, such as PEG, can protect SLNs against hydrolytic enzymes and agglomeration. PEG coating increased the oral administration of medicines via SLNs, according to several studies. A hydrophilic coating on SLNs can potentially be used to administer vaccines via the nasal mucosa (Bhatt et al., 2015). A lipid-based nanoparticle formulation successfully administered pilocarpine, a standard glaucoma therapy, through the ocular pathway.

## 6.5   LIPID-BASED NANOPARTICLES APPLICATION FOR NUCLEIC ACID AND DRUG DELIVERY

In the pharmaceutical industry, lipid nanoparticles (LNPs) have shown to be a reliable delivery system for a variety of medicines. LNPs play a crucial role in

---

**TABLE 6.4**
**Several Administration Route with Examples**

| Route of Administration | Drug Name | Uses | References |
|---|---|---|---|
| Oral drug delivery | Cyclosporine | Immunosuppressant | Guada et al. (2015) |
| | Rifampicin | Tuberculosis treatment | Sarathchandiran (2012) |
| | Hydrochlorothiazide | Hypertension treatment | Cirri et al. (2018) |
| | Lovastatin | Antihyperlipidemic | Chen et al. (2010) |
| Nasal delivery | Galantamine | Alzheimer's disease treatment | Misra et al. (2016) |
| | Ropinirole HCL | Parkinson's disease treatment | Pardesho et al. (2013) |
| Parenteral injection | Apomorphine | Parkinson's disease treatment | Pfeiffer (2016) |
| | Bromocriptine | Parkinson's disease treatment | Esposito et al. (2008) |
| Ocular | Moxifloxacin | Endophthalmitis treatment | |
| | Mangiferin | Cataract treatment | Liu et al. (2012) |
| | Pilocarpine | Glaucoma treatment | |
| Topical applications | Vitamin E | Anti-inflammatory agent | |
| | Octylmethoxycinnamate | UVB absorber, sunscreen | |
| | Clotrimazole | Antifungal | |

maintaining and delivering mRNA to cells. They are now in the spotlight as a significant Covid-19 mRNA vaccine component (Tenchov et al., 2021). Numerous cationic lipid amphiphiles have been synthesized and investigated for use as nucleic acid carriers. The most often used nonviral delivery route for nucleic acid therapy is cationic LNPs, which are stable complexes of synthetic cationic lipids and anionic nucleic acids. The development of LNP-based mRNA vaccines against the Zika virus, Cytomegalovirus (CMV), tuberculosis, and influenza has started in clinical trials (Wang et al., 2021).

### 6.5.1 LIPID-BASED NANOPARTICLES TARGETING BRAIN DRUG DELIVERY

One of the medication delivery methods that has received the most attention is brain pharmacological delivery. Because the central nervous system has some features that make it an extremely isolated system and prevent therapeutic entry, such as the BBBs and the blood–cerebrospinal fluid barrier that provide physiochemical protection, the immune-privileged environment, and the lack of tissue regeneration properties, CNS disorders are particularly difficult to diagnose, evaluate, monitor, and treat (Abbott, 2013). CNS illnesses are frequently linked to neurodegenerative disorders, which can range from minor impacts to more lasting neuron loss. Brain injuries, strokes, trauma, infections, mental problems, autoimmune illnesses, and neurodegenerative diseases include, for instance, CNS tumours. They are all very difficult to treat since the majority of medicines have a hard time getting through the barriers, and because these medications are not site specific, they may also be rather dangerous (Witika et al., 2022). In addition to being able to penetrate the blood–brain barrier, accumulate in the CNS, and potentially avoid the significant health risks associated with the majority of drugs used to treat CNS problems, lipid-based drug delivery systems provide a variety of advantages for treating brain disorders (see Table 6.5).

### 6.5.2 LIPID-BASED NANOPARTICLES APPLIED TO VARIOUS CNS DISEASES TREATMENT

#### 6.5.2.1 Alzheimer's Disease

Alzheimer's disease (AD) is a long-term chronic illness. It is the seventh greatest cause of death (Haddad et al., 2022). It's a type of dementia that becomes worse over time. Various illnesses that affect memory, thinking, and personality are together referred to as dementia. These changes have a significant influence on a person's day-to-day living, and Alzheimer's disease is responsible for 60% to 80% of dementia cases. Everyone has the potential to acquire Alzheimer's disease, although certain individuals over 65 and those with a family history of the condition are more vulnerable than others. Extracellular amyloidal protein deposits and intracellular neurofibrillary accumulate as senile plaques, which are pathological features. Amyloid plaques accumulate in the brain hippocampus (which plays a very critical function in the control of emotional reactions) and cerebral blood vessels, resulting in congophilic angiopathy, which is also called cerebral amyloid angiopathy (CAA) (Thakur et al., 2018). With hyperphosphorylated tau proteins, NFTs developed paired helical filaments. These NFTs are marked

**TABLE 6.5**
**Summary of Nanocarrier Application in CNS Diseases**

| Disease | API | Nanocarrier Type | Lipid Composition | Route of Administration | In Vivo Results | |
|---|---|---|---|---|---|---|
| Ischaemic stroke | Vinpocetine | Cyclodextrin—NLC | Compritol® 888 ATO and Miglyol® 812N | Oral administration | Increase bioavailability | Lin, et al. (2014) |
| | Baicalein | NLC | Tripalmitin, vitamin E, phospholipids, and poloxamer 188 | Intravenous administration | Higher accumulation | |
| | Asialoerythropoietin | PEGylated liposomes | Distearoyl phosphatidylethanolamine (DSPE)-PEG-N-hydroxysuccinimide (NHS) | Intravenous administration | Increased the accumulation of AEPO | |
| | Dexamethasone phosphate and FK506 (Tacrolimus) | Liposomes | Dipalmitoylphosphatidyl-choline, distearoylphosphatidyle-thanolamine | Intravenous administration | Counteracted the inflammation in ischaemic stroke | |
| Alzheimer's disease | DPL | SLN | Glyceryl monostearate, blend of Tween® 80 and poloxamer 188 (1:1) as surfactant | Intranasal route | Good brain target delivery | Sood et al. (2013) |
| | Curcumin and donepezil | NLC | Soya lecithin, Tween 80, Precirol (solid lipid), capmul MCM (liquid lipid) | Intranasal route | Reduced the levels of oxidative stress, improved memory | |
| | Growth hormone | SLN | Glyceryl behenate, Pluronic® F-127 as surfactant, and Tween® 80 as co-surfactant | Oral administration | Prolonged drug release | |

*(Continued)*

**TABLE 6.5 (Continued)**
**Summary of Nanocarrier Application in CNS Diseases**

| Disease | API | Nanocarrier Type | Lipid Composition | Route of Administration | In Vivo Results | |
|---|---|---|---|---|---|---|
| Parkinson's disease | Ropinirole hydrochloride | SLN | Tripalmitin, Pluronic R-68, and stearylamine | Intranasal delivery route | Decrease in symptoms | Pardeshi et al. (2013) |
| | Piribedil | SLN | Palmitic acid (PA) and polyvinyl alcohol (PVA), Tristearin, stearic triglyceride (tristearin) | Intranasal delivery route | Increased nose-to-brain transport and a four-fold increase in area under the time–versus-concentration curve | |
| | Fibroblast growth factor (bFGF) | NLC | Nonionic copolymer poloxamer 188 | Intranasal administration | Preserved nigrostriatal dopamine neurons and encouraged their regeneration | |
| | Glial cell-line derived neurotrophic factor | Chitosan (CS)-coated NLC | Glyceryl distearate, caprylic/capric triglyceride, Tween 80 and Lutrol. Chitosan, Triton X-110 | Intranasal administration | Rapid motor recovery | |
| | Vitexin | SLN | | | Increase in reactive antioxidant activity | |

by neuronal and synaptic loss, as well as some specific lesions (Thakur et al., 2018). Currently, there are no proper treatments for Alzheimer's disease, although a few medications are available on the market that target the Aβ and τ proteins to improve cognitive symptoms and halt the deterioration process. Acetylcholinesterase inhibitors such as galantamine, rivastigmine, tacrine, donepezil, and N-methyl D-aspartate (NMDA) receptor antagonists are among the medicines available for the disorder (Twohig et al., 2019). Memantine is a strong NMDA-receptor antagonist with fast on/off kinetics and intermediate affinity. Memantine is most frequently prescribed to those with mild to severe Alzheimer's disease because it is neuroprotective, reducing tau hyperphosphorylation and NMDA- and glutamate-induced cell death in cultured experimental model cerebellar, cortical, mesencephalic, and hippocampal neurons. (Robinson et al., 2006). Several drugs are in clinical trials for the treatment of Alzheimer's disease, including Tarenfl urbil, Bapineuzumab (a monoclonal antibody to Aβ that undertook a phase III clinical trial in patients with mild to moderate Alzheimer's disease from 2007 to 2012), solanezumab, lithium, and valproic acid aiming Tau-targeted therapy (Briggs et al., 2016). However, it is estimated that 40% of these trials were abandoned owing to a lack of clinical relevance, and only 12% progressed to Phase III, with just a few being authorized for the treatment of Alzheimer's disease. The difficulty of these antibodies or drugs to pass the BBB is primarily responsible for their lack of clinical importance. These small particles are unable to pass through the BBB (Witika et al., 2022). Novel drug delivery methods that can move potentially therapeutic substances over the blood–brain barrier are therefore demanded. Nanocarriers, particularly those based on lipidic transporting medium, have been demonstrated to have optimal qualities for drug transport across the BBB because of their attraction and interaction with endothelial cells, capacity to be absorbed into minute blood vessels, and surface-modifying abilities with biologically active proteins and cell penetrating peptides (CPP). They are also biodegradable, non-toxic, and non-immunogenic. Donepezil is a water-loving tiny molecule licensed by the US FDA for the treatment of AD but with limited BBB permeability, higher dosing frequency, and significant cholinergic adverse effects. In comparison to DPL-Sol, investigations demonstrated that the improved DPL-SLN formulation was more stable and was able to overcome the aforementioned problems (Yasir et al., 2018). When compared to DPL-Sol, pharmacokinetic and brain targeting experiments in mice model indicated much improved brain target delivery and high drug concentration in the brain after intranasal administration of DP-SLNs. Another example of drug-loaded SLN is in comparison to sol GH, which released 81% of the drug in the early hour (first one hour); GH-SLN demonstrated a sustained drug distribution throughout 24 hours, with over 99% of the drug released (Yasir et al., 2018). This suggests that combining these therapeutically potent compounds with nanolipid-based drug delivery methods delivered intranasally or orally might boost the success rate of novel molecules for improved therapeutic results.

### 6.5.2.2   Ischaemic Stroke

The most prevalent form of stroke is ischaemic stroke. It is generally caused by a blood clot in the brain that stops or plugs a blood artery. It is a neurological condition caused by a stenosis or blockage of the cerebrovascular system, in which blood flow to the brain is stopped and haemorrhagic (Durukan & Tatlisumak, 2007). This disruption causes pH, carbohydrates, oxygen, minerals, and other nutrients to become unbalanced, resulting in the death of brain cells. Ferulic acid (FA) has been shown to help in ischaemic stroke. Using a high-pressure homogenization method, FA was introduced into NLCs. FA-NLCs at concentrations of 80 and 100 lM were reported to be beneficial in lowering neurobehavioural impairments while limiting oxidative stress, cytotoxicity, inflammation, and cellular dysfunction in ischaemic experimental mice models (Hassanzadeh et al., 2018). Other nanocarriers have also been developed. As compared to baicalin solution, baicalin-containing PEGylated cationic SLNs coated with OX26 antibody, for example, were shown to be effective in the treatment of cerebral ischaemia (Tsai et al., 2012). Asialo-erythropoietin (AEPO) is a potent agent for ischaemia region targeting. It attaches to the EPO receptors present on neuronal cells. It also has a significant cytoprotective impact against the induction of apoptosis. Therefore, it appears that AEPO-modified liposomes have both a nerve cell targeting and a neuroprotective effect on cerebral ischaemia impairment (Ishii et al., 2012).

### 6.5.2.3  Parkinson's Disease

Older persons are very much susceptible to Parkinson's disease (PD), which is also known as paralysis agitans. In terms of disability, it affects 2% to 4% of senior people and holds the second position only to Alzheimer's disease. Parkinson's disease is a multi-system disease involving motor, non-motor, and psychiatric conditions (Marras et al., 2012). Tremor, stiffening, postural disorders, slowness of movement, and loss of coordination are all motor signs. Non-motor symptoms include constipation, sensory issues, sleeping disorders, and autonomic dysfunction, while neurological issues include dementia, all of which have a significant influence on the patient's lifestyle (Pfeiffer et al., 2016). PD is characterized pathologically by the slow death of dopaminergic neurons in the midbrain region's substantia nigra, which causes dopamine deficiency in the basal ganglia and the production of protein clumps called α synuclein. (De Lau et al., 2006). PD, like Alzheimer's disease, is currently incurable, and no medication exists to halt or decrease the illness's course. The major therapy methods offered primarily ease symptoms. The treatment focuses on restoring dopamine, improving mitochondrial function, and lowering neuroinflammation by lowering oxidative stress. Levodopa is the first-line therapy for Parkinson's disease, with the ability to cross the BBB but at insufficient amounts. It can also cause dyskinesia (Witika et al., 2022). This issue was solved by employing microemulsion technology to incorporate LD into SLN. Experimental animals with Parkinson's disease were used to evaluate vitexin-loaded SLN. Studies on mice showed a substantial rise in overall reactive antioxidant activity. Additionally, it was shown that SLN loaded with vitexin decreased striatal ROS. In comparison to the marketed oral version of Ropinirole, Ropinirole-loaded SLN showed no adverse effects and had much superior therapeutic effectiveness (dopaminergic agonist) (Rahman et al., 2019).

## 6.6  LIPID-BASED NANOPARTICLES FOR MEDICAL IMAGING

The use of molecular and diagnostic imaging in the treatment of a variety of illnesses has increasingly been a hot topic of research. Magnetic resonance imaging, radionuclide imaging (PET and SPECT), ultrasonic molecular imaging, computed axial tomography, positron emission tomography, and diffusion optical tomography are among some of the medical imaging modalities with distinct strengths (Naseri et al., 2018). Among these, the two main technologies now employed for diagnostic imaging are MRI and CT. To get a proper picture with the optimum quality, technologies such as MRI require particular contrast agents. Contrast agents are pharmaceuticals that affect the intrinsic characteristics of tissues, influencing the underlying mechanisms of contrast, to enhance the sensitivity and specificity of diagnostic imaging. Some approved contrasting agents include gadopentetate dimeglumine (Gd-DTPA or gadolinium diethylenetriamine pentaacetic acid gadobenate dimeglumine) (Kuhn et al., 2007). But there are significant drawbacks to adopting these techniques. For example, the imaging agents used in MRI require high doses because of their rapid dispersion qualities and poor sensitivity, nonspecificity, and a variety of unpleasant side effects in patients. Lipid-based nanoparticles have been presented as a solution to these issues. Liposomes encasing paramagnetic compounds in the aqueous lumen were the first reported type of liposomal that works as an MRI contrast agent (Mirahadi et al., 2018). This sort of contrast agent has been shown to upgrade the method for the identification of cancers in the liver of experimental mice with hepatic metastases. The paramagnetic molecule is carried in the lipid bilayer by a second class of liposomal contrast agents, making amphiphilic paramagnetic complexes an intrinsic component of the liposomal surface (Mulder et al., 2006). When compared to the way of enclosing the paramagnetic molecules in the aqueous interior, this method improves the metal's ionic relaxivity (Naseri et al., 2018). Other lipid-based nanoparticles, such as solid lipid nanoparticles and nanostructured lipid carriers, can also be used as imaging carriers in addition to liposomes. The hepatic and spleen macrophages had a poor grasp of SLNs. Furthermore, the increased absorption in the brain was a fascinating finding.

### 6.6.1 MAGNETIC RESONANCE IMAGING (MRI)

MRI is a non-invasive type of medical imaging technique that was clinically developed in the 1980s. It is a flexible approach that may be applied for both research and diagnostic purposes. Spins dissipate the energy they absorbed from a radiofrequency pulse during the process of relaxation (Thakor et al., 2016). A number of variables, including different forms of relaxation, affect the MRI signal: T1 (longitudinal relaxation) and T2 (transverse relaxation) (Naseri et al., 2018). Contrasting agents that shorten the longitudinal and transverse relaxation durations can change the tissue's relaxation timings. The ability to see the vascular system, irritated joints or tissue associated with arthritis, and arterial plaques is considerably improved by combining MRI with contrast agents (Naseri et al., 2018). Since they are biodegradable, biocompatible, and less toxic, have a long retention time and controlled delivery of therapeutic agents, can cross the blood–brain barrier, are target specific, and have the potential to insulate active pharmaceuticals, liposomes, a type of nanoparticulate colloid, have recently found use in the developing field of cellular and molecular MR imaging. There are several uses for nanoparticles in contrast-enhanced MRI. They can also be utilized as blood pool agents for magnetic resonance angiography (MRA) with prolonged circulation periods. PEGylated liposomes have been employed as a blood pool agent, for lymph node identification, and to provide long-term malignancy contrast enhancement. Gd–DTPA and Mn–DTPA are two examples (Mulder et al., 2006). *In vivo* MRI findings from SLN-FeB reveal that the slow blood clearance of SLN allows for prolonged brain absorption. Because SLNs may cross the blood–brain barrier, they are good candidates for use as a central nervous system MRI contrast agent. Recently, theranostics for treating cancer have been proposed utilizing SLNs loaded with quantum dots (QDs) (Mirahadi et al., 2018) (Green, 2004). On stimulation with a light source, semiconductor nanocrystals called quantum dots generate fluorescence (Green, 2004). The high illumination of QDs within SLNs permitted *in situ* representations of SLNs intracellular absorption into tumour tissue or cells, implying that this multipurpose and visually detectable nanocarrier might be employed as efficient antitumour diagnostics and therapeutics. The use of SLNs as a pharmacological tool is being studied for a variety of diagnostic agents, including superparamagnetic iron oxide and technetium-99. The findings showed that the recommended ideas have great promise for the future of medication and screening, since they allow for tailored distribution of therapeutic drugs while also allowing *in vivo* imaging (Mirahadi et al., 2018).

### 6.6.2 COMPUTED TOMOGRAPHY SCANNING

A computed tomography scan, often known as a CT scan or CAT scan, is a medical imaging procedure that uses X-rays to produce comprehensive inside pictures of the body without causing any discomfort to the patient. It is also not as expensive as other techniques (Lusic & Grinstaff, 2013). Because each kind of tissue in the body has a distinct atomic number, electron density, and therefore X-ray attenuation coefficient, the CT is able to differentiate between them in order to provide pictures of the body's tissues and structures. In order to increase the sensitivity of the image, many CT scans are presently performed with intravenously administered iodine-based (iohexol and iopromide) or oral barium sulphate solution contrast agents (Lusic & Grinstaff, 2013; He et al., 2015). However, clinical operations show that these compounds are not ideal due to a number of drawbacks, including a brief blood half-life, non-specific biodistribution, rapid clearance, stomach cramps, heart failure, a small risk of renal toxicity, hypersensitivity, and poor contrast in individuals who are obese (Lee et al., 2013). To get over this restriction, nanoparticulate contrast agents for CT have been developed in recent years. Positive CT imaging was made possible by the development of novel iodinated lipid-based contrast agent. Compared to standard water-soluble contrast agents, they are able to stay in the blood pool for prolonged periods of time (Weichert et al., 1998).

### 6.7 CONCLUSION AND FUTURE ASPECTS

In this chapter, we discussed lipid-based nanoplatforms that are being looked at for the creation of theranostic agents. As was previously mentioned, these nanoparticles can have distinctive properties and have been successfully modified and functionalized. A wide range of neurodegenerative illnesses have been treated with LBNPs, a

**TABLE 6.6**

**Imaging Application of Lipid Based Nanoparticles**

| Type of Nanocarriers | Diagnosing Agents | Modality | Uses | References |
|---|---|---|---|---|
| SLN | Superparamagnetic iron oxide | MRI | Tissues containing iron oxides appear hypointense (dark) relative to surrounding tissues on MR images | Mirahadi et al. (2018) Naseri et al. (2018) |
| | Technetium-99 | SPECT imaging | Higher accumulation, in foam cells | Mirahadi et al. (2018) |
| | 64Cu | PET | To track the biodistribution of radiolabel SLNs | Andreozzi et al. (2011) |
| | Quantum dots | Fluorescence imaging and MRI | *In situ* fluorescence imaging | Mirahadi et al. (2018) |
| | Gadolinium (Gd) | MRI | Improve the clarity of the images | Morel et al., (1998) |
| Thermosensitive liposomal | Gd-DTPA | MRI | Facilitating *in vivo* fluorescence imaging of tumour tissue | Li et al. (2014) |

broad and varied group of chemical substances. LNPs are now considered a standard in the area of drug delivery since their usage in clinical applications has significantly reduced the off-target toxicity of many drugs. Medical imaging with LNPs is projected to have the most impact by aiding in the early diagnosis of illness. The benefits of using lipid-based carriers that have been stated underline the need to speed up their introduction into clinical practice, and technological advances must be made to enable adequate manageability and the creation of cost-effective, inexpensive pharmaceutical formulations.

## REFERENCES

Abbott, N. J. (2013). Blood-brain barrier structure and function and the challenges for CNS drug delivery. *Journal of Inherited Metabolic Disease, 36*(3), 437–449.

Ahmed, K. S., Hussein, S. A., Ali, A. H., Korma, S. A., Lipeng, Q., & Jinghua, C. (2019). Liposome: Composition, characterisation, preparation, and recent innovation in clinical applications. *Journal of Drug Targeting, 27*(7), 742–761.

Adler-Moore, J. P., & Proffitt, R. T. (2008). Amphotericin B lipid preparations: what are the differences?. *Clinical Microbiology and Infection, 14*, 25–36.

Andreozzi, E., Seo, J. W., Ferrara, K., & Louie, A. (2011). Novel method to label solid lipid nanoparticles with 64cu for positron emission tomography imaging. *Bioconjugate chemistry, 22*(4), 808–818.

AZoNano.com. (2022, May 19). *Real-Time droplet size monitoring of nano emulsions during high pressure homogenization.* https://www.azonano.com/article.aspx?ArticleID=5679

Bangham, A. D. (1998). Artificial lung expanding compound (ALEC™). In *Elsevier eBooks* (pp. 455–472). Elsevier Science BV.

Barenholz, Y. C. (2012). Doxil®—The first FDA-approved nano-drug: Lessons learned. *Journal of controlled release, 160*(2), 117–134.

Bhatt, R., Singh, D., Prakash, A., & Mishra, N. (2015). Development, characterization and nasal delivery of rosmarinic acid-loaded solid lipid nanoparticles for the effective management of Huntington's disease. *Drug Delivery, 22*(7), 931–939.

Briggs, R., Kennelly, S. P., & O'Neill, D. (2016). Drug treatments in Alzheimer's disease. *Clinical medicine, 16*(3), 247.

Bulbake, U., Doppalapudi, S., Kommineni, N., & Khan, W. (2017). Liposomal formulations in clinical use: An updated review. *Pharmaceutics, 9*(2), 12.

Buse, J., & El-Aneed, A. (2010). Properties, engineering and applications of lipid-based nanoparticle drug-delivery systems: Current research and advances. *Nanomedicine, 5*(8), 1237–1260.

Chen, C. C., Tsai, T. H., Huang, Z. R., Fang, J. Y. (2010). Effects of lipophilic emulsifiers on the oral administration of lovastatin from nanostructured lipid carriers: Physicochemical characterization and pharmacokinetics. *Eur. J. Pharm. Biopharm, 74*, 474–482.

Chen, F., Ehlerding, E. B., & Cai, W. (2014). Theranostic nanoparticles. *Journal of Nuclear Medicine, 55*(12), 1919–1922.

Cirri, M., Maestrini, L., Maestrelli, F., Mennini, N., Mura, P., Ghelardini, C., Di Cesare Mannelli, L. (2018). Design, characterization and in vivo evaluation of nanostructured lipid carriers (NLC) as a new drug delivery system for hydrochlorothiazide oral administration in pediatric therapy. *Drug Deliv, 25*, 1910–1921.

Cormode, D. P., Skajaa, T., Fayad, Z. A., & Mulder, W. J. (2009). Nanotechnology in medical imaging: Probe design and applications. *Arteriosclerosis, Thrombosis, and Vascular Biology, 29*(7), 992–1000.

Cross, R. (2021). Without these lipid shells, there would be no mRNA vaccines for COVID-19. *Chem Eng News.*

De Lau, L., & Breteler, M. M. (2006). Epidemiology of Parkinson's disease. *Lancet Neurology, 5(6)*, 525–535.

Durukan, A., & Tatlisumak, T. (2007). Acute ischemic stroke: overview of major experimental rodent models, pathophysiology, and therapy of focal cerebral ischemia. *Pharmacology Biochemistry and Behavior, 87*(1), 179–197.

Esposito, E., Fantin, M., Marti, M., Drechsler, M., Paccamiccio, L., Mariani, P., Sivieri, E., Lain, F., Menegatti, E., Morari, M., & Cortesi, R. (2008). Solid lipid nanoparticles as delivery systems for bromocriptine. *Pharmaceutical Research, 25*(7), 1521–1530.

Fernandes, R. S., Silva, J. O., Monteiro, L. O., Leite, E. A., Cassali, G. D., Rubello, D., & de Barros, A. L. (2016). Doxorubicin-loaded nanocarriers: A comparative study of liposome and nanostructured lipid carrier as alternatives for cancer therapy. *Biomedicine & Pharmacotherapy , 84*, 252–257.

García-Pinel, B., Porras-Alcalá, C., Ortega-Rodríguez, A., Sarabia, F., Prados, J., Melguizo, C., & López-Romero, J. M. (2019). Lipid-based nanoparticles: application and recent advances in cancer treatment. *Nanomaterials, 9*(4), 638.

Gaba, B., Fazil, M., Ali, A., Baboota, S., Sahni, J. K., & Ali, J. (2015). Nanostructured lipid (NLCs) carriers as a bioavailability enhancement tool for oral administration. *Drug Delivery, 22*(6), 691–700.

Gadhave, D. G., Tagalpallewar, A. A., & Kokare, C. R. (2019). Agranulocytosis-protective olanzapine-loaded nanostructured lipid carriers engineered for CNS delivery: Optimization and hematological toxicity studies. *AAPS PharmSciTech, 20*(1), 1–15.

Green, M. (2004). Semiconductor quantum dots as biological imaging agents. *Angewandte Chemie International Edition, 43*(32), 4129–4131.

Guada, M., Sebastián, V., Irusta, S., Feijoó, E., del Carmen Dios-Viéitez, M., & Blanco-Prieto, M. J. (2015). Lipid nanoparticles for cyclosporine a administration: Development, characterization, and in vitro evaluation of their immunosuppression activity. *International Journal of Nanomedicine, 10*, 6541.

Haddad, H., Malone, G. W., Comardelle, N. J., Degueure, A. E., Poliwoda, S., Kaye, R., Murnane, K. S., Kaye, A. M., & Kaye, A. D. (2022). Aduhelm, a novel anti-amyloid monoclonal antibody, for the treatment of Alzheimer's Disease: A comprehensive review. *Health Psychology Research, 10*(2).

Haider, M., Abdin, S. M., Kamal, L., & Orive, G. (2020). Nanostructured lipid carriers for delivery of chemotherapeutics: A review. *Pharmaceutics, 12*(3), 288.

Haron, A. S., Syed Alwi, S. S., Saiful Yazan, L., Abd Razak, R., Ong, Y. S., Zakarial Ansar, F. H., & Roshini Alexander, H. (2018). Cytotoxic effect of thymoquinone-loaded nanostructured lipid carrier (TQ-NLC) on liver cancer cell integrated with hepatitis B genome, Hep3B. *Evidence-Based Complementary and Alternative Medicine, 2018*.

Hassanzadeh, P., Arbabi, E., Atyabi, F., & Dinarvand, R. (2018). Ferulic acid-loaded nanostructured lipid carriers: A promising nanoformulation against the ischemic neural injuries. *Life Sciences, 193*, 64–76.

He, W., Ai, K., & Lu, L. (2015). Nanoparticulate X-ray CT contrast agents. *Science China-chemistry, 58*(5), 753–760.

Huwyler, J., Wu, D., & Pardridge, W. M. (1996). Brain drug delivery of small molecules using immunoliposomes. *Proceedings of the National Academy of Sciences, 93*(24), 14164–14169.

Ishii, T., Asai, T., Fukuta, T., Oyama, D., Yasuda, N., Agato, Y., . . . & Oku, N. (2012). A single injection of liposomal asialo-erythropoietin improves motor function deficit caused by cerebral ischemia/reperfusion. *International Journal of Pharmaceutics, 439*(1–2), 269–274.

Jenning, V., Lippacher, A., & Gohla, S. H. (2002). Medium scale production of solid lipid nanoparticles (SLN) by high pressure homogenization. *Journal of Microencapsulation, 19*(1), 1–10.

Jores, K., Mehnert, W., Drechsler, M., Bunjes, H., Johann, C., & Mäder, K. (2004). Investigations on the structure of solid lipid nanoparticles (SLN) and oil-loaded solid lipid nanoparticles by photon correlation spectroscopy, field-flow fractionation and transmission electron microscopy. *Journal of Controlled Release, 95*(2), 217–227.

Karami, Z., Zanjani, M. R. S., & Hamidi, M. (2019). Nanoemulsions in CNS drug delivery: Recent developments, impacts and challenges. *Drug Discovery Today, 24*(5), 1104–1115.

Kasongo, K. W., Müller, R. H., & Walker, R. B. (2012). The use of hot and cold high pressure homogenization to enhance the loading capacity and encapsulation efficiency of nanostructured lipid carriers for the hydrophilic antiretroviral drug, didanosine for potential administration to paediatric patients. *Pharmaceutical Development and Technology, 17*(3), 353–362.

Kuhn, M. J., Picozzi, P., Maldjian, J. A., Schmalfuss, I. M., Maravilla, K. R., Bowen, B. C., . . . & Anzalone, N. (2007). Evaluation of intraaxial enhancing brain tumors on magnetic resonance imaging: Intraindividual crossover comparison of gadobenate dimeglumine and gadopentetate dimeglumine for visualization and assessment, and implications for surgical intervention. *Journal of Neurosurgery, 106*(4), 557–566.

Kumar, R. (2019). Lipid-based nanoparticles for drug-delivery systems. In *Nanocarriers for drug delivery* (pp. 249–284). Elsevier.

Large, D. E., Abdelmessih, R. G., Fink, E. A., & Auguste, D. T. (2021). Liposome composition in drug delivery design, synthesis, characterization, and clinical application. *Advanced Drug Delivery Reviews, 176*, 113851.

Lee, N., Choi, S. H., & Hyeon, T. (2013). Nano-sized CT contrast agents. *Advanced Materials, 25*(19), 2641-2660.

Li, X., Anton, N., Zuber, G., & Vandamme, T. F. (2014). Contrast agents for preclinical targeted X-ray imaging. *Advanced Drug Delivery Reviews, 76*, 116–133.

Li, X., Jia, X., & Niu, H. (2018). Nanostructured lipid carriers co-delivering lapachone and doxorubicin for overcoming multidrug resistance in breast cancer therapy. *International Journal of Nanomedicine, 13*, 4107.

Li, X., Sui, Z., Li, X., Xu, W., Guo, Q., Sun, J., & Jing, F. (2018). Perfluorooctylbromide nanoparticles for ultrasound imaging and drug delivery. *International Journal of Nanomedicine*, 3053–3067.

Lin, C., Chen, F., Ye, T., Zhang, L., Zhang, W., Liu, D., Xiong, W., Yang, X. (2014). Pan, W. A novel oral delivery system consisting in "drug-in cyclodextrin-in nanostructured lipid carriers" for poorly water-soluble drug: Vinpocetine. *Int. J. Pharm*, *465*, 90–96.

Liu, R., Liu, Z., Zhang, C., Zhang, B. (2012). Nanostructured lipid carriers as novel ophthalmic delivery system for mangiferin: Improving in vivo ocular bioavailability. *J. Pharm. Sci.*, *101*, 3833–3844.

Lu, Y. M., Huang, J. Y., Wang, H., Lou, X. F., Liao, M. H., Hong, L. J., ... & Han, F. (2014). Targeted therapy of brain ischaemia using Fas ligand antibody conjugated PEG-lipid nanoparticles. *Biomaterials*, *35*(1), 530–537.

Lusic, H., & Grinstaff, M. W. (2013). X-ray-computed tomography contrast agents. *Chemical reviews, 113*(3), 1641–1666.

Marras, C., & Tanner, C. M. (2012). Chapter 10: Epidemiology of Parkinson's disease. In Watts, R. L., Standaert, D. G. & Obeso, J. A. (eds.) *Movement disorders*. The McGraw-Hill Companies, New York, NY, USA.

Mirahadi, M., Ghanbarzadeh, S., Ghorbani, M., Gholizadeh, A., & Hamishehkar, H. (2018). A review on the role of lipid-based nanoparticles in medical diagnosis and imaging. *Therapeutic Delivery*, *9*(8), 557–569.

Misra, S., Chopra, K., Sinha, V. R., & Medhi, B. (2016). Galantamine-loaded solid–lipid nanoparticles for enhanced brain delivery: Preparation, characterization, in vitro and in vivo evaluations. *Drug delivery, 23*(4), 1434–1443.

Morel, S., Terreno, E., Ugazio, E., Aime, S., & Gasco, M. R. (1998). NMR relaxometric investigations of solid lipid nanoparticles (SLN) containing gadolinium(III) complexes. *European Journal of Pharmaceutics and Biopharmaceutics, 45*(2), 157–163.

Mulder, W. J., Strijkers, G. J., van Tilborg, G. A., Griffioen, A. W., & Nicolay, K. (2006). Lipid-based nanoparticles for contrast-enhanced MRI and molecular imaging. *NMR in Biomedicine: An International Journal Devoted to the Development and Application of Magnetic Resonance In Vivo*, *19*(1), 142–164.

Muller, R. H., Shegokar, R., & Keck, C. M. (2011). 20 years of lipid nanoparticles (SLN & NLC): Present state of development & industrial applications. *Current Drug Discovery Technologies*, *8*(3), 207–227.

Naik, A. B., & Selukar, N. B. (2009). Role of nanotechnology in medicine. *Everyman's Science*, *44*(3), 151–153.

Nair, R., Kumar, K. A., Vishnupriya, K., Badivaddin, T. M., & Sevukarajan, M. (2011). Preparation and characterization of rizatriptan loaded solid lipid nanoparticles (SLNs). *J. Biomed. Sci. Res, 3*(2), 392–396.

Naseri, N., Ajorlou, E., Asghari, F., & Pilehvar-Soltanahmadi, Y. (2018). An update on nanoparticle-based contrast agents in medical imaging. *Artificial Cells, Nanomedicine, and Biotechnology*, *46*(6), 1111–1121.

Pardeshi, C. V., Rajput, P. V., Belgamwar, V. S., Tekade, A. R., & Surana, S. J. (2013). Novel surface modified solid lipid nanoparticles as intranasal carriers for ropinirole hydrochloride: Application of factorial design approach. *Drug Deliv, 20*, 47–56.

Participants, V. R. (2005). Guidelines for using verteporfin (Visudyne) in photodynamic therapy for choroidal neovascularization due to age-related macular degeneration and other causes: Update. *Retina , 25*(2), 119–134.

Pfeiffer, R. F. (2016). Non-motor symptoms in Parkinson's disease. *Parkinsonism & related disorders, 22*, S119–S122.

Pindiprolu, S. K. S., Chintamaneni, P. K., Krishnamurthy, P. T., & Ratna Sree Ganapathineedi, K. (2019). Formulation-optimization of solid lipid nanocarrier system of STAT3 inhibitor to improve its activity in triple negative breast cancer cells. *Drug Development and Industrial Pharmacy*, *45*(2), 304–313.

Qu, C. Y., Zhou, M., Chen, Y. W., Chen, M. M., Shen, F., & Xu, L. M. (2015). Engineering of lipid prodrug-based, hyaluronic acid-decorated nanostructured lipid carriers platform for 5-fluorouracil and cisplatin combination gastric cancer therapy. *International Journal of Nanomedicine, 10,* 3911.

Rahiminejad, A., Dinarvand, R., Johari, B., Nodooshan, S. J., Rashti, A., Rismani, E., . . . & Khosravani, M. (2019). Preparation and investigation of indirubin-loaded SLN nanoparticles and their anti-cancer effects on human glioblastoma U87MG cells. *Cell Biology International, 43*(1), 2–11.

Rahimpour, Y., & Hamishehkar, H. (2012). Liposomes in cosmeceutics. *Expert Opinion on Drug Delivery, 9*(4), 443–455.

Rahman, M., & Kumar, V. (2019). Fabrication of solid lipid nanoparticles containing vitexin protects dopaminergic neurons against 6-hydroxydopamine induced Parkinson's disease model via altered the genetic backgrounds. *Journal of the Neurological Sciences, 405,* 248.

Rawat, M., Singh, D., Saraf, S. A. S. S., & Saraf, S. (2006). Nanocarriers: Promising vehicle for bioactive drugs. *Biological and Pharmaceutical Bulletin, 29*(9), 1790–1798.

Real, D., Formica, M. L., Picchio, M. L., & Paredes, A. J. (2022). Manufacturing techniques for nanoparticles in drug delivery. In Shahzad, Y., Rizvi, S.A.A., Yousaf, A.M., Hussain, T. (eds.), *Drug delivery using nanomaterials,* CRC Press: London, UK, pp. 23–48 CRC Press.

Robinson, D. M., & Keating, G. M. (2006). Memantine. *Drugs, 66*(11), 1515–1534.

Sarathchandiran, I. (2012). A review on nanotechnology in solid lipid nanoparticles. *Int J Pharm Dev Technol, 2*(1), 45–61.

Saupe, A., & Rades, T. (2006). Solid lipid nanoparticles. In *Nanocarrier technologies* (pp. 41–50). Springer, Dordrecht.

Sharma, A., Anghore, D., Awasthi, R., Kosey, S., Jindal, S., Gupta, N., . . . & Sood, R. (2015). A review on current carbon nanomaterials and other nanoparticles technology and their applications in biomedicine. *World Journal Pharmacy and Pharmaceutical Science, 4*(12), 1088–1113.

Sood, S., Jain, K., & Gowthamarajan, K. (2013). Curcumin-donepezil-loaded nanostructured lipid carriers for intranasal delivery in an Alzheimer's disease model. *Alzheimer's Dement, 9,* P299.

Souto, E. B., & Müller, R. H. (2006). Investigation of the factors influencing the incorporation of clotrimazole in SLN and NLC prepared by hot high-pressure homogenization. *Journal of Microencapsulation, 23*(4), 377–388.

Sriraman, S. K., & Torchilin, V. P. (2014). Recent advances with liposomes as drug carriers. *Advanced Biomaterials and Biodevices , 79–119.*

Takeuchi, H., Yamamoto, H., Niwa, T., Hino, T., & Kawashima, Y. (1996). Enteral absorption of insulin in rats from mucoadhesive chitosan-coated liposomes. *Pharmaceutical Research, 13*(6), 896–901.

Tenchov, R., Bird, R., Curtze, A. E., & Zhou, Q. (2021). Lipid nanoparticles: From liposomes to mRNA vaccine delivery, a landscape of research diversity and advancement. *ACS nano, 15*(11), 16982–17015.

Thakor, A. S., Jokerst, J. V., Ghanouni, P., Campbell, J. L., Mittra, E., & Gambhir, S. S. (2016). Clinically approved nanoparticle imaging agents. *Journal of Nuclear Medicine, 57*(12), 1833–1837.

Thakur, A. K., Kamboj, P., Goswami, K., & Ahuja, K. (2018). Pathophysiology and management of Alzheimer's disease: An overview. *J. Anal. Pharm. Res, 7*(1), 226–235.

Torchilin, V. P. (1994). Immunoliposomes and PEGylated immunoliposomes: Possible use for targeted delivery of imaging agents. *Immunomethods, 4*(3), 244–258.

Tsai, M. J.; Wu, P. C., Huang, Y. B., Chang, J. S., Lin, C. L., Tsai, Y. H., Fang, J. Y. (2012). Baicalein loaded in tocol nanostructured lipid carriers (tocol NLCs) for enhanced stability and brain targeting. *Int. J. Pharm, 423,* 461–470.

Turánek, J., Miller, A. D., Kauerová, Z., Lukáč, R., Mašek, J., Koudelka, Š., & Raška, M. (2015). Lipid-based nanoparticles and microbubbles-multifunctional lipid-based biocompatible particles for in vivo imaging and theranostics. *Advances in Bioengineering*, 79–116.

Twohig, D., & Nielsen, H. M. (2019). α-synuclein in the pathophysiology of Alzheimer's disease. *Mol. Neurodegener*, *14*, 23.

Üner, M. (2006). Preparation, characterization and physico-chemical properties of solid lipid nanoparticles (SLN) and nanostructured lipid carriers (NLC): Their benefits as colloidal drug carrier systems. *Die Pharmazie-an International journal of Pharmaceutical Sciences*, *61*(5), 375–386.

Wang, J., Zhu, R., Sun, X., Zhu, Y., Liu, H., & Wang, S. L. (2014). Intracellular uptake of etoposide-loaded solid lipid nanoparticles induces an enhancing inhibitory effect on gastric cancer through mitochondria-mediated apoptosis pathway. *International Journal of Nanomedicine*, *9*, 3987.

Wang, Y., Zhang, Z., Luo, J., Han, X., Wei, Y., & Wei, X. (2021). mRNA vaccine: A potential therapeutic strategy. *Molecular Cancer*, *20*(1), 1–23.

Weichert, J. P., Lee Jr, F. T., Longino, M. A., Chosy, S. G., & Counsell, R. E. (1998). Lipid-based blood-pool CT imaging of the liver. *Academic Radiology*, 5, S16–S19.

Witika, B. A., Poka, M. S., Demana, P. H., Matafwali, S. K., Melamane, S., Malungelo Khamanga, S. M., & Makoni, P. A. (2022). Lipid-based nanocarriers for neurological disorders: A review of the state-of-the-art and therapeutic success to date. *Pharmaceutics*, *14*(4), 836.

Yang, S. C., Lu, L. F., Cai, Y., Zhu, J. B., Liang, B. W., & Yang, C. Z. (1999). Body distribution in mice of intravenously injected camptothecin solid lipid nanoparticles and targeting effect on brain. *Journal of Controlled Release*, *59*(3), 299–307.

Yasir, M., Sara, U. V. S., Chauhan, I., Gaur, P. K., Singh, A. P., Puri, D., & Ameeduzzafar, A. (2018). Solid lipid nanoparticles for nose to brain delivery of donepezil: Formulation, optimization by box-behnken design, in vitro and in vivo evaluation. *Artif. Cells Nanomed. Biotechnol*, *46*, 1838–1851.

Zhang, X., Zhang, H., Wu, Z., Wang, Z., Niu, H., & Li, C. (2008). Nasal absorption enhancement of insulin using PEG-grafted chitosan nanoparticles. *European Journal of Pharmaceutics and Biopharmaceutics, 68*(3), 526–534.

Zara, G. P., Cavalli, R., Bargoni, A., Fundarò, A., Vighetto, D., & Gasco, M. R. (2002). Intravenous administration to rabbits of non-stealth and stealth doxorubicin-loaded solid lipid nanoparticles at increasing concentrations of stealth agent: Pharmacokinetics and distribution of doxorubicin in brain and other tissues. *Journal of Drug Targeting*, *10*(4), 327–335.

# 7 Amino Acid Functionalized Inorganic Nanoparticles in Diagnosis

*Anindita De*

## 7.1 INTRODUCTION

Since their inception, nanomaterials have attracted a tremendous amount of attention from researchers all over the globe due to their exceptional chemical and physical properties. Application of nanomaterials ranges from catalysis [1] to energy storage [2] and therapy/diagnosis [3] to fabrication of biomedical devices [4]. One of the main difficulties encountered in synthesizing and handling inorganic nanomaterials is their tendency to aggregate. Therefore, suitable capping agents are required to stop this aggregation process. Most of the time, the capping agents are organic molecules containing functional groups which impart a positive or negative charge on the surface of the nanomaterials. Steric repulsion between the like charges prohibits the aggregation process. Moreover, organic molecules also regulate the shape and size of the nanomaterials by selective binding to particular crystalline phases during the growth phase [5]. Inclusion of organic molecules onto the nanoparticle surface brings about surface modification, which is often different from the nanoparticle composition and thus makes it suitable for certain applications [5, 6]. Molecules containing functional groups create a monolayer on the nanomaterial surface and affect properties such as hydrophilicity/hydrophobicity, self-assembly, and adsorption behaviour depending on the types of functional group present [7, 8]. Molecules such as long chain alkane thiols, long chain carboxylic acid and terminally functionalized thiols have been used for functionalization. In the last example, the thiol group anchors on the nanoparticle surface while the terminal functionality, such as amine or carboxylic acid, is responsible for further reactivity. However, synthesis of multifunctional molecules is challenging and requires multi-step organic synthesis. Besides, the presence of terminal, organic functional groups also increases the overall polarity and makes it unstable in different solvents as well as in mediums of different pH [8]. Biological molecules like sugars, lipids and protein have been used for functionalization purposes but could not bypass the problem of aggregation. Amino groups containing surfactants have also been used by some researchers [9], but their uses are severely limited as the presence of surfactants makes the nanoparticles hydrophobic

DOI: 10.1201/9781003316398-7

due to the presence of hydrocarbon chains. Instead of using biomolecules or surfactants, one can use amino acids because they contain multiple functional groups in their structure. Amino acids are easily available and easy to purify compared to other types of biomolecules, and they are non-toxic, versatile and highly stable. Amino acids have low molecular weight and contain both acidic and basic groups (positive and negative) and themselves exert some useful biological functions. Moreover, capping of amino acids makes the surface polar, increases the water dispersibility and biocompatibility and thus makes the nanomaterials more suitable for biological applications. The chirality, optical properties of functional group containing amino acids in combination with optical, electrical and magnetic properties of the nanoparticle, has inspired their novel applications in various fields of therapy and diagnosis such as biosensing, bioreactors, biofuel cells, drug delivery, bioimaging, as MRI contrast agents, as antimicrobial agents and in cancer therapy [10] (Figure 7.1).

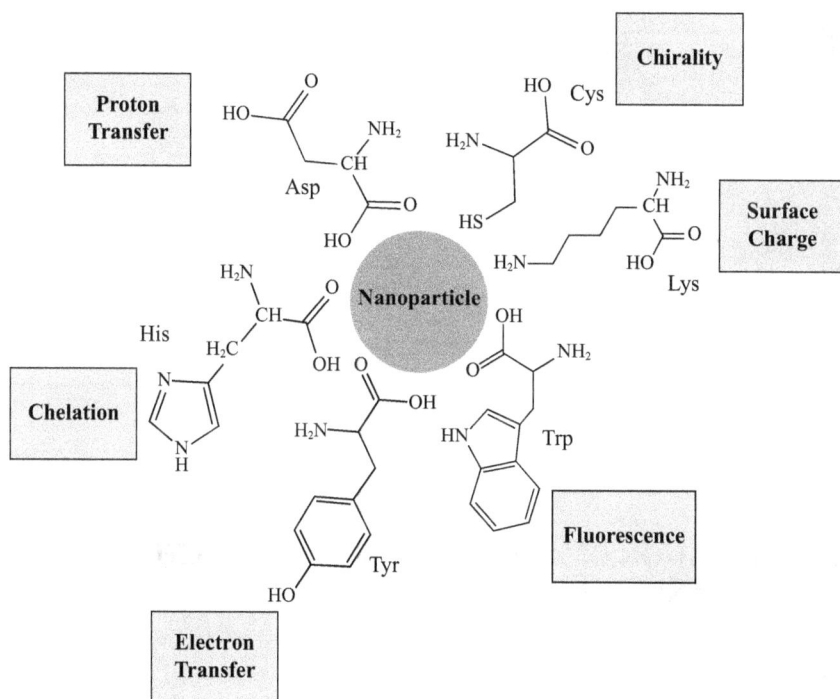

**FIGURE 7.1**   Various properties of amino acid functionalized inorganic nanoparticles [9].

Amino acids are the elementary unit of which peptides and proteins are made. They can be classified into three distinct groups: nonpolar, polar and electrically charged. Positively charged amino acids are basic in nature while negatively charged amino acids are acidic (Figure 7.2). In this chapter amino acid functionalized inorganic nanoparticles have been discussed in detail in terms of their synthesis, properties and applications, with special emphasis on diagnostic applications.

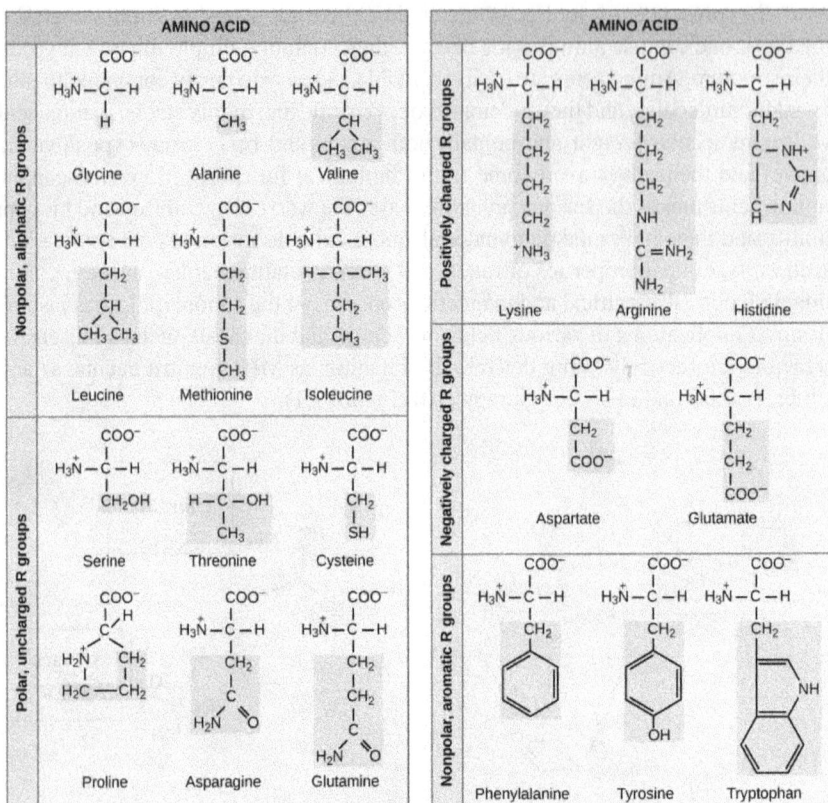

**FIGURE 7.2**     Different types of amino acids and their structure. Image source: Wikimedia commons (CC by 4.0).

## 7.2   FABRICATION OF AMINO ACID FUNCTIONALIZED INORGANIC NANOPARTICLES

The synthetic procedure can be broadly classified into two types: use of amino acid during nanoparticle generation and growth. Here, amino acid itself acts as a reducing agent, and commonly no reducing agent is used. The second strategy is post-synthesis functionalization, where amino acids are incorporated on the already prepared nanoparticle synthesis. In the following subsections, these two synthetic protocols are discussed in detail.

### 7.2.1   FUNCTIONALIZATION OF NANOPARTICLES USING AMINO ACIDS DURING SYNTHESIS

This is a direct method of synthesis where the amino acids are attached onto the nanoparticle surface during the formation stage from the precursor material. Inorganic nanoparticles are synthesized by reduction of their respective metal salts using a chemical reducing agent in the presence of capping agents. Traditionally used reducing agents include trisodium citrate, sodium borohydrate, ascorbic acid and hydrazine. The unreacted reducing agents and the by-products can contaminate the nanoparticle surface, thereby limiting its efficiency. Amino acids can serve the dual purpose of serving both as reducing agent and as functionalizing agent. For instance, tyrosine, tryptophan and aspartic acid attached stable gold and silver nanoparticles were synthesized by Selvakanna et al. and Daima et al. [11–13]. Amino acids such as aspartic acid, lysine, tyrosine and tryptophan have one extra functional group in addition to amino and carboxylic acid functionality, which makes them more useful. Moreover, they are highly stable and retain their water dispersibility property even after completely drying out. Anisotropic growth of gold nanoparticles on using lysine on a gold seed has been studied [14] and reveals that the amount of lysine and the gold seed dictate the anisotropic growth of the gold nanoparticles (Figure 7.3).

Facile synthesis of bluish-green fluorescent gold nanoclusters have been reported using histidine amino acid [15]. In the last examples, amino acid itself acts as a reducing agent, and no other reducing agent is required to reduce the metal salt precursor (Figure 7.4).

## 7.2.2 Functionalization of Nanoparticles Using Amino Acid Post Synthesis

Some amino acids such as phenylalanine and cysteine are unsuitable reducing agents for the metal ions. Instead, they can be used to functionalize already synthesized nanoparticles synthesized using some other type of capping agent, for example citrate

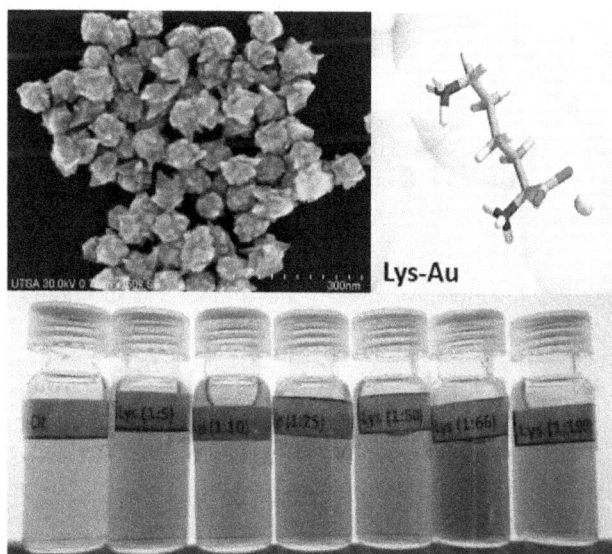

**FIGURE 7.3**  Lysine amino acid functionalized gold nanoparticles [14]. Reprinted with permission from Plascencia-Villa et al. (2015). Copyright (2015) American Chemical Society.

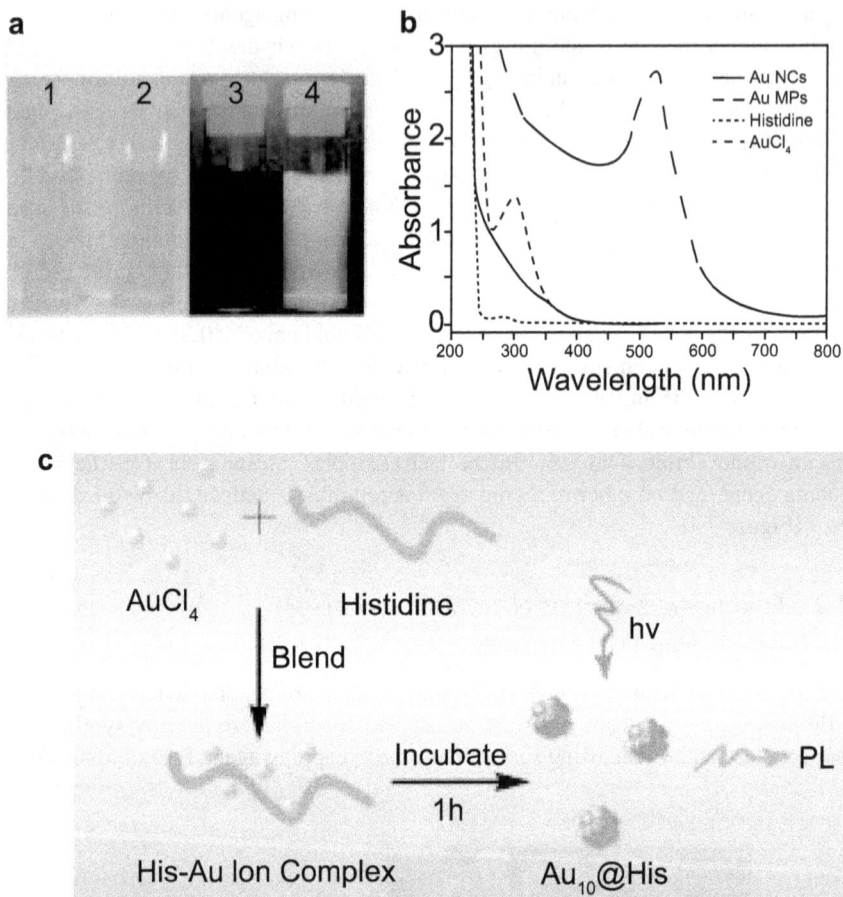

**FIGURE 7.4** (a) HAuCl4 solution in daylight (1) and under 365nm UV radiation (2). (b) Lysine-capped gold nanocluster in daylight (3) and under and under 365 nm UV radiation (4). (c) Schematic diagram of lysine-capped gold nanocluster synthesis [15]. Reprinted with permission from Yang et al. (2010). Copyright (2010) Royal Society of Chemistry.

ions. During the functionalization process, the amino acid substitutes the capping agents from the nanoparticle surface. Amino acids can bind with the nanoparticle surface either by thiol functionality (-SH), amine (-NH$_2$) or by carboxylic functionality (-COOH) [16–18]. Basic amino acids are those which are easily protonated in water to assume a positive charge at physiological pH. Cationic amino acids promote silica hydrolysis and hence have been used for the synthesis of mesoporous silica. When amino acids are removed, mesoporous silica is generated. In some instances, presence of chirality in the amino acid is retained in the nanomaterial structure [19, 20]. Use of arginine during synthesis causes tetraethyl orthosilicate hydrolysis leading to the formation of uniformly shaped silica nanoparticles [21]. The SEM image and the size distribution for arginine-capped silica nanoparticles are shown in Figure 7.5. In another example, polyacrylic acid functionalized CeO$_2$ nanoparticles were

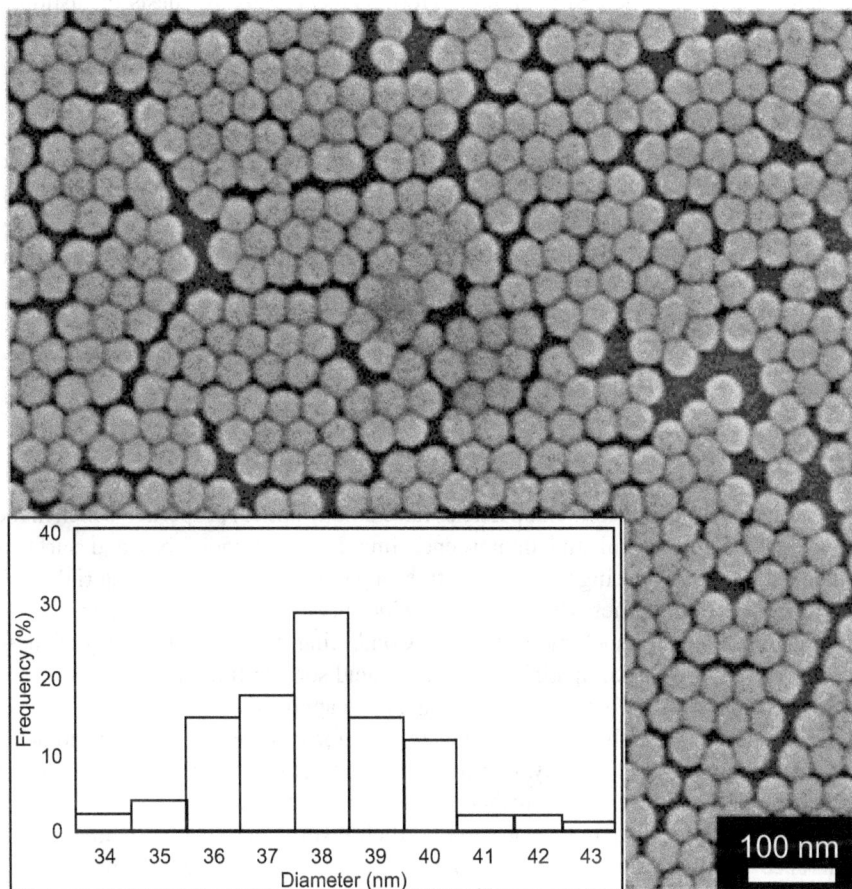

**FIGURE 7.5**   SEM image of arginine-capped silica nanoparticles. Inset: size distribution data obtained from DLS experiment [21]. Reprinted with permission from Wang et al. (2010). Copyright (2010) American Chemical Society.

subjected to amidation, which leads to the formation of D and L phenylalanine functionalized $CeO_2$ nanoparticles. FTIR spectroscopy can prove the presence of amino acid attached to the surface [22]. Synthesis of arginine functionalized CuO nanoparticles have been reported by El-Trass et al. [23].

Quantum dots and semiconductor nanoparticles are generally synthesized in nonpolar organic solvent, and as in polar solvent the fluorescence is quenched. Moreover, semiconductor nanoparticles have very poor dispersibility in polar solvent such as water, which further limits their applications. However, amino acid functionalized quantum dots and semiconductor nanoparticles have the ability to disperse in water without losing fluorescence property [24]. CdS, ZnS, ZnO and $TiO_2$ nanoparticles are well-known semiconductor nanoparticles which can be synthesized in the presence of amino acids from their precursor metal salt. Histidine amino acids can interact strongly with both zinc and cadmium. Thus, sonochemical synthesis of histidine-functionalized ZnS and CdS nanoparticles with good fluorescence properties have been reported. The amino acid–nanoparticle interaction is very important as it dictates the kinetics of formation as well as their stability [25].

Magnetite nanomaterials have widespread application in the biomedical field but they can cause toxicity. For stabilization and for making them biocompatible, various types of biomolecules including amino acids have been used. Nosrati et al. have studied the one-pot synthesis of iron oxide nanoparticles functionalized with different amino acids. The iron nanoparticles were generated from FeCl3 and FeCl2 and subsequently treated with glycine, L-lysine, L-phenylalanine, L-arginine and L-tyrosine. It was reported that amino acid–coated iron nanoparticles have significantly less toxicity and more biocompatibility compared to the bare nanoparticles when tested against the HFF2 cell line [26].

## 7.3    OPTICAL AND CHIRO-OPTICAL PROPERTIES

Metallic and semiconductor nanoparticles have unique optical properties which make them suitable for optical and fluorescence imaging, immunoprobes and surface-enhanced Raman scattering (SERS) spectroscopy [27–30]. These nanomaterials have tuneable optical properties which lie over a wide spectral range, and compared to the organic dyes they are much more efficient. Combining the optical property of these nanomaterials with biocompatibility of amino acid surface functionalization makes them useful for various bioimaging and diagnostic applications.

In addition to that, chiral amino acids (except for glycine, all other amino acids are chiral) show optical activity. Chirality in combination with their other types of optical properties present can further extend their applications. Chiral molecules, including amino acids, can exhibit circular dichroism which can be defined as the difference in absorption for the left and right circularly polarized light. Circular dichroism for chiral compounds falls in the UV region, whereas absorption and emission for metallic nanoparticles fall in the visible region. Conjugation of amino acid onto the nanoparticle surface makes it chiral, and the presence of surface plasmon resonance and fluorescence can move the circular dichroism adsorption towards the visible region of the spectrum. Thus, by controlling the chirality (number of functional groups/amino acids on the nanoparticle surface) one can fine-tune the chiro-optical properties for use in optical devices, for fabricating chiral catalysts as well as for enantiospecific separation. Two main phenomena responsible for optical properties exhibited by metallic and semiconductor nanoparticles are described next.

### 7.3.1    Surface Plasmon Resonance

Surface plasmon resonance (SPR) can be defined as the collective oscillation of conduction band electrons in the presence of some electromagnetic radiation. When the frequency of oscillation matches with that of the electromagnetic radiation, energy transfer occurs. Due to this phenomenon, metallic nanoparticles have intense absorption in the visible or in the near infrared range. Surface plasmon resonance is highly susceptible to shape, size, chemical composition, dispersing medium and state of aggregation of the nanoparticles. Surface plasmon resonance absorption has a high extinction coefficient ($\varepsilon = 10^8$ M$^{-1}$ cm$^{-1}$) compared to absorption cross section for commonly used organic dyes in bioimaging studies. As the SPR is dependent on the surface groups, any change in surface functionalization can lead to SPR absorbance and as a result visual change in colour. Some of the amino acids interact with the nanoparticles strongly leading to the charge transfer and resulting change in SPR absorption. Even when the nanoparticle composition is the same, different types of amino acid functionalization can show different SPR absorption. Sethi et al. have studied the effect of different amino acids on SPR behaviour of the gold nanoparticles by partially introducing arginine (Figure 7.6), alanine, cysteine, and histidine amino acid on the already synthesized, citrate-capped gold nanoparticle surface [31]. Treating the gold nanoparticles with amino acid exchanges the citrate ions from the surface. The concentration of the amino acids and their interaction with the gold nanoparticle dictates the self-assembly process. It was observed that except for cysteine, the binding of all other amino acids to the gold nanoparticle surface is reversible. The authors concluded that amino acid binding to the nanoparticle surface is subjected to the type of functional group present, which in turn also dictates the aggregation and optical properties. Anisotropic nanoparticles have two different SPR absorption—one transverse and another longitudinal. Use of amino acids during synthesis can lead to the formation of anisotropic gold nanoparticles. Plascencia-Villa reported the synthesis of star-shaped gold nanoparticle synthesis using lysine, a type of basic amino acid [14] (Figure 7.3). The optical property is tuneable from visible to near infrared range by changing the gold salt to lysine volume ratio. The authors reported that the amino acid lysine interacts with the gold nanoparticle via the -COOH group and not by the positively charged side chain.

## 7.3.2 Fluorescence Properties

Semiconductor nanoparticles demonstrate fluorescence behaviour and, similar to SPR, fluorescence properties also depend on size, shape, composition, state of aggregation and surface functionalization [32]. Consequently, semiconductor nanoparticles are extensively used in bioimaging applications. However, these nanoparticles are toxic in nature and have poor dispersibility in water. Functionalization of the semiconductor nanoparticles with amino acid can lead to better biocompatibility and better water dispersion without disrupting its fluorescence behaviour. Amino acid functionalized nanoparticles can bind with another substrate or biomolecule and cause enhancement or quenching of fluorescence emission [33–35]. Tryptophan amino acid has a strong absorption in the UV region of the spectrum which tends to change in the presence of nanoparticles. Kasture et al. have reported the synthesis of white light emitting, CTAB-capped, anisotropic gold nanoparticles in the presence of tryptophan as reducing agent [36] (Figure 7.7). In general, the fluorescence is quenched in the presence of metallic nanoparticles. However, in this case the fluorescence increases.

In another work, fluorescence of tryptophan bound to zinc oxide nanoparticles have been studied. It was observed that fluorescence resulting from the amino acid is quenched when it is bound to zinc oxide nanoparticles, probably due to strong

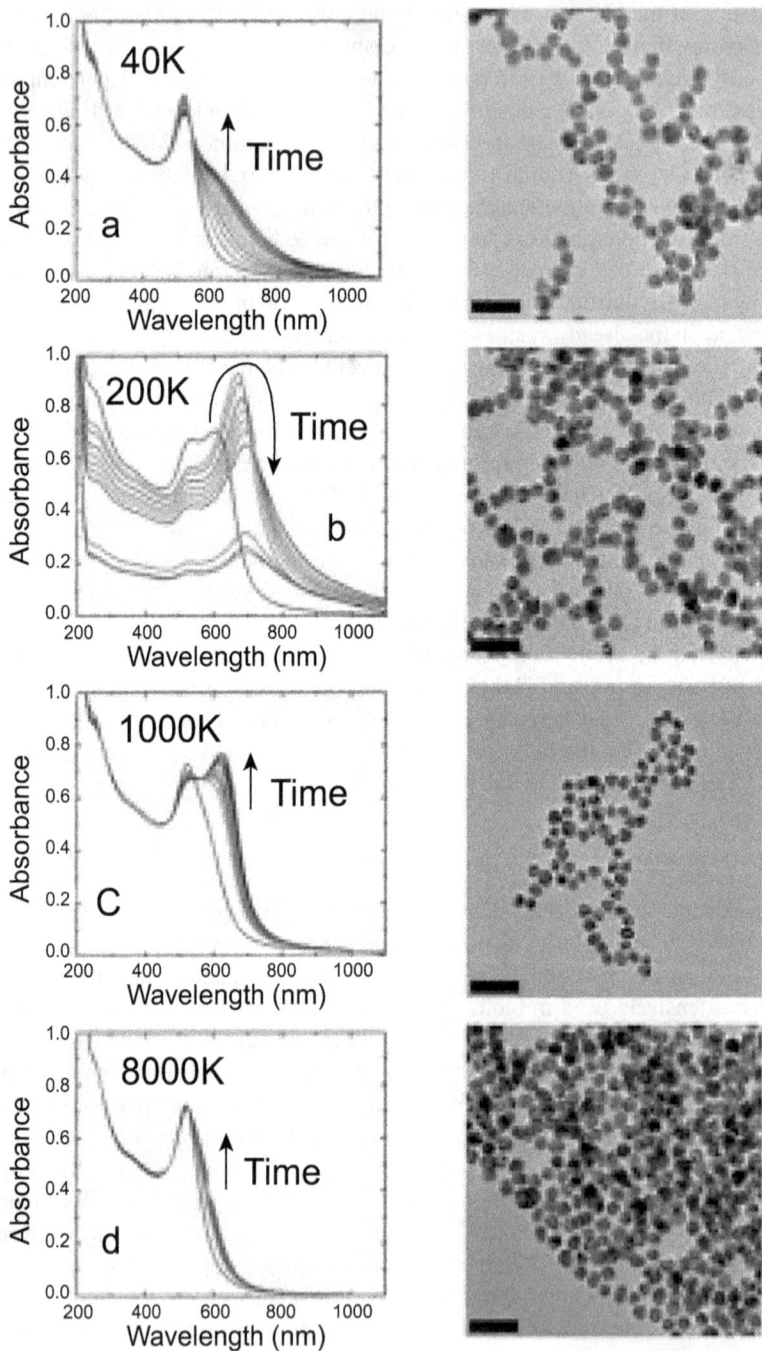

**FIGURE 7.6** UV-visible spectra and TEM images of arginine-capped Au nanoparticles with amino acid: nanoparticle ratio of (a) 40K, (b) 200K, (c) 1000K and (d) 8000K at room temperature and studied over a period of 6 hours [31]. Reprinted with permission from Sethi et al. (2010). Copyright (2010) Royal Society of Chemistry.

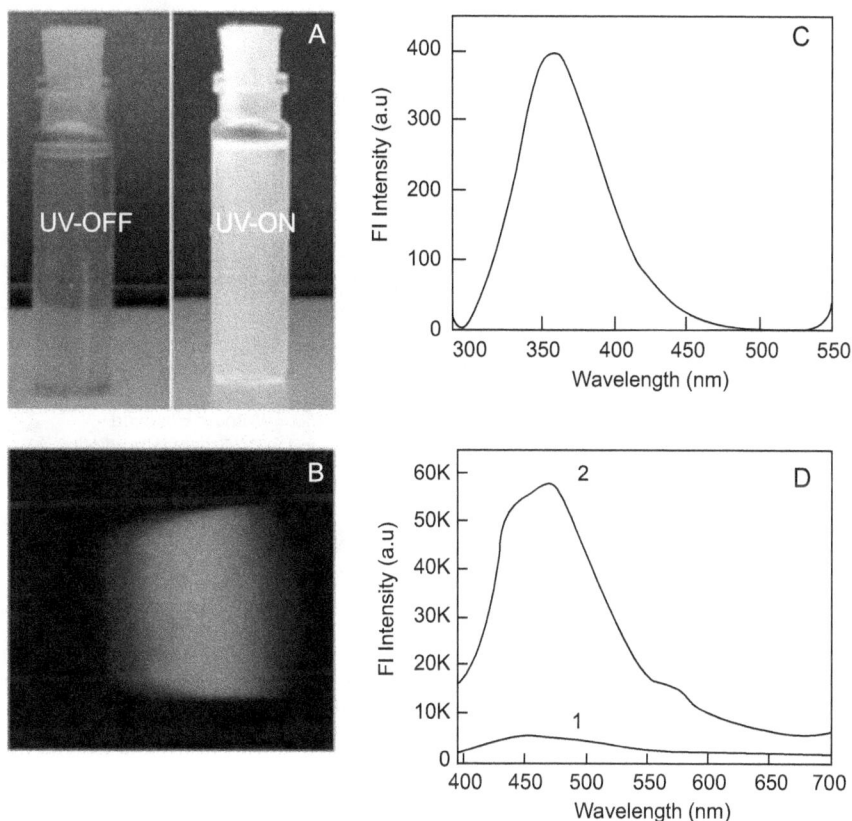

**FIGURE 7.7** (a) Generation of fluorescence. (b) Continuous spectrum obtained when while light is passed through a prism. (c) Fluorescence spectrum of tryptophan excited with 365 nm radiation. (d) Fluorescence spectrum of tryptophan-capped gold nanoparticles excited with 365 nm radiation [36].

electrostatic interaction between the ZnO nanoparticle and the tryptophan [37]. Histidine-bound cadmium sulphide nanoparticles have also been synthesized by a sonochemical method. The presence of amino acid controls the size as it has a strong affinity for cadmium. These nanoparticles were found to be highly fluorescent due to the confinement effect [38] (Figure 7.8).

## 7.4 BINDING OF AMINO ACID ON NANOPARTICLE SURFACE: THERMODYNAMIC ASPECT

To use the amino acid–bound metallic nanoparticles for various applications, one should have a thorough understanding of how the amino acid is interacting with the metallic surface. In addition to that, this interaction is also dictated by various physicochemical factors such as temperature, ionic strength and pH. Thus, the study of the thermodynamics of how amino acids interact with metallic nanoparticles is of prime

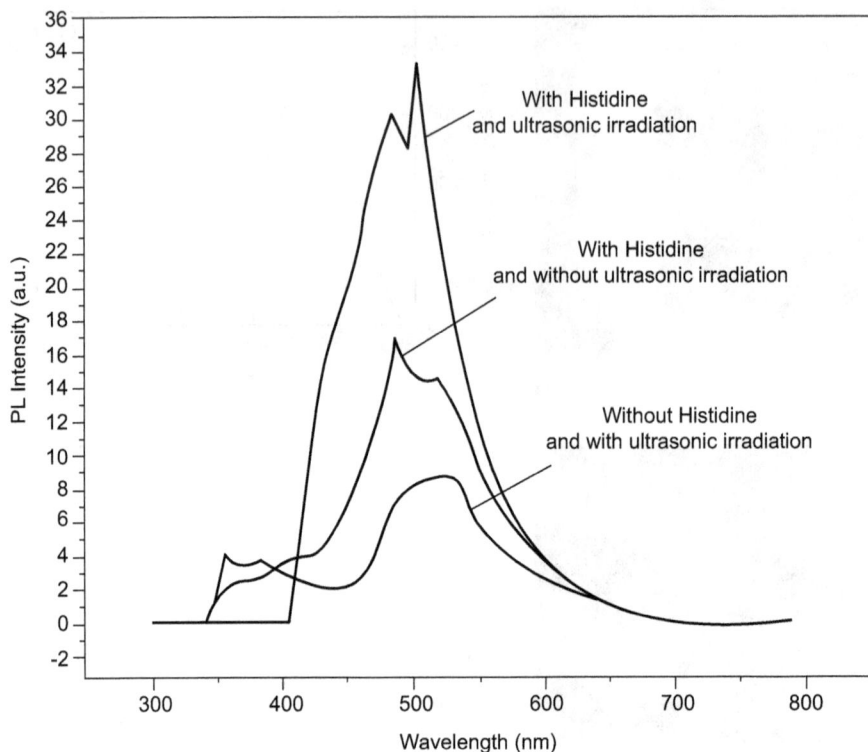

**FIGURE 7.8** Photoluminescence spectrum of CdS nanoparticles synthesized under different conditions [38]. Reprinted with permission from Yadav et al. (2010). Copyright (2010) Elsevier.

importance, as it will help us to understand and predict the behaviour of the nanoparticles under different conditions. Interaction of the amino acid with the metallic nanoparticle surface could be covalent or noncovalent such as electrostatic, pi-pi, Van der Waals, hydrogen bonding and hydrophobic. Furthermore, amino acids contain multiple functional groups which can interact differently and individually with the metallic surface, thus increasing the complexity. It is therefore very important to experimentally establish how the metallic nanoparticles are interacting with amino acids. One such experimental technique is isothermal titration calorimetry (ITC), which can provide information about the binding energy by measuring the heat of the reaction. This technique has been widely employed to study various types of biological interaction such as protein-protein interactions, protein-lipid interactions, protein-metal ion interactions, protein-DNA interactions as well as drug-enzyme interactions [39]. Joshi et al. have studied the heat of interaction of adding two different amino acids, aspartic acid and lysine, in aqueous dispersion of gold nanoparticles. The heat of interaction was measured at a function of pH in physiological and alkaline pH range. As more and more amino acid molecules occupy the vacant site on the gold nanoparticle surface, exothermic interaction decreases and eventually saturates. At neutral pH, aspartic acid, which is an acidic amino acid, binds strongly while at basic pH, lysine binds strongly. The authors concluded that both amino acids can bind with the gold nanoparticles if the -NH$_2$ group remains deprotonated. In the case of aspartic acid, deprotonation does not occur at physiological pH, and for lysine in higher pH the -NH$_2$ group is not protonated [40]. NMR spectroscopy and UV-visible spectroscopy are also used to study amino acid–nanoparticle interaction.

Selvakannan et al. have studied the effect of the gradual addition of lysine to gold nanoparticle dispersion by UV-visible spectrophotometer. It was observed that with continuous addition, surface plasmon resonance absorption becomes broad and is moved towards higher wavelength regions, indicating slight aggregation of nanoparticles. However, the overall stability of the nanoparticles was high, and lysine-capped nanoparticles are stable for months. Most importantly, lysine-capped gold nanoparticles could be isolated in dry form and can be easily redispersed in water. Thermogravimetric analysis as well as proton NMR studies were also performed by the authors. Proton NMR studies on the lysine amino acid and lysine-capped gold nanoparticles were performed in D$_2$O to understand the interaction at the molecular level and reveal that the NMR spectrum for both is not same. Major changes occur for the proton of the alpha carbon of the amino acid and the proton present in the carbon next to the second amino acid. Peak from the terminal -NH$_2$ is also broadened, indicating the presence of hydrogen bonding with the -COOH group of the lysine bound to the neighbouring gold nanoparticles. From these results the authors could conclude that binding of the amino acid with the nanoparticle surface is via amino group and the second amino acid is responsible for the hydrogen bonding [41]. Same authors have studied the capping of tyrosine amino acid on silver nanoparticles, which also reveals the profound effect that pH of the medium plays in binding of different groups to nanoparticles of silver. The nanoparticles are highly stable and can be easily dispersed on water. The authors extended their work by using gold nanoparticle-surface bound tyrosine molecules to reduce Ag(I) ions to generate silver nanoparticles onto the surface of the gold nanoparticles. To understand the binding, the authors utilized the ITC technique. They gradually added tyrosine to pre-synthesized gold nanoparticles and monitored the heat of the reaction. The process is highly exothermic, indicating the strong affinity tyrosine has towards the gold nanoparticle surface. By using the ICT technique along with UV-visible spectroscopy, the authors concluded that reduction of Ag(I) took place only in basic pH. This is because, at higher pH, the phenolic side chain of tyrosine amino acid would be deprotonated and the reduction process accompanied by possible formation of a semiquinone type moiety (Figure 7.9) [42].

## 7.5  APPLICATIONS

Proteins, hormones, pathogens and other biomarkers are routinely diagnosed in pathological laboratories. However, the standard methods are time consuming and expensive and require extensive use of chemicals and skills to interpret the data. Development of analytical tools that can detect biomolecules easily with a high accuracy level is one of the forefronts of biomedical research nowadays. The necessary quality that an analytical probe possesses is easy sample preparation, portable instrument, low cost, rapid analysis, easy to comprehend results as well as the specificity

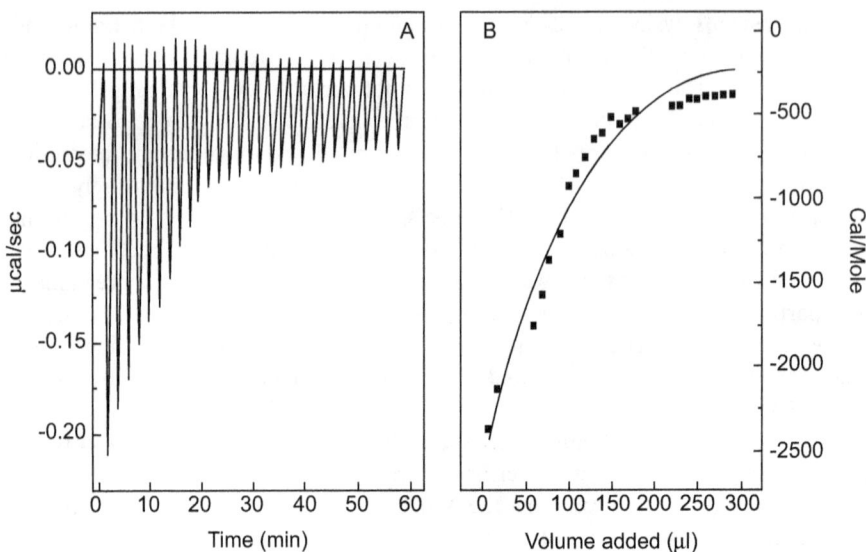

**FIGURE 7.9** (A) ITC data obtained on gradual addition of tyrosine solution to borohydride reduced gold nanoparticle. (B) Binding isotherm [42]. Reprinted with permission from Selvakannan et al. (2004). Copyright (2004) American Chemical Society.

and accuracy of the probe [43]. Amino acids are bi/multifunctional and thus one of the functional groups can bind with the metallic nanoparticle while the other interacts with the analyte to cause a signalling effect. There are 20 amino acids, each with different functional groups. On the other hand, inorganic nanoparticles could be of different types. Thus, various types of sensing probes are designed and fabricated. Inorganic nanoparticles have interesting optical properties (section 7.4) and give fast optical response (change in colour or change in fluorescence) in the presence of analyte. Hence, amino acid functionalized inorganic nanoparticle-based probes can be an alternative to clinical assays or other instrument-based techniques. The analytical probes can be of two types—fluorescence sensor and colorimetric sensor.

### 7.5.1 TOWARDS SENSING FOR DIAGNOSTIC APPLICATIONS

Levothyroxine or L-thyroxine is an artificial thyroid enzyme that is used to treat abnormality in thyroid gland function. Cysteine-functionalized zinc sulphide quantum dots can quantify L-thyroxin up to a level of $9.5 \times 10^{-7}$ mol/L. The process is based on quenching photoluminescence in the presence of the quantum dots. The authors observed that on addition of L-thyroxin, the photoluminescence is quenched (Figure 7.10) and the efficiency of the process further increases on addition of a surfactant such as cetyltrimethyl ammonium bromide (CTAB). The authors explained the mechanism that the cationic surfactant can neutralize the negative charge present on the quantum dots surface due to the presence of anionic cysteine amino acid. As a result, L-thyroxin can now come closer to the zinc and interact with it causing

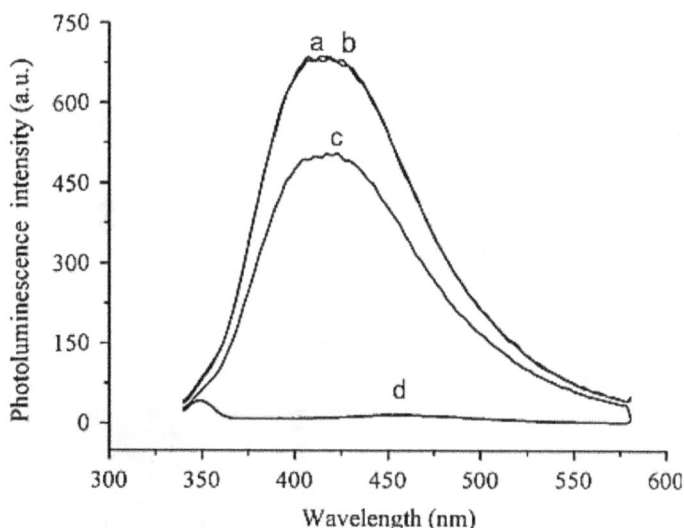

FIGURE 7.10   (a) Photoluminescence spectrum of aqueous dispersion of cysteine-function-alized ZnS quantum dots. (b) Photoluminescence spectrum of saliva resuspended on aqueous dispersion of cysteine-functionalized ZnS quantum dots. (c) Photoluminescence spectrum of saliva containing L-thyroxine resuspended on aqueous dispersion of cysteine-functionalized ZnS quantum dots. (d) Emission spectrum of saliva resuspended in water [44]. Reprinted with permission from Khan et al. (2014). Copyright (2014) Elsevier.

quenching. Thyroxine withdrawal electron density from the conduction band of the quantum dot results in static quenching of the photoluminescence. The probe was used to detect L-thyroxine in drugs that contain the artificial thyroxine hormone as active constituent and in human saliva containing the analyte. In human saliva samples, the limit of detection was found to be $4.0 \times 10^{-6}$ mol/L [44].

In another example, ascorbic acid was detected by a biosensor based on N-acetyl cysteine bound CdTe/CdS/ZnS quantum dots. The limit of detection is 1.8 nm in human urine samples.

The process requires the addition of $KMnO_4$ in the dispersion of nanomaterial in tris-HCl buffer solution. On addition of $KMnO_4$, the fluorescence is quenched as it binds to the surface of the nanoparticles and electron transfer occurs. When ascorbic acid is introduced in the system, a redox reaction occurs between $KMnO_4$ and ascorbic acid which eliminates the molecule of $KMnO_4$ from the nanoparticle surface and, consequently, the fluorescence starts again (Figure 7.11).

Except proteinuria, the effects of other compounds present in the urine sample such as salt, nucleotide and amino acid were also considered. The exact benefit of using N-acetyl substituted amino acid is not clear; however, the most probable reason is the electron withdrawing effect of the substituent, which helps in the removal of electrons from the conduction band of zinc sulphide shells [45]. Glutathione is an important biomarker that is used to detect several diseases such as HIV, Parkinson's disease and Alzheimer's disease. Glutathione detection by a turn on/off fluorescence

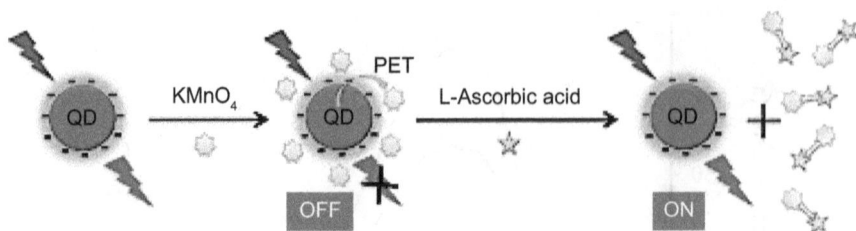

**FIGURE 7.11** Schematic representation of N-acetyl cysteine bound CdTe/CdS/ZnS quantum dots based ascorbic acid fluorescent biosensor [45]. CC by 2004.

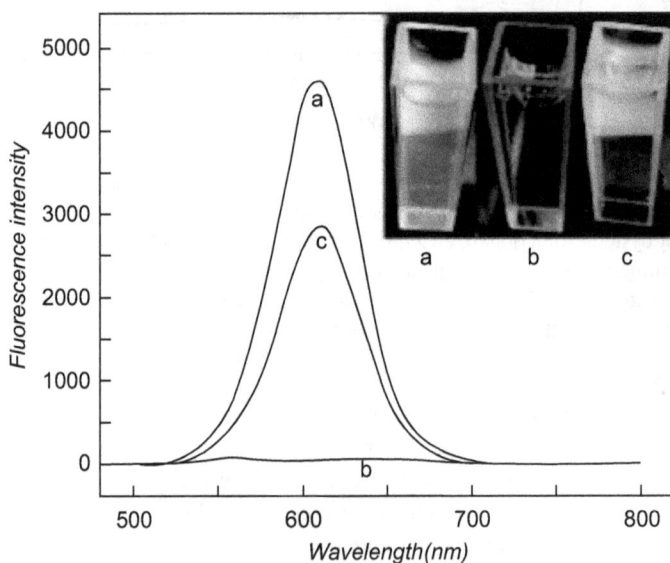

**FIGURE 7.12** The fluorescence spectra of (a) NALC-functionalized CdTe quantum dots; (b) NALC-functionalized CdTe quantum dots on addition of Hg(II); (c) NALC-functionalized CdTe quantum dots + Hg(II) + GSH. Inset: (a) NALC-functionalized CdTe quantum dots; (b) NALC-functionalized CdTe quantum dots on addition of Hg(II); (c) NALC-functionalized CdTe quantum dots + Hg(II) + GSH under UV illumination [46].

biosensor was reported by an N-acetyl cysteine-capped CdTe quantum dot. The fluorescence of the nanomaterial is quenched when Hg(II) is added because it binds with substituted amino acid present on the nanoparticle surface (Figure 7.12).

On addition of glutathione, $Hg^{2+}$ ion binds strongly with it and releases the nanoparticle, and as a result fluorescence is restored. The biosensor is highly selective and can detect glutathione up to a limit of 2.49 ng/mL. The authors reported that the nanoparticles were able to determine the human serum glutathione level and hence can be used clinically. However, the presence of cysteine can hamper the accuracy of the detection [46]. Asparagine is a non-essential amino acid in human beings. In one more reported work, asparagine-capped gold nanoparticles were used as immune sensors when they are combined with anti-CA125 antibody. A commonly used antibody-bound immunosensor is EDS-NHS activated gold nanoparticles which is used to modify the electrode for sensing. However, according to this report, asparagine-modified gold nanoparticles are even better and have about 2.5 times greater specificity than the EDS-NHS modified gold nanoparticles [47]. In another work, cysteine amino acid-capped ZnS semiconductor nanoparticles having a synchronous fluorescence spectrum at $\lambda = 267$ nm were synthesized and used as fluorescent probes for deoxyribonucleic acid (DNA). It was observed that in the presence of DNA, the intensity of the synchronous fluorescence spectra decreases greatly, although the position of the emission maxima remains in the same position. The interaction between the DNA and the nanoparticle is non-intercalative because of a size mismatch and the non-planar structure of the cysteine-capped ZnS nanoparticle [48]. For these examples, it is evident that amino acid–capped inorganic nanoparticles can be a very powerful diagnostic tool which is simple, non-toxic and economical. Although the examples cited in this section are all in the laboratory scale, clinical use is also possible.

## 7.5.2 Bioimaging with Emphasis on MRI

Iron oxide nanoparticles are widely used as MRI contrast agents as they are biodegradable and biocompatible. Amino acid functionalized iron oxide nanoparticles are extensively used to image tutor cells. Both normal cells and tutor cells have a negative charge due to the presence of phospholipid such as phosphatidylserine on the cell membrane. On the other hand, poly-amino acids such as poly-aspartic acid–capped iron oxide nanoparticles have a large positive surface charge and thus have a high colloidal stability. These nanoparticles can be taken by the tutor cells at acidic pH and retained inside the cell for about three days without losing any contrasting signal. Despite having a very low concentration dose, accumulation of nanoparticles inside the cells is not desired as it may lead to chronic cytotoxicity. Therefore, to monitor their biodistribution and excretion from the body is of vital importance [49]. Yang et al. have prepared superparamagnetic, water-soluble iron oxide nanoparticles functionalized with poly-aspartic acid. Anti-HER/neu2 antibodies were then conjugated onto this poly-aspartic acid–coated nanoparticle to produce a negative contrast enhancement agent to image breast cancer cells. The nanoparticles were ~31 nm in diameter and showed much less cytotoxicity even in high doses. Moreover, the nanoparticles are excellent for enhancing the signal intensity of the T2 relaxed MRI image [50] (Figure 7.13).

Poly-aspartic acid–coated manganese oxide nanoparticles have been used to image liver cells. The nanoparticles were synthesized by a thermal decomposition method in three different particle sizes. Incorporation of poly-amino acid changes the nanoparticles from hydrophobic to hydrophilic and makes it useful to be used in the MRI technique. *In vivo* liver imaging established that smaller size particles were more effective T1 contrasting agents. Moreover, the authors mentioned that the nanoparticles are highly stable as indicated by high negative zeta potential value and have very low cytotoxicity [51] (Figure 7.14).

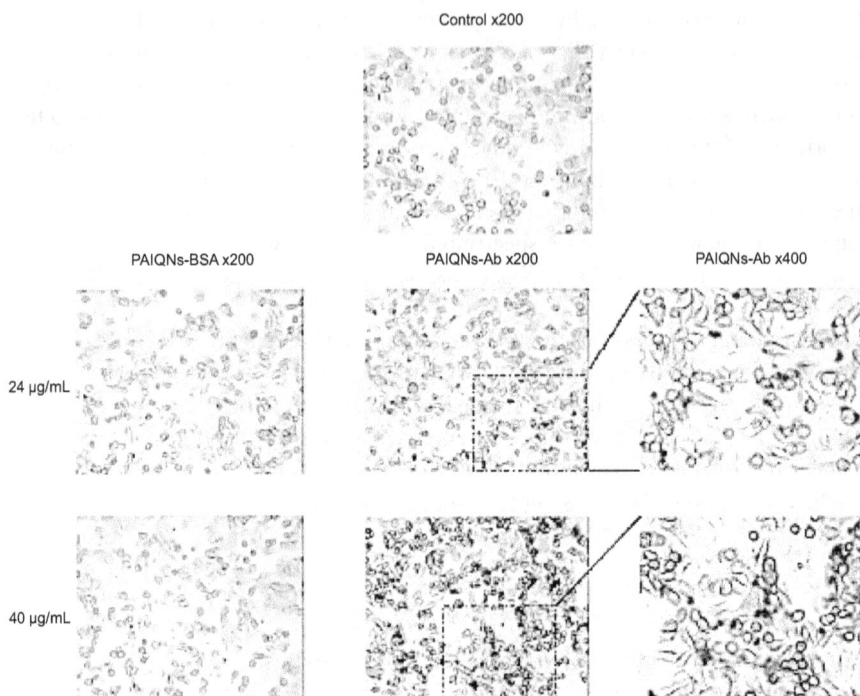

**FIGURE 7.13**    T2 relaxed MRI image of SKBR-3 cells strained with Prussian blue for poly-aspartic acid–capped iron oxide nanoparticles combined with BSA (as negative control) and poly-aspartic acid–coated iron oxide nanoparticles combined with antibody [50]. Reprinted with permission from Yang et al. (2010). Copyright (2010) American Chemical Society.

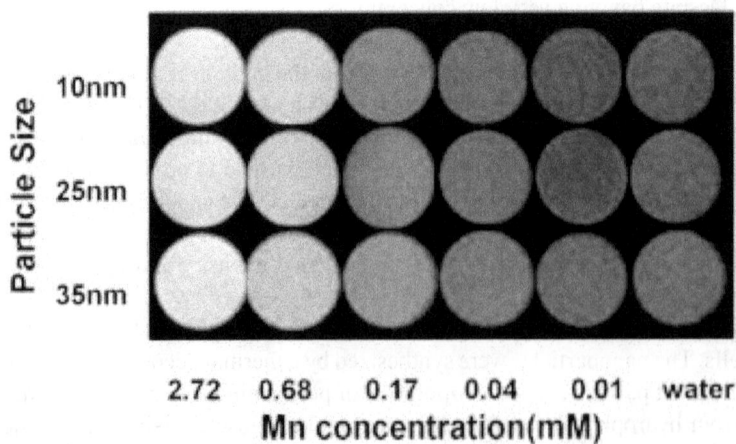

**FIGURE 7.14**    Changes in liver cell image contrasting with particle size [51]. Reprinted with permission from Xing et al. (2011). Copyright (2011) Royal Society of Chemistry.

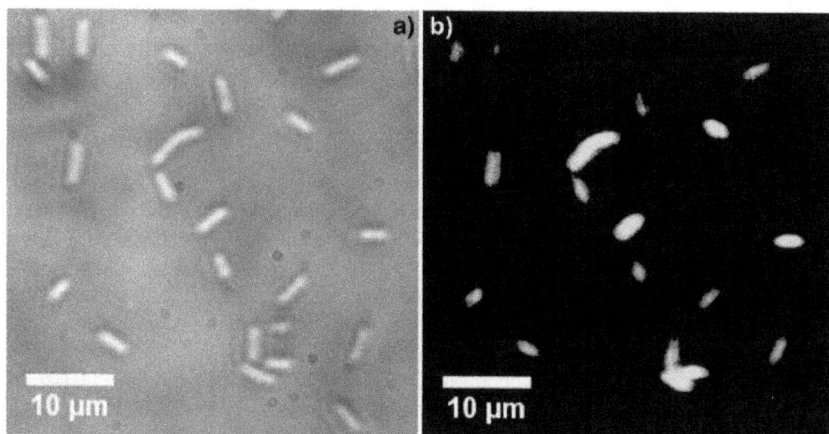

**FIGURE 7.15** (a) Bright-field (b) fluorescent imaging of *E. coli* bacteria cells incubated with tryptophan-capped gold nanoparticle sample [53]. Reprinted with permission from Pajovic et al. (2015). Copyright (2015) Elsevier.

In addition to this, to increase image quality, the fluorescent properties of the nanoparticle can be combined with the magnetic properties. Such particles generally used iron oxide or gold core capped with amino acid, which in turn was connected to a fluorophore. For example, poly D-glutamic acid and poly D-lysine coated iron oxide and Au nanoparticles were synthesized. Then, they were combined with commercial fluorophores available, namely either Alexa Fluor 430 succinimide ester or Alexa Fluor 350 hydrazide. The nanoparticles were stable and exhibited excellent fluorescence and magnetic behaviour and thus indicated the possibility of using it as an MRI probe. However, the problem is, the fluorophore may get cleaved, giving rise to non-specific signalling [52]. Tryptophan-capped gold nanoparticles have been used as UV-violet fluorescent probes to image bacteria cells. Tryptophan amino acid can absorb and emit in the UV-visible region. The emission for tryptophan-functionalized Au nanoparticles appeared at 340 nm. The authors incubated *E. coli* cells with tryptophan-functionalized Au nanoparticles and then studied the fluorescence spectrum by deep UV imaging. Bacteria autofluorescence was also observed at the 452–486 nm region, whereas that in the 327–353 nm region is for the nanoparticles. So, it was possible to distinguish the two different types of fluorescent emission. The autofluorescence from bacteria and that from the nanostructure can be easily distinguished. Moreover, the authors also inferred that nanoparticles are mainly concentrated near the middle part of the *E. coli* cells [53] (Figure 7.15).

## 7.6  CONCLUSION

Amino acid functionalized inorganic nanoparticles are highly promising candidates for diagnostic as well as environmental application. Amino acids are biocompatible, non-toxic, economic, easily available and multifunctional molecules. Due to their versatile nature, they can interact with a wide range of analyte, both inorganic and organic biomolecules. The functionalized nanoparticles are highly selective and ensure a high degree of sensitivity towards various types of analytes.

## REFERENCES

[1] Li, S., Zhao, Z., Zhao, J., Zhang, Z., Li, X. and Zhang, J., 2020. Recent advances of ferro-, piezo-, and pyroelectric nanomaterials for catalytic applications. *ACS Applied Nano Materials*, *3*(2), pp. 1063–1079.

[2] Sonawane, J.M., Yadav, A., Ghosh, P.C. and Adeloju, S.B., 2017. Recent advances in the development and utilization of modern anode materials for high performance microbial fuel cells. *Biosensors and Bioelectronics*, *90*, pp. 558–576.

[3] Barreto, J.A., O'Malley, W., Kubeil, M., Graham, B., Stephan, H. and Spiccia, L., 2011. Nanomaterials: Applications in cancer imaging and therapy. *Advanced Materials*, *23*(12), pp. H18–H40.

[4] Chimene, D., Alge, D.L. and Gaharwar, A.K., 2015. Two-dimensional nanomaterials for biomedical applications: Emerging trends and future prospects. *Advanced Materials*, *27*(45), pp. 7261–7284.

[5] Thomas, K.G. and Kamat, P.V., 2003. Chromophore-functionalized gold nanoparticles. *Accounts of Chemical Research*, *36*(12), pp. 888–898.

[6] Shenhar, R. and Rotello, V.M., 2003. Nanoparticles: Scaffolds and building blocks. *Accounts of Chemical Research*, *36*(7), pp. 549–561.

[7] Templeton, A.C., Wuelfing, W.P. and Murray, R.W., 2000. Monolayer-protected cluster molecules. *Acc Chem Res*, *33*(1), pp. 27–36.

[8] Sastry, M., Rao, M. and Ganesh, K.N., 2002. Electrostatic assembly of nanoparticles and biomacromolecules. *Acc Chem Res*, *35*(10), pp. 847–855.

[9] Periasamy, S., Dumbre, D., Babu, L., Madapusi, S., Soni, S.K., Daima, H.K. and Bhargava, S.K., 2021. Amino acids functionalized inorganic metal nanoparticles: Synthetic nano-zymes for target specific binding, sensing and catalytic applications. In *Nanozymes for Environmental Engineering* (pp. 1–33). Springer, Cham.

[10] Chakraborty, A., Boer, J.C., Selomulya, C. and Plebanski, M., 2017. Amino acid func-tionalized inorganic nanoparticles as cutting-edge therapeutic and diagnostic agents. *Bioconjugate Chemistry*, *29*(3), pp. 657–671.

[11] Daima, H.K., Selvakannan, P.R., Shukla, R., Bhargava, S.K. and Bansal, V., 2013. Fine-tuning the antimicrobial profile of biocompatible gold nanoparticles by sequential sur-face functionalization using polyoxometalates and lysine. *PloS One*, *8*(10), p. e79676.

[12] Daima, H.K., Selvakannan, P.R., Kandjani, A.E., Shukla, R., Bhargava, S.K. and Bansal, V., 2014. Synergistic influence of polyoxometalate surface corona towards enhancing the antibacterial performance of tyrosine-capped Ag nanoparticles. *Nanoscale*, *6*(2), pp. 758–765.

[13] Selvakannan, P.R., Ramanathan, R., Plowman, B.J., Sabri, Y.M., Daima, H.K., O'Mullane, A.P., Bansal, V. and Bhargava, S.K., 2013. Probing the effect of charge trans-fer enhancement in off resonance mode SERS via conjugation of the probe dye between silver nanoparticles and metal substrates. *Physical Chemistry Chemical Physics*, *15*(31), pp. 12920–12929.

[14] Plascencia-Villa, G., Torrente, D., Marucho, M. and Jose-Yacaman, M., 2015. Biodirected synthesis and nanostructural characterization of anisotropic gold nanopar-ticles. *Langmuir*, *31*(11), pp. 3527–3536.

[15] Yang, X., Shi, M., Zhou, R., Chen, X. and Chen, H., 2011. Blending of HAuCl 4 and histidine in aqueous solution: A simple approach to the Au 10 cluster. *Nanoscale*, *3*(6), pp. 2596–2601.

[16] Maity, A., De, S.K. and Chakraborty, A., 2021. Interaction of aromatic amino acid-functionalized gold nanoparticles with lipid bilayers: Insight into the emergence of novel lipid corona formation. *The Journal of Physical Chemistry B*, *125*(8), pp. 2113–2123.

[17] Zhu, Z., Liu, W., Li, Z., Han, B., Zhou, Y., Gao, Y. and Tang, Z., 2012. Manipulation of collective optical activity in one-dimensional plasmonic assembly. *ACS Nano*, *6*(3), pp. 2326–2332.

[18] Ma, J., Qiu, J. and Wang, S., 2020. Nanozymes for catalytic cancer immunotherapy. *ACS Applied Nano Materials*, *3*(6), pp. 4925–4943.

[19] Selvakannan, P.R., Mantri, K., Tardio, J. and Bhargava, S.K., 2013. High surface area Au-SBA-15 and Au-MCM-41 materials synthesis: Tryptophan amino acid mediated confinement of gold nanostructures within the mesoporous silica pore walls. *Journal of Colloid and Interface Science*, *394*, pp. 475–484.

[20] Lacasta, S., Sebastián, V., Casado, C., Mayoral, Á., Romero, P., Larrea, Á., Vispe, E., López-Ram-de-Viu, P., Uriel, S. and Coronas, J., 2011. Chiral imprinting with amino acids of ordered mesoporous silica exhibiting enantioselectivity after calcination. *Chemistry of Materials*, *23*(5), pp. 1280–1287.

[21] Wang, J., Sugawara, A., Shimojima, A. and Okubo, T., 2010. Preparation of anisotropic silica nanoparticles via controlled assembly of presynthesized spherical seeds. *Langmuir*, *26*(23), pp. 18491–18498.

[22] Sun, Y., Zhao, C., Gao, N., Ren, J. and Qu, X., 2017. Stereoselective nanozyme based on ceria nanoparticles engineered with amino acids. *Chemistry: A European Journal*, *23*(72), pp. 18146–18150.

[23] El-Trass, A., ElShamy, H., El-Mehasseb, I. and El-Kemary, M., 2012. CuO nanoparticles: Synthesis, characterization, optical properties and interaction with amino acids. *Applied Surface Science*, *258*(7), pp. 2997–3001.

[24] Jaiswal, J.K. and Simon, S.M., 2004. Potentials and pitfalls of fluorescent quantum dots for biological imaging. *Trends in Cell Biology*, *14*(9), pp. 497–504.

[25] Yadav, R.S., Mishra, P., Mishra, R., Kumar, M. and Pandey, A.C., 2010. Growth mechanism and optical property of CdS nanoparticles synthesized using amino-acid histidine as chelating agent under sonochemical process. *Ultrasonics Sonochemistry*, *17*(1), pp. 116–122.

[26] Nosrati, H., Salehiabar, M., Attari, E., Davaran, S., Danafar, H. and Manjili, H.K., 2018. Green and one-pot surface coating of iron oxide magnetic nanoparticles with natural amino acids and biocompatibility investigation. *Applied Organometallic Chemistry*, *32*(2), p. e4069.

[27] Jain, P.K., Huang, X., El-Sayed, I.H. and El-Sayed, M.A., 2008. Noble metals on the nanoscale: Optical and photothermal properties and some applications in imaging, sensing, biology, and medicine. *Accounts of Chemical Research*, *41*(12), pp. 1578–1586.

[28] Green, M., 2004. Semiconductor quantum dots as biological imaging agents. *Angewandte Chemie International Edition*, *43*(32), pp. 4129–4131.

[29] Jaiswal, J.K. and Simon, S.M., 2004. Potentials and pitfalls of fluorescent quantum dots for biological imaging. *Trends in Cell Biology*, *14*(9), pp. 497–504.

[30] Xiao, L. and Yeung, E.S., 2014. Optical imaging of individual plasmonic nanoparticles in biological samples. *Annual Review of Analytical Chemistry*, *7*, pp. 89–111.

[31] Sethi, M., Law, W.C., Fennell, W.A., Prasad, P.N. and Knecht, M.R., 2011. Employing materials assembly to elucidate surface interactions of amino acids with Au nanoparticles. *Soft Matter*, *7*(14), pp. 6532–6541.

[32] Trindade, T., O'Brien, P. and Pickett, N.L., 2001. Nanocrystalline semiconductors: Synthesis, properties, and perspectives. *Chemistry of Materials*, *13*(11), pp. 3843–3858.

[33] Green, M., 2004. Semiconductor quantum dots as biological imaging agents. *Angewandte Chemie International Edition*, *43*(32), pp. 4129–4131.

[34] Jaiswal, J.K. and Simon, S.M., 2004. Potentials and pitfalls of fluorescent quantum dots for biological imaging. *Trends in Cell Biology*, *14*(9), pp. 497–504.

[35] Ahmad, S., Tripathy, D.B. and Mishra, A., 2016. Sustainable nanomaterials. In *Sustainable Inorganic Chemistry, Wiley and Sons Ltd. UK*, (p. 205).

[36] Kasture, M., Sastry, M. and Prasad, B.L.V., 2010. Halide ion-controlled shape dependent gold nanoparticle synthesis with tryptophan as reducing agent: Enhanced fluorescent properties and white light emission. *Chemical Physics Letters*, *484*(4–6), pp. 271–275.

[37] Mandal, G., Bhattacharya, S. and Ganguly, T., 2009. Nature of interactions of tryptophan with zinc oxide nanoparticles and L-aspartic acid: A spectroscopic approach. *Chemical Physics Letters*, *472*(1–3), pp. 128–133.

[38] Yadav, R.S., Mishra, P., Mishra, R., Kumar, M. and Pandey, A.C., 2010. Growth mechanism and optical property of CdS nanoparticles synthesized using amino-acid histidine as chelating agent under sonochemical process. *Ultrasonics Sonochemistry*, *17*(1), pp. 116–122.

[39] Ladbury, J.E. and Chowdhry, B.Z., 1996. Sensing the heat: The application of isothermal titration calorimetry to thermodynamic studies of biomolecular interactions. *Chemistry & Biology*, *3*(10), pp. 791–801.

[40] Joshi, H., Shirude, P.S., Bansal, V., Ganesh, K.N. and Sastry, M., 2004. Isothermal titration calorimetry studies on the binding of amino acids to gold nanoparticles. *The Journal of Physical Chemistry B*, *108*(31), pp. 11535–11540.

[41] Selvakannan, P.R., Mandal, S., Phadtare, S., Pasricha, R. and Sastry, M., 2003. Capping of gold nanoparticles by the amino acid lysine renders them water-dispersible. *Langmuir*, *19*(8), pp. 3545–3549.

[42] Selvakannan, P.R., Swami, A., Srisathiyanarayanan, D., Shirude, P.S., Pasricha, R., Mandale, A.B. and Sastry, M., 2004. Synthesis of aqueous Au core-Ag shell nanoparticles using tyrosine as a pH-dependent reducing agent and assembling phase-transferred silver nanoparticles at the air-water interface. *Langmuir*, *20*(18), pp. 7825–7836.

[43] Song, W., Zhao, B., Wang, C., Ozaki, Y. and Lu, X., 2019. Functional nanomaterials with unique enzyme-like characteristics for sensing applications. *Journal of Materials Chemistry B*, *7*(6), pp. 850–875.

[44] Khan, S., Carneiro, L.S., Romani, E.C., Larrudé, D.G. and Aucelio, R.Q., 2014. Quantification of thyroxine by the selective photoluminescence quenching of L-cysteine-ZnS quantum dots in aqueous solution containing hexadecyltrimethylammonium bromide. *Journal of Luminescence*, *156*, pp. 16–24.

[45] Huang, S., Zhu, F., Xiao, Q., Su, W., Sheng, J., Huang, C. and Hu, B., 2014. A CdTe/CdS/ZnS core/shell/shell QDs-based "off-on" fluorescent biosensor for sensitive and specific determination of L-ascorbic acid. *RSC Advances*, *4*(87), pp. 46751–46761.

[46] Tan, X., Yang, J., Li, Q. and Yang, Q., 2015. Detection of glutathione with an "off-on" fluorescent biosensor based on N-acetyl-L-cysteine capped CdTe quantum dots. *Analyst*, *140*(19), pp. 6748–6757.

[47] Raghav, R. and Srivastava, S., 2016. Immobilization strategy for enhancing sensitivity of immunosensors: L-Asparagine-AuNPs as a promising alternative of EDC-NHS activated citrate-AuNPs for antibody immobilization. *Biosensors and Bioelectronics*, *78*, pp. 396–403.

[48] Li, Y., Chen, J., Zhu, C., Wang, L., Zhao, D., Zhuo, S. and Wu, Y., 2004. Preparation and application of cysteine-capped ZnS nanoparticles as fluorescence probe in the determination of nucleic acids. *Spectrochimica Acta Part A: Molecular and Biomolecular Spectroscopy*, *60*(8–9), pp. 1719–1724.

[49] Chakraborty, A., Boer, J.C., Selomulya, C. and Plebanski, M., 2017. Amino acid functionalized inorganic nanoparticles as cutting-edge therapeutic and diagnostic agents. *Bioconjugate Chemistry*, *29*(3), pp. 657–671.

[50] Yang, H.M., Park, C.W., Woo, M.A., Kim, M.I., Jo, Y.M., Park, H.G. and Kim, J.D., 2010. HER2/neu antibody conjugated poly (amino acid)-coated iron oxide nanoparticles for breast cancer MR imaging. *Biomacromolecules*, *11*(11), pp. 2866–2872.

[51] Xing, R., Zhang, F., Xie, J., Aronova, M., Zhang, G., Guo, N., Huang, X., Sun, X., Liu, G., Bryant, L.H. and Bhirde, A., 2011. Polyaspartic acid coated manganese oxide nanoparticles for efficient liver MRI. *Nanoscale*, *3*(12), pp. 4943–4945.

[52] Perego, D., Masciocchi, N., Guagliardi, A., Domínguez-Vera, J.M. and Gálvez, N., 2013. Poly (amino acid) functionalized maghemite and gold nanoparticles. *Nanotechnology*, *24*(7), p. 075102.

[53] Pajović, J.D., Dojčilović, R., Božanić, D.K., Kaščáková, S., Refregiers, M., Dimitrijević-Branković, S., Vodnik, V.V., Milosavljević, A.R., Piscopiello, E., Luyt, A.S. and Djoković, V., 2015. Tryptophan-functionalized gold nanoparticles for deep UV imaging of microbial cells. *Colloids and Surfaces B: Biointerfaces*, *135*, pp. 742–750.

# 8 Hybrid Nanocomposites
## *Eco-Friendly Future Material*

*Kiran T., Chethana M. V., Pratibha
Singh, Jyoti Joshi, and Preeti Jain*

## 8.1 INTRODUCTION

A composite is a structural substance made up of two or more combined parts
that are not soluble in each other and are mixed at a macroscopic level. The
reinforcing phase is one of the constituents, and the matrix is the one in which it
is embedded. For the reinforcing phase, materials like fibres, particles, and flakes
can be used. In most cases, matrix phase materials are continuous. The words
"hybrid" and "composite" are combined in a hybrid composite material. Hybrid
composites are also created by blending two or more kinds of fibres into a single
matrix. The two or more forms of fibre complement one other while also filling
in for each other's inadequacies. Because the two materials are mixed together,
hybrid composites vary from ordinary composites. Due to that, they show either
entirely new properties or their properties lie between those of the original material. The factors which influence hybrid composites are fibre content, orientation,
length of distinct fibres, bonding of fibre to matrix, and arrangement of sequence
in both fibres.

The composite matrix is made up of thermoplastic and thermosetting polymers.
Thermosets are crosslinked polymer chains that harden after curing, resulting in
a stiff product that cannot be moulded. Thermosets have the benefit of being able
to withstand high temperatures while maintaining structural stiffness. Polyester,
epoxy, vinyl ester, polyurethane, phenolic, cyanate ester, polyimide, and bisma-
leimide are examples of thermosetting polymer matrices. Thermoplastics, unlike
thermosets, may be reformed into a new product by applying heat and pressure,
and hence are more widely recycled. Polyethylene (PE), high-density polyethylene (HDPE), polypropylene (PP), and polyvinyl chloride (PVC) are common
thermoplastic polymers used as matrix. The purpose of a matrix is to help you
figure out what you want to do with your life. The fibres are protected from the
environment by the matrix, which also serves as a load transfer medium between
the fibre and the matrix [1]. The mechanical properties of composites are determined by the interfacial connection between the fibre and the matrix. Curing
improves fibre-matrix adhesion, resulting in improved mechanical properties of
the composites.

Natural fibres are divided into numerous categories based on their source of origin, as illustrated in Figure 8.2.

DOI: 10.1201/9781003316398-8

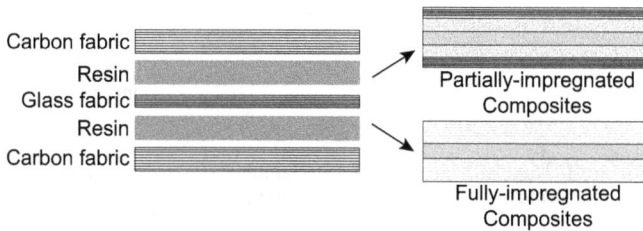

FIGURE 8.1   Structure of hybrid composites.

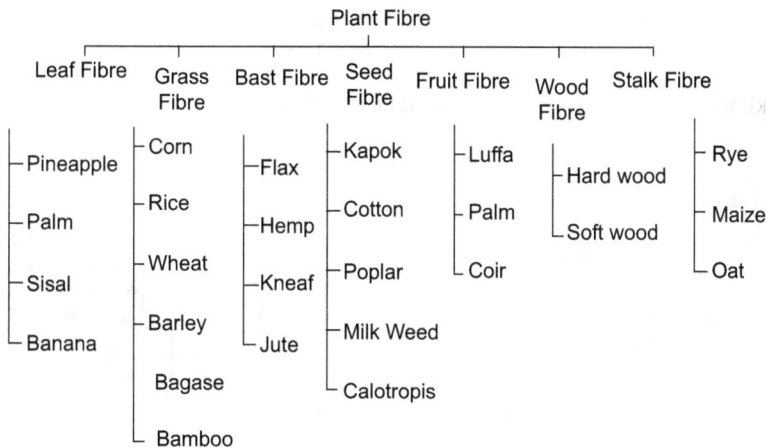

FIGURE 8.2   Classification of plant fibres present in hybrid composites.

Natural or synthetic fibres can be utilized to make composite materials (Figure 8.3). The cellulose content of plant fibres is high. Alternative fibres derived from nature for composite reinforcement include kenaf, flax, hemp, jute, sisal, and banana. Natural fibres provide a number of advantages, including low manufacturing costs, eco-friendliness, light weight, and optimal mechanical qualities. Natural fibres have two major disadvantages: moisture absorption and matrix adherence. Various chemical–physical and modified chemical strategies can be used to circumvent these restrictions. Jute fibres, low fibre–matrix adhesion, hydrophilic character, and poor thermal properties can all be improved using these modification procedures. Different surface-modification methods, such as alkali and radiation, have also been proven to be effective in improving the mechanical capabilities of natural fibres while limiting their drawbacks. Carbon, glass, and aramid are some of the synthetic fibres employed. Initially, composites were made with synthetic fibres, primarily glass fibres, since they are less expensive than other synthetic fibres and have better mechanical qualities. Synthetic fibre has the disadvantages of increased density, higher cost, non-biodegradable nature, higher energy consumption, and hazard to human health, among others. Better mechanical qualities are

```
                              Synthetic Fibres
                                    │
                  ┌─────────────────┴─────────────────┐
                  ↓                                   ↓
               Organic                            Inorganic
                  │                                   │
        ┌─────────┴─────────┐                         ↓
        ↓                   ↓                      Boron
  From Natural        From Synthetic               Glass
   Polymers             Polymers                   Carbon
        │                   │                      Metal
        ↓                   ↓                      Ceramic
    Acetate             Aramid
    Triacetate          Polyester
    Lyocell             Polyamide
    Rubber              Fluorofibre
    Modal               Polyethylene
                        Polypropylene
```

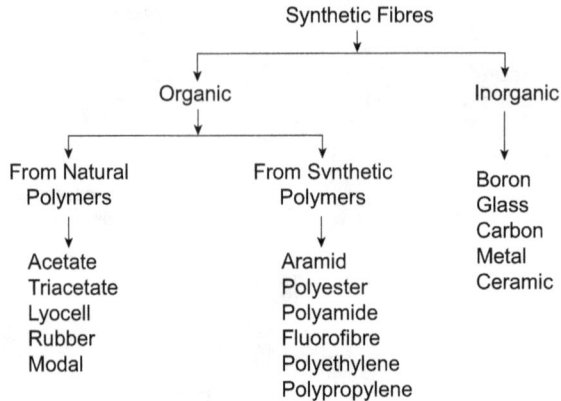

**FIGURE 8.3**    Types of synthetic fibres used for hybrid composites.

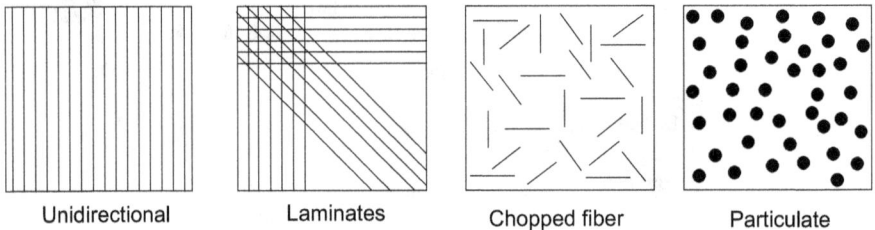

    Unidirectional      Laminates      Chopped fiber      Particulate

**FIGURE 8.4**    Various geometries of the fibre.

due to the connection between the fibre and the matrix. The composite should have good mechanical characteristics if the bond is strong. Surface treatment is one of the strategies for improving the connection between two surfaces.

The mechanical characteristics of composites are influenced by the fibre shape. The stacking sequence is a sequence in which items are stacked one on top of the other. Superior mechanical qualities are produced by a better stacking sequence of fibres and matrix [2]. Mechanical qualities should be improved when fibre content is raised up to a specified limit less than the weight percentage of the matrix. The mechanical characteristics of the composite are enhanced by increasing the fibre content [3, 4]. Due to the moisture absorption properties of natural fibres, composites reinforced with them cannot be utilized in outdoor applications. The mechanical qualities of natural fibres are lower than those of synthetic fibres. One technique to improve the mechanical characteristics of composites reinforced with natural fibres is to hybridize them.

## 8.2   CLASSIFICATION OF HYBRID COMPOSITES

The resultant materials of a hybrid exhibit new properties after mixing the various components, and these qualities may be controlled by the precise chemical and physical properties of individual components, as well as the structure and interfaces between distinct components. Hybrid materials are categorized based on the potential interactions between organic and inorganic species contained in the substance.

Class I materials exhibit weak (van der Waals, hydrogen bonding, or weak electrostatic) interactions between the two components.

Class II materials show strong chemical (i.e. covalent bonding) interactions between the two components.

## 8.3   SYNTHESIS OF HYBRID COMPOSITE MATERIALS

Mainly two varied strategies can be employed for the development of hybrid materials.

1. Functionally distinct prefabricated building pieces react with one another to generate a hybrid substance in that the predecessors retain some of their original property.
2. Precursors are used to create both structural components, which then form a new structure, i.e. network. It is critical that the interface between inorganic and organic materials be adjusted in order to overcome substantial issues in the development of hybrid materials (Figure 8.5).

Various building blocks and procedures can be utilized for preparation, and these have to be tailored to link the differences of inorganic and organic materials. Various approaches are summarized as follows.

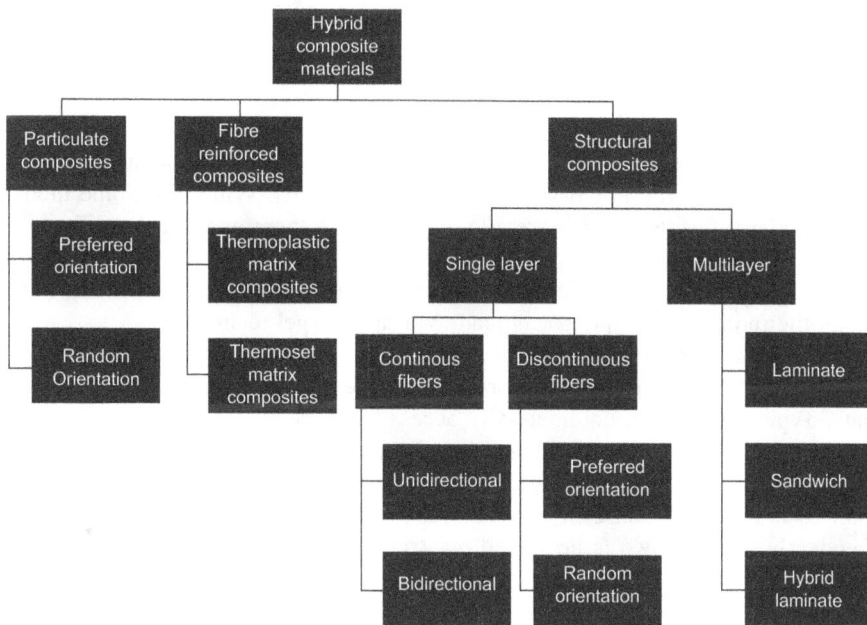

FIGURE 8.5   Classification of hybrid composite materials.

### 8.3.1 BUILDING BLOCK APPROACH

The molecular dependability of building blocks is maintained at least considerably throughout the material configuration, implying that structural units discovered in these sources for material creation may also be found in the finished material. Nanoparticles with reactive organic groups attached are good examples of such defined building components. Cluster compounds have a single functional group that interacts with an organic matrix, such as by copolymerization. These building blocks can adapt an organic matrix with one functional group or produce semi- or fully crosslinked materials, depending on the amount of groups that can interrelate. Two reactive groups, for example, can control the growth of chain structures. If these blocks have at minimum three reactive groups, they can be employed to create a crosslinked material without the use of additional molecules. Nanoscale building components, such as nanoparticles and nanorods, can also be employed to create nanocomposites.

The fact that the minimum one structural unit of the building block is unique and does not experience structural changes during matrix development is a significant benefit of this strategy. Predictions of structural properties that are more accurate are also encouraging. Furthermore, these blocks can be designed to provide the best representation in the materials' production, such as high solubility of inorganic compounds due to surface groups with comparable polar nature to the monomers.

A large number of building blocks were employed in the manufacture of hybrid materials in recent years, and they may be built on a molecular basis using sophisticated technologies. This design is also one of the key motivations for the hybrid materials building block method.

#### 8.3.1.1 *In situ* formation

The *in situ* preparation of the hybrid materials is dependent on the chemical transformation of the precursors throughout materials' synthesis. Some methods of *in situ* formation of inorganic materials are mentioned in the following figure.

It is usually when organic polymers are created, but it can also be the case when the inorganic component is made via the sol–gel technique. Various molecules are changed into three-dimensional structures in these circumstances, which often have features that are radically different from the original antecedents. Typically, simple, commercially accessible molecules are helpful, and the resulting material's internal structure is influenced not only by the content of these precursors, but also by the reaction conditions. As a result, maintaining circumstances is a crucial phase in the process. A change in a single parameter may frequently result in two quite varied materials. When the inorganic species is a derivative of silica produced by the sol–gel method, switching from base to acid catalyst creates a significant impact, as a basic catalyst produces a more particle-like microstructure whereas acid catalysis produces a polymer-like microstructure. As a result, the resultant materials' final result is highly dependent on their process and optimization.

Many traditional inorganic solid materials are made with solid precursors and high-temperature methods, typically incompatible with the presence of organic groups due to high-temperature degradation. As a result, these high-temperature methods are ineffective for forming *in situ* hybrid materials. The nature of the reactions should be more like that of a traditional covalent bond arrangement in solutions. The sol–gel method is one of the most well-known processes that meets these requirements. However, such low-temperature procedures typically result in kinetic products rather than the thermodynamically most stable structure, which has ramifications for the structures formed. Low-temperature materials are frequently amorphous, with crystalline structures only visible at the nanoscale scale. The arrangement of nanoparticles of metals in inorganic or organic matrices by reducing metal salts or organometallic precursors is an example of the latter.

## 8.4   PREPARATION OF ORGANIC POLYMERS (WITH PREFORMED INORGANIC MATERIALS)

The inorganic material may have no surface functionalization, exposed material surface, non-reactive organic groups (e.g. alkyl chains), or reactive surface groups (e.g. polymerizable functionalities). To surpass the incompatibility of the two species, one must discern between many potentials if organic polymerization happens in an inorganic substance to generate the hybrid material. The material can be pre-processed based on these requirements; for example, a pure inorganic surface can be treated with surfactants and coupling agents to make it compatible with organic monomers, or functional monomers can be given that react with the inorganic material's surface. The final substance following organic polymerization is a blend when an inorganic component contains non-reactive organic groups present on its surface which is dissolved in a monomer and then polymerized.

The inorganic component interacts only weakly with the organic polymer in this example, resulting in a class I material. Only if the aggregation of the inorganic components in the organic is retained can homogeneous materials be formed, which can be accomplished if the interactions present in the inorganic components and the monomers are enhanced or equal. However, due to diffusion effects in the emerging material, the long-term stability of a once homogenous material is unknown if no strong chemical connections are generated. The final material is more stable when the respective interaction between the components is stronger. The strongest interaction is obtained if class II materials are produced, for instance with covalent interactions.

## 8.4.1 APPROACH OF HYBRID MATERIALS BY SIMULTANEOUS FORMATION OF BOTH COMPONENTS

The most homogenous sort of interpenetrating networks may be created by developing inorganic and organic polymers at the same time. Precursors for the sol–gel process are typically combined with monomers for polymerization, and both processes can take place at the same time, with or without the use of a solvent. Three processes are pitted against each other using this strategy:

- the hydrolysis and condensation kinetics which form the inorganic phase,
- the polymerization process of the organic phase, and
- phase separation between the two phases and their thermodynamics.

By optimizing the two polymerizations that occur concurrently, phase separation is prevented or minimized. To avoid phase separation, additional characteristics like attractive interactions present in two moieties might be applied. Because of the sensitivity of many polymerization processes, sol–gel conditions come from the simultaneous creation of both networks. Ionic polymerizations, for example, typically interact with the sol–gel process's precursors or intermediates. As a result, they are seldom beneficial in these responses.

### 8.4.1.1 Fabrication of Hybrid Composites by Using Natural Fibre

#### 8.4.1.1.1 Fibre Synthesis

Selection of fibre and its extraction are important aspects in this process. Synthetic fibres must be washed with water and dried to remove any remaining water. The fibres must then be separated somewhat by hand. Individual fibre strands might be separated utilizing a mechanical combing procedure on the fibre laminates during the hand-retting process. After sorting them, the fibres are examined many times with a cotton-checking tool to isolate the filaments. After that, the fibres are allowed to dry at ambient temperature to eliminate any remaining moisture. After that, all of the fibres' weight and length are determined according to the required measurements. Natural banana, flax, hemp, and pineapple strengthening ingredients are employed to prepare the samples.

### 8.4.1.2 Weight Fraction of Reinforcement and Matrix Materials

Based on the hybridization process, the weight fraction of the constituents used for the fabrication process can be measured.

### 8.4.1.3 Preparation of Matrix Material

As a matrix material, epoxy resin or other types of matrix materials, coupled with hardener, can be utilized. The proportions of both resin and hardener in terms of weight should be measured for the optimum resin solution. In a plastic container, combine the epoxy resin and hardener and stir for two to three minutes with a plastic stirrer at room temperature to obtain a uniform colour mixture. The solution is then swirled for 30 seconds more to clear the container's sides and bottom. The resin solution can be used on fibres after adequate mixing of both hardener and epoxy.

#### 8.4.1.4 Composite Fabrication Procedure

The hand layup technique is used to finish the manufacturing process. The coir fibres must first be carefully washed and dried in the shade. The coir mould must be cleaned and stored in a dust-free area before production begins. The mould surface must next be coated with a coating of polyvinyl alcohol. After that, combine the epoxy resins (Araldite LY554) and the hardener (amino hydrocarbon)(in a ratio of 10:1. The resin should then be applied to the mould using a brush, and the coir threads should be arranged horizontally on top. Care must be given to avoid fibre overlapping, since this might result in a variance in the thickness of the finished composite. The roller is adjusted to set the coir fibres firmly over the antecedent layer of resin and shuns.

After that, using a brush, a double coating of resin must be applied to the coir layer. Over the resin layer, put the same quantity of coir fibres in a vertical manner once again. Because transmutation in orientation has the power to protect the material from the proliferation of applied stresses, this may be done to raise the overall vigour of the composite. After that, the laminate must be allowed to improve for one hour. To prevent the formation of air bubbles on the composite's surface, a force must be applied.

Finally, verify that the resin has cured, and then carefully remove from mould and cut into the desired measurements. All of the samples must be processed in the same way. As illustrated in Figure 8.6, three samples should be made using the same process but with various fibre orientations.

### 8.5 TESTING OF HYBRID COMPOSITES

#### 8.5.1 TENSILE TEST

The tensile test is one of the most often used by specialists to determine flexible qualities. The adaptability test should be carried out using test models developed by the American Society for Testing and Materials (ASTM). The mouldable testing machine in the current assessment is Associated Scientific Eng. Works, FIE social-occasion of India, which is equipped with gear turn-speeds of 1.25, 1.5, and 2.5mm/run and will, in general, be pursued for a most amazing stack at 5 tonnes. Tests may be performed on models manufactured from three different materials: coir, hair, and a hybrid of coir and hair. By securing the model in the Universat Testing Machine (UTM), pliable force is applied at the discontinuities. Every model's pliant-breaking load is carefully measured. Furthermore, the variety is adaptable.

#### 8.5.2 COMPRESSIVE TEST

A typical piece of overlay should be cut in accordance with ASTM standards for the compressive test. It is done in a JE social gathering in India, in a controlled environment.

(a)

(b)

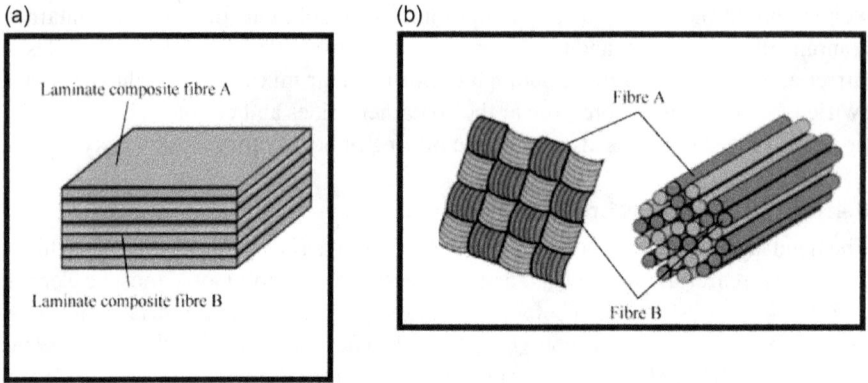

**FIGURE 8.6** (a) Hybrid composite interlayer or layer by layer; (b) hybrid composite interlayer or yarn by yarn, and hybrid composite intrayarn or fibre by fibre [5].

Compressive force is applied at the terminuses by cutting the model in the UTM, and the tests are repeated. Every model is grown by the compressive-breaking load. The uniqueness of compressive pressing factor with strain can also be plotted.

### 8.5.3 FLEXURAL STRENGTH TEST

Flexural testing is carried out to determine the composite materials' mechanical characteristics. This test is designed to figure out what level of stress is optimal for external weight extension.

### 8.5.4 FLEXURAL MODULUS TEST

The flexural-modulus test of composite material is portrayed as the capacity for that composite to wind. It is resolved from the inclination of the pressing factor and uprooting twist. It is also known as modulus of elasticity and tangent modulus.

### 8.5.5 IMPACT TEST

The material receives a lot of energy from the effect test. Furthermore, this impact test should be possible in an ASTM Charpy impact machine. The models for this test should be able to go around the typical estimates. The model takes into account the weight that is passed up the sledge, and the impact energy is determined by the hit that it is unable to bear. Tests should be repeated on models derived from the three types of models: coir, hair, and a mixture of coir and hair.

### 8.5.6 INTERLAMINAR SHEAR STRENGTH

Preliminary shear tests of various sorts are used extensively in the polymer business to discover the strength of the fortification to matrix materials. The interlaminar shear strength (ILSS) of fiber reinforced composites describes the shear strength between laminate planes of composites.

### 8.5.7 Moisture Absorption Test

It is used to determine how quickly the composite consumes water (dampness) when exposed to a wet media. The composite example is placed in a holder filled with refined water to determine the ingestion rate. The instances were put to the test in accordance with ASTM D-570.

% Weight acquire = (Final weight − Initial weight/Initial weight) × 100

### 8.5.8 Hardness Test

The hardness test determines the inflexibility of polymer-based materials in a vacuum. The test is carried out in accordance with ASTM guidelines. The sample should be indented using a solidified steel indenter with sufficient power and precision.

## 8.6 FUNDAMENTAL PROPERTIES

The properties of density, tensile strength, modulus, percentage of elongation, and moisture content are shown in Table 8.1.

The mechanical and wear characteristics of aloe vera, kenaf, and jute supported epoxy composites were studied by Ashwin et al. [6]. The composites were assembled via pressing factor moulding. Jute, kenaf, and aloe vera were used to arrange the composite elements into distinct stacking groups. Given that they were obtained through early analysis, it is clear that the stacking progression has an impact on

**TABLE 8.1**
**Natural Fibres' Properties**

| fibre | Density (G/CM3) | Tensile Strength (MPA) | Young's Modulus (GPA) | Elongation at Break (%) | Moisture Content (wt. %) |
|---|---|---|---|---|---|
| Flax | 1.5 | 345–1100 | 27.6 | 2.7–3.2 | 10.0 |
| Hemp | 1.47 | 690 | 30–60 | 1.6 | 10.8 |
| Jute | 1.3–1.4 | 393–773 | 13.0–26.5 | 1.2–1.5 | 12.6 |
| Kenaf | 1.3 | 930 | 53 | 1.6 | – |
| Ramie | 1.5 | 400–938 | 61.4–128 | 1.2–3.8 | – |
| Abaca | 1.5 | 400 | 12.0 | 3.0–10.0 | 15 |
| Pineapple | 1.3 | 413–1627 | 34.5–82.7 | 1.6 | 11.8 |
| Sisal | 1.4 | 468–640 | 9.4–22.0 | 3.0–7.0 | 11.0 |
| Coir | 1.1 | 131–175 | 4.0–6.0 | 15–40 | 8.0 |
| Cotton | 1.5–1.6 | 287–800 | 5.5–12.6 | 7.0–8.0 | - |

*Source*: Fictitious data, for illustration purposes only.

composite characteristics. The mechanical characteristics of epoxy and polyester composites supported with standard strands were examined by Indira et al. [7]. Glass strands, jute, and pineapple leaf were used. The composites were made using a hand layup method. Epoxy pitch composites have been discovered to perform better than polyester composites. The researchers found that these composites are light in weight and have better mechanical properties. They also found that polyester and epoxy contribute to composites for improving the mechanical properties of vehicle parts, and packaging materials.

Mechanical parameters like elasticity and flexural strength have an impact on the strength and hardness of typical fibre-based composites. Similarly, the surface morphology of strands was considered. The stiffness and impact strength of composites improve as the percentage of fibres rises. The flexural strength of composites grows to a certain point and then declines due to fibre loading. Surface treatment of strands enhances grid and fibre holding strength, resulting in improved characteristics. Hybridization decreases the use of designed strands while improving mechanical qualities. Because ordinary strands absorb a lot of moisture and have a low fire resistance, they cannot be utilized in areas where there's a lot of water and fire.

These composites have the ability to displace the manufactured strands due to the great mechanical qualities of conventional fibre composites. However, because of its superior qualities, the complete replacement of planned strands is nonsensical.

In comparison to metallic materials and engineered strands, natural fibres have an extraordinary malleable modulus and a lower thickness [8]. By combining trademark strands with these materials, the use of oil-based and designed ingredients may be reduced. Recently, researchers have looked at the impact on the qualities and variety of materials, including composites.

Understanding the features of fundamental materials for a particular applications is one of the most important considerations while deciding its composition.

Safri investigated the impact injury of glass fibre reinforced polymers (GFRPs) employing low velocity [9, 10] and quick influence tests [11, 12]. According to the two tests, composites with a larger number of fibre layers can withstand higher impact energy; as a result, they are appropriate for basic applications.

## 8.7    FACTORS INFLUENCING THE MECHANICAL PROPERTIES

### 8.7.1    Fibre Content

The proportion of fibre present in the composite by weight or volume percentage is called Fibre content. Properties like rigidity, flexural strength, and hardness increase in proportion to the amount of fibre in the composite. The best assessments of the properties are found at the 40 percent of weight part of fibre [13].

### 8.7.2    Chemical Treatment

The unbending nature and flexural strength of composites are improved by compound treatment or surface treatment of fibres prior to composite arrangement. The contamination existing on the surface of strands are removed by the surface treatment which expands the bonding between cross section and fibre, resulting in improvement of its characteristics. The impact strength of composites, on the other hand, falls when the strength of the fibre weakens following treatment.

### 8.7.3 Bonding between the Matrix and Fibre

Bonding between matrix and fibre has a huge impact on composite qualities. The qualities of composites are superior because they have a greater grip on the network and fibre. It might also be cultivated by surface treatment. Surface treatment creates a few spaces outward of the fibre, which are piled with organization and lead to an improvement in the holding between the framework and the fibre.

### 8.7.4 Fibre Form

Long fibres, short strands, and fibre in powder form are all examples of fibre types. Because larger strands contain more energy, impact strength improves as the length of the fibres grows. In terms of malleable and flexural strength, the fibre length has little effect. The ideal strand size provides superior malleability and flexural strength.

### 8.7.5 Water Absorption

The direct intake of water by strands reduces the holding force between the grid and the fibres. Similarly, it causes composite characteristics to deteriorate. As a result, fibre-reinforced composites can be employed in areas where there is less water traction.

### 8.7.6 Stacking Sequence

In certain composites more than one fibre is used. The arrangement of fibres additionally impacts the properties. By picking the best stacking course of action of fibres, the properties of composites can be improved.

### 8.7.7 Hybridization

Hybridization is just the help of both standard and designed strands with the matrix. In hybrid composites more than one fibre is used. Hybridization prop decreases the proportion of use of designed strands and improves the properties of these composites. Blend composites properties are superior to basic fibre-based composites.

## 8.8 ADVANTAGES OF USAGE OF NATURAL FIBRE

- Density of strands is less and it lessens the substantialness of the composites.
- Natural fibres are less expensive than manufactured fibres. Natural strands are environmentally safer.
- Ordinary fibre-reinforced composites have a safer manufacturing process than manufactured fibre composites.

- Waste generated during the collection of basic fibre composites can be recycled. Particles have a higher fuel efficiency in automobiles since composites are lighter.
- Regular polymer lattices might include inorganic bundles or nanoparticles with unmistakable optical, electrical, or attractive qualities.
- In contrast to pure solid state inorganic materials, which demand a high temperature treatment for planning, hybrid materials exhibit a more polymer-like behaviour as a result of their massive normal substance or as a result of the formation of cross-linked inorganic associations from small nuclear precursors, which is similar to polymerization reactions.
- By avoiding light dissipation in homogenous cross variety materials, optical straight imposition of the material may be prevented.

## 8.9 NOVEL APPLICATION OF HYBRID POLYMER COMPOSITE

An expected increase in the number of applications for fibre-reinforced polymers is seen. Figure 8.7 depicts a segment of these systems that are important, regardless, in extremely specific conditions where limitations, such as enormous cost, are seen as aids to qualities such as low thickness, high inflexibility, and high strength in applications where weight is not as important.

*Nanotechnology*: In the realm of nanotechnology, composite materials have a varied range of applications. The nanocomposite is a unique start, with nanometre-scale fillers dispersed throughout the master. With the addition of just an infinitesimal amount of fillers, fire retardance and stiffness of the sap increases throughout. Polymer nanofibres are employed in therapeutic applications for the repair of burns or wounds on humanoid skin, as well as organized haemostatic frameworks and contraptions with unique and prohibitive properties.

*Wind power generation*: Because it saves resources and is typically secure, wind power planning is a need of the energy age. All the blades which power the world's growing number of wind turbines are made of composite materials. The edges, for example, were created using a cross variety of strands (carbon, glass).

*Application of a hybrid thermoplastic*: For a long time, thermoplastic composite joints have held promise to produce high performance lightweight structures. These head protectors, known as PASGT coverings, are composed of a thermoset lattice with a composite-containing aramid surface. The innovation should also be able to survive the wear and tear of an officer's duties over time, as well as provide enhanced ballistic assurance.

### 8.9.1 Aviation Applications

Because of their increased strength, durability, use resistance, and damage obstruction features, fibre-reinforced composites have become a conceptually appealing alternative to traditional metals for several plane components. A high-level regular jet planned to meet a pair of power and security proportions. Glass and carbon

**FIGURE 8.7** Applications of composites and hybrid materials.

maintained half-and-half composites are the most recommended materials due to significant innovation that has gone beyond the arrangement and application.

### 8.9.2 SMART MEMORY HYBRID COMPOSITES

Smart composites can be created by hybridizing or combining shape-memory materials with other functional or structural materials. These composites can take advantage of the distinctive functions or properties of the individual bulk materials to achieve a variety of responses and ideal properties, or they can be tuned to adjust to environmental changes. Shape-memory elements can be found in the matrix or the reinforcement of a range of shape-memory hybrid composites that have been developed and produced. The creation of new paradigms for material-structural interactions has a great deal of potential with hybrid composites.

*Chemical industry*: Composite materials are among the most well-known materials utilized in the manufacturing sector for a variety of reasons, including their low weight, improved fire resistance, and ability to protect against synthetic compounds. In the subsistence sector, composites are widely employed in the assembly of basic platforms, exhaust stacks and blowers, storage tanks, areas, syphons, reactors, and so on for acidic and vital conditions.

*Orthopaedic aids for prosthetics and orthotics*: iBy equipping them with artificial support, you can help persons who have acquired insufficiency or were born with disfigurements. A prosthesis is a fictitious replacement for a missing component of today's society.

Despite the possible benefits and varied inclinations of less weight and high strength resulting from usage resistance, advanced composites are rarely employed in automotive applications. On an overall level, some effort should be made to make advanced composites material engaging for widespread use in trucks, automobiles, and so on.

### 8.9.3 BRIDGE CONSTRUCTION

The creative effort of completely creamer FRP-structures in main planning has been made in a few nations during the last decade. Combination fibre pontification plastics were used to make all of the necessary pieces (GFRP-glass fiber-reinforced plastic and CFRP-carbon fiber-reinforced plastic). Because of the platform's highly disastrous environment, the all-HFRP-hybrid fiber-reinforced plastic approach was chosen. The pultruded composite elements of the augmentations provide important mechanical qualities while being substantially lighter than steel. The West Mill Bridge (River Cole near Shrivenham in Oxfordshire, UK in 2002) spans a distance of 10 metres and is 6.8 metres wide. The advancement's total heap is 37 t, with a load passing through breaking point of 46 t for vehicles up to 46 t and a turn pile of I 3.St.

## 8.10 CHALLENGES, OPPORTUNITIES, AND FUTURE TRENDS

To increase the planned use of hybrid composites, a couple of issues must be overcome. These difficulties will be solved by planning, analysis, and product enhancement efforts, as well as business development capabilities. In this endeavour, there is a critical necessity to address current concerns. The examination of fundamental treatment of hybrids should be prioritized, particularly issues affecting microstructural authenticity. In hybrid composites, there is a need to increase damage-tolerant qualities, notably break strength and flexibility. Work should be done to portray strongholds of high type and simplicity that are free of mechanical waste and outcomes. Hybrids that are reliant on non-standard fibres and organization should be improved. There is a greater necessity to define distinct hybrid assessments based on property profile and accumulating cost. To measure distortions in hybrid composites, there is a critical need for basic, practical, and useful non-ruinous packs.

A few challenges must be overcome in order to improve the design and use of hybrid composites. To overcome these challenges, configuration, investigation, and item development efforts, as well as business development skills, are required. There is a fundamental necessity to handle the related challenges in this endeavour.

- The essential handling of mixtures, particularly issues impacting microstructural respectability, should be viewed more broadly.
- It is critical to increase the harm-tolerant features of hybrid composites, particularly break durability and flexibility.
- Efforts must be made to increase high-quality, low-cost fortifications from mechanical waste and results.
- Efforts must be undertaken to develop hybrids that rely on non-standard filaments and grids.
- A greater requirement to aggregate various hybrid assessments based on property characteristics and assembly cost [14].
- Basic, inexpensive, and small non-ruinous equipment to analyse irritating flaws in hybrid composites are urgently needed.
- Basic knowledge on how to begin producing strategies for hybrid composites should be thoroughly reviewed, particularly the highlights affecting the underlying tiny dependability. In addition, there is a genuine need to increase the fracture resistance properties of new hybrid composites, such as break sturdiness and pliability.
- Efforts should be made to produce high-quality fortifications with lower costs from farming and mechanical subsidiaries and deposits.

Hybrid composite is one of the newest types of unique materials to emerge, and it has spurred a slew of current modifications. A single lattice with at least two fortifications (biobased/biobased, biobased/engineered, and miniature/nano) results in a unique hybrid composite with a wide range of material conduct and characteristics. Hybridization of two different types of strands has shown to be the most effective way to develop innovative materials that meet certain requirements. The collection of selected properties that develop as a result of the specific arrangement and interactions between the various components opens up a world of possibilities for cutting-edge material advancements.

## 8.11 CONCLUSION

The current chapter summarizes previous research articles by displaying several hybrid composite materials and their mechanical and impact characteristics. Although several study papers have been committed to increasing the performance of hybrid composites, further research is needed to evaluate the exhibition of nano hybrid composites, which also possess remarkable underlying features. Hybrid composite materials are increasingly being employed in a variety of design applications because they have improved qualities and different focus points from traditional composite materials. The mechanical characteristics of hybrid composites are made up of (n > 2) interconnected working phases, which are extremely important. As a result, the mechanical characteristics of hybrid composites are demonstrated via a direct linkage of mathematical recreation models, as previously mentioned. However, the mechanical behaviour of crossover composites is influenced not by the personality of the framework and fortifications, but by the characteristics of the interface between these segments and the network, which should be considered when expressing mechanical features mathematically. In addition, while mathematically exhibiting hybrid composite materials, the influence of ecological maturation should be taken into account. Composite materials have various applications in various fields such as auto, aviation, wind energy, electrical, sports, common development, clinical-compound ventures, and so on. Composite materials have an exceptional chance of being used in designs that are primarily oppressed by compressive-burdens.

## REFERENCES

[1] Wambua, P., Ivens, J., Verpoest, I. Natural fibres: Can they replace glass in fibre reinforced plastics? *Composites Science and Technology* 2003;63:1259–1264.

[2] Sailesha, A., Arunkumar, R., Saravanan, S. Mechanical properties and wear properties of kenaf-aloe vera-jute fibre reinforced natural fibre composites. *Materials Today: Proceedings* 2018;5:7184–7190.

[3] Indra Reddya, M., Prasad Varmaa, U.R., Ajit Kumara, I., Manikantha, V., Kumar Raju, P.V. Comparative evaluation on mechanical properties of jute, pineapple leaf fibre and glass fibre reinforced composites with polyester and epoxy resin matrices. *Materials Today: Proceedings* 2018;5:5649–5654.

[4] Indra Reddy, M., Anil Kumar, M., Rama Bhadri Raju, Ch. Tensile and flexural properties of jute, pineapple leaf and glass fibre reinforced polymer matrix hybrid composites. *Materials Today: Proceedings* 2018;5:458–462.

[5] Swolfs, Y., Gorbatikh, L., Verpoest, I. Fibre hybridisation in polymer composites: A review. *Compos Part A Appl Sci Manuf* 2014; 67:181–200.

[6] Sailesha, A., Arunkumar, R., Saravanan, S. Mechanical properties and wear properties of kenaf-aloe vera-jute fibre reinforced natural fibre composites. *Materials Today: Proceedings* 2018;5:7184–7190.

[7] Indra Reddya, M., Prasad Varmaa, U.R., Ajit Kumara, I., Manikantha, V., Kumar Raju, P.V. Comparative evaluation on mechanical properties of jute, pineapple leaf fibre and glass fibre reinforced composites with polyester and epoxy resin matrices. *Materials Today: Proceedings* 2018;5:5649–5654.

[8] Ahmad, F., Choi, H.S., Park, M.K. A review: Naturalfibre composites selection inview of mechanical, light weight, and economic properties. *Macromol MaterEng* 2015;300(1):10e24.

[9] Safri, S.N.A., Sultan, M.T.H., Aminanda, Y. Impact characterisation of glassfibrereinforced polymer (gfrp) type C-600 and E-800 using a drop weight ma-chine. *Appl Mech Mater* 2014;629:461e6.

[10] Safri, S.N.A., Sultan, M.T.H., Cardona, F. Impact damage evaluation of glass-fibrereinforced polymer (gfrp) using the drop test rigean experimental basedapproach. *ARPN J Eng Appl Sci* 2015;10(20):9916e28.

[11] Safri, S.N.A., Sultan, M.T.H., Razali, N., Basri, S., Yidris, N., Mustapha, F. The effect oflayers and bullet type on impact properties of glassfibre reinforced polymer(gfrp) using a single stage gas gun (ssgg). *Appl Mech Mater* 2014;564:428e33.

[12] Safri, S.N.A., Sultan, M.T.H., Cardona, F. Impact characterisation of glassfibre-reinforced polymer (gfrp) type C-600 and E-800 using a single stage gasgun (ssgg). *Pertanika Journals Sci Technol* 2017;25(1):303e16.

[13] Kumre, A., Rana, R.S., Purohit, R. A review on mechanical property of sisal glass fibre reinforced polymer composites. *Materials Today: Proceedings* 2017;4:3466–3476.

[14] Wood, A.S. Hybrid reinforcement: Low cost route to premium performance. *Mod Plastics* 1978 March;55(3):48–50.

# 9 Design Strategies and Drug Delivery Applications of Multifunctional Mesoporous Silica Nanoparticles

*Umar Farooq, Ravi Shekhar, Pooja Sharma, and Syed Salman Ali*

## 9.1 INTRODUCTION

With continuous development of human civilizations and technology, there is a continuous threat of both evolution of already occurring diseases and development of new diseases. Continuous efforts are being made in the field of medical sciences and drug designing technology to eradicate currently occurring diseases and to stop new diseases [1]. The designing and development of new drug formulations have gained huge research interest to stop the threat posed by different diseases to the present human civilization. However, drugs already designed to cure a particular disease are associated with some serious side effects [2]. The main important reason for the side effects is nonspecific cell or tissue biodistribution of these drugs. Therefore, to reduce this serious issue of nonspecific biodistribution, significant effort has been made and considerable success has already been achieved [3–5]. Out of these efforts/developments, encapsulation is considered a promising way to minimize the side effects of developed new/already existing drugs [6–7]. By the encapsulation strategy, drugs are delivered to the target cell/tissue where the release of the drugs is controlled according to the demand. In this respect, nanotechnology has gained immense interest because nanomaterials show different properties from that of their bulk counterparts [8–10]. Nanoparticles have different electronic, optical and magnetic properties which can be used to guide the drug-loaded nanoparticles to the desired site, where the drugs can be released in a controlled fashion [11–13]. Also, these nanoparticles have an immensely high surface-to-volume ratio due to the large amount of drugs that can be loaded and delivered to the targeted sites [14, 15]. The use of nanoparticles/nanomedicines has significantly improved the already existing drugs/treatments

DOI: 10.1201/9781003316398-9

by changing their biodistribution profiles and pharmacokinetics. For improved therapeutic applications of different nanomedicines, researchers have explored the nanoparticles (NPs) showing response to the surrounding environment; that is, when NPs are subjected to external stimuli, physicochemical changes are produced that favour the controlled release of the drug at a specific target site. The external stimuli to which nanodimensional formulation responds include (i) physical stimuli such as electric field, magnetic field, temperature and pressure and (ii) chemical stimuli such as surrounding pH, redox potential, enzymes and ionic strength. Different nanoparticles have been employed as carriers for delivering developed drugs to the targeted sites [16–18]. Among different nanostructured materials pristine, end capped, and surface-functionalized mesoporous nanoparticles of silica have attracted great interest as a potential target-specific drug-delivering agent because of their unique structure which facilitates the stimuli-responsive and -controlled drug/protein/gene release [19–21]. The unique properties of mesoporous silica nanoparticles include narrow straight channels, a high volume of these channels and large surface area which allow adsorption of drugs deep into the structure of silica nanoparticles. These structural properties, particularly the presence of straight channels in mesoporous silica nanoparticles (MSNs), allow MSNs to show a controlled and targeted release of adsorbed drugs around the targeted tissues [22, 23]. The targeted, controlled and sustainable release by MSNs is controlled by various characteristic features which involve the nature of loaded drugs, nature of the targeted medium/surroundings at which the drug is to be released, surface functionalization, pore size/surface area and morphology. The presence of straight channels in MSNs offers a property of a reservoir for loaded drugs, and the drug-releasing properties (opening and closing of these channels) can be controlled by polymers, photoactive derivatives, nanocrystals and more. To improve the targeted drug delivery of silica nanoparticles different modification strategies like functionalization and capping with different agents such as small molecules and polymers are being employed [24]. The main aim of functionalization, the use of a capping agent on MSNs, is to improve the physicochemical characteristics for better drug adsorption and efficient drug delivery with sustainable release. The dosage and controllable release of a specific drug at its target site can be controlled efficiently by varying one of the following parameters: composition, structure of matrix, modification of pore entrance, pore size and chemical interaction between host (MSNs) and guest (drugs) [25–27]. Continuous research is proceeding in the field of developing drug delivery vehicles, and numerous reports are available in which MSNs have been used in target-specific delivery of important pharmacologically important drugs. The use of MSNs for delivery of different important drugs reduces their cytotoxicity and improves efficiency. For example, a drug (ramipril) is used to treat myocardial infarction, but its use is associated with severe side effects like reduction in glucose level and neutropenia (abnormality involving a low count of white blood cells). Researchers have used MSNs for the targeted delivery of ramipril where MSNs show controlled and sustained release and side effects were considerably reduced [28]. Similarly, nanorods of silica were investigated for their bactericidal activity due to the release of nitric oxide. From this investigation it was observed that particle shape and size play a significant role in the drug-release property of silica nanorods.

In this chapter we discuss progress made in the field of target-specific responsive drug delivery using mesoporous silica nanoparticles, different modifications of MSNs for improved and efficient drug delivery, issues like cytotoxicity and biocompatibility arising due to modifications of silica nanoparticles and challenges that need to be addressed.

## 9.2   ADVANTAGES OF TARGETED DRUG DELIVERY

Herein, various advantages of using different carriers for target-specific therapeutic applications are discussed. The response of each human being to a particular drug varies considerably from one person to another. In everyone the distribution and metabolism of a drug, especially for those drugs that act and show response only at a particular receptor, differ considerably according to the demand of each individual [29]. Therefore, monitoring the progress of any disease and responding to the change in chemistry (chemical changes) taking place in the body of any individual provide an unprecedented opportunity for delivering personalized medical care. For individualized medical care the analysis of markers as an indication for different stages of a disease provides an efficient way to vary the dose and type of drug which would be most appropriate at a particular stage. Hence, for different therapeutic regimes the use of chemical biosensors for determining the level of various biomarkers may act as a guide for safe and efficient drug delivery [30]. The development of individualized therapeutic ways to control disease is a long-held aim of researchers. In this regard considerable progress has been made, but it remains an unattainable goal. Therefore, personalized medical care advances like targeted drug delivery have been made. The main advantages and advances in targeted drug delivery are discussed as follows:

### 9.2.1   CONTROLLED DRUG RELEASE

Currently most of the available drug delivery systems function on the principle of controlled release. Controlled drug release signifies release of a certain amount of drug over a specific period and that this release can be easily predicted. As the diseases of every person are unpredictable the use of controlled drug delivery is limited, as these systems cannot respond as per individual pharmacologic demand. Nevertheless, the use of controlled drug delivery systems is responsible for low therapeutic index with improved compliance of an individual patient [31]. By employing the controlled drug delivery approach the variation of concentration of drug in blood plasma is small, thereby limiting the risk of under- and overdosing drugs in patients. This problem of individual variance becomes challenging when the therapeutic treatment involves more drugs, and this problem is more complicated by certain medical conditions which are unique in different persons. For example, the development of breast cancer in patients varies based on different mutant genes. Similarly, there may be different mutant genes of cancer present in a single individual which makes controlled release more formidable [32]. Sometimes medical conditions result in an unpredictable change, and the observed symptoms do not provide sufficient lead for accurate and timely dosing. In these conditions physicians and patients may opt for a preventive approach, employing controlled release which maintains the constant presence of drug in the patient's body. However, this approach may result in the development of tolerance towards the drug and side effects due to constant exposure to the drug.

### 9.2.2 STIMULI-RESPONSIVE DRUG DELIVERY

To manage the aforementioned unpredictability associated with controlled drug delivery, alternative ways are being explored. One way is to put the patient on constant alert to the warnings associated with arising medical crises. Such an approach is mostly used in diabetic patients who must constantly be alert for the onset of hypoglycaemia, which is a common complication associated with such patients. To have constant vigil on glucose level, the patient must undergo constant stress along with continuous finger pricking which causes pain and is not considered a convenient approach. To avoid such stressful conditions and pain, a stimuli-responsive drug delivery is considered a convenient approach. In a stimuli-responsive drug delivery approach, the drug delivery system is inserted in the body of a patient and continuously senses the different chemical markers representing the onset of crises, and as soon as the markers associated with any disease is detected a prompt response with countermeasures are made by the drug delivery system without disrupting the daily routine/life of a patient.

At present, advances in the personalized therapeutic regimes involve either *ex vivo* detection of different markers with *in vivo* response or *in vivo* detection of markers with *ex vivo* drug delivery. Both approaches have their advantages and disadvantages. However, currently there are limited systems which employ *in vivo* detection and response systems that could make stimuli-responsive drug delivery a reality for individualized treatment. An important example of personalized responsive therapy is an implanted cardiac defibrillator (ICD). Although this is not a drug delivery system, it is a suitable example explaining how it is possible to make an advance and development in personalized health care. In ICD heartbeat is monitored continuously; if any irregularities are detected, electric pulses are sent to the heart [33]. This technique is used for the treatment of ventricular fibrillation. The stimuli-responsive delivery system should be able to detect a minute concentration of markers with high precision in the body. For this purpose (detection and delivery of drug) a single technique should be developed. The analytical technique used for *in vivo* detection and delivery should be small and biocompatible with accurate and reproducible performance.

### 9.2.3 SENSING AND BIOSENSORS

To monitor small amounts of body fluids or chemicals produced during any infection/disease with great accuracy, sensitivity and selectivity is a prime requirement for designing and developing target-specific responsive delivery systems. For this purpose, the design of molecular bioreagents using molecular biology and protein engineering is an important strategy. These bioreagents could be employed as biosensors in different biological sensing systems having high sensitivity, selectivity, low detection limits and instant response. In addition, these biosensors could be used for detection of hormones and neurotransmitters [34–37]. These biosensors could be integrated with stimuli-responsive delivery systems for development of target-specific individualized healthcare response. It is also observed that these *in vivo* biosensors are considerably affected by physiological changes of the body and are associated with a short life span. Therefore, for efficient drug delivery systems with stimuli-responding delivering properties, considerable efforts have been already made and much more investigation is needed to have an optimal delivery system with high stability, selectivity, specificity and quick response. Table 9.1 represents various developed drug delivery systems (DDS).

## 9.3  SILICA NANOPARTICLES AS DRUG DELIVERY SYSTEMS

The first mesoporous silica materials were reported back in the 1990s by Japanese scientist Kuroda [44, 45]. For the development of bulk mesoporous silica materials, surface active molecules (surfactants) are employed as templates for the condensation of precursors of silica around them. After carrying out the synthesis, the template is removed and material with network cavities is developed. Homogeneous pore size distribution with pore size ranging from 2 to 20 nm high surface area, density of surface silanol groups favouring the diverse functionalization with different groups and high pore volume are important characteristics of as-developed mesoporous silica [46]. These properties make mesoporous silica a prime material for application in processes where adsorption is a prerequisite condition, like for drug delivery systems. Based on different significant applications like bioactive behaviour, bone tissue engineering of bulk mesoporous silica, the translation to nanoscale could result

**TABLE 9.1**

**Different Types of Drug Delivery Systems**

| Type of DDS | Examples | Main Features | References |
|---|---|---|---|
| Commercial | (a) Implanted fusion pumps | Delivery by preprogramming of release rate through a tube at specific target inside the body. | [38] |
| | (b) Noninvasive reversible iontophoresis devices | | [37] |
| | | Extraction of analyte from the skin using electric current across the skin. | |
| Responsive | (a) Closed loop | Delivery in response to sensing. | [39] |
| | (b) Responsive polymers | Release in response to a stimulus using smart polymer. | [40] |
| Controlled release | Duros implant technology | Used for pain medication with continuous release, tailored as per individual demand by employing osmotic gradient. | [41] |
| Micro and small systems | (a) Sacrificial valves (microfabrication) | Sealed reservoirs containing nano/low microlitre volumes of drugs; sealed reservoirs are opened electrochemically. | [42] |
| | (b) Miniaturized valves ("artificial muscle") | Hydrogel/polymer blends which behave like/mimic natural muscles. | [43] |

in the improvement of biomedical application. Therefore, the improved properties of nanodimensional mesoporous silica have triggered great interest in biomedical and biotechnological applications. With the continuous development of nanotechnology, continuous modifications are being incorporated into MSNs to have better intimate interaction between the developed materials and the biological systems where nanoparticles are being used. In the coming section, we will discuss developments made in the synthesis strategies for preparation of mesoporous silica nanoparticles and advances in their application as systems for targeted, stimuli-responsive drug delivery technologies.

### 9.3.1   SYNTHESIS OF MESOPOROUS SILICA NANOPARTICLES

Since the discovery of MSNs, different methods have been employed for their synthesis. The most used method involves the use of template-assisted methods using cationic surfactant templates, the sol–gel method. Other methods have been employed but in general the synthesis approaches involve three processes like production of colloidal amorphous metal oxides (silica) via the sol–gel approach, preparation of mesoporous material using surface-directing agents (surfactants) and development of spherical nanoparticles by modification of the Stober process [47, 48]. The tailoring of morphologies (fibres, hollow spheres, spheres etc.), mesostructures (wormlike, cubic, hexagonal, lamellar etc.) and dimensions (nm and cm scale) can be easily controlled by using different morphologies and liquid crystal morphologies of assembled surfactant molecules and by controlling the reaction conditions such as silica precursor, surfactants and pH. In different research articles the investigations regarding the preparation of specific morphologies of silica were carried out by analysing the interactions taking place between silica species and surfactant molecules. From these investigations it was concluded that the spontaneous assembling of silica-surfactant nanocomposites takes place due to the matching interaction between inorganic and organic components present on silica and surfactants respectively. In addition to this matching interaction, thermodynamics of silica-surfactant assemblies' kinetics of the sol–gel process has a significant role in controlling the morphologies and dimensions of final mesoporous silica nanoparticles [49, 50].

Conventionally for synthesis of silica nanoparticles the sol–gel hydrolysis process is used in which silicon alkoxide undergoes hydrolysis and condensation process under basic and acidic catalytic conditions. During the typical reaction progress polycondensation reaction takes place around the template molecules (surfactant), and the silica precursor present in the reaction system results in the formation of an oxide network colloidal solution (sol), which results in the formation of gel/discrete particles with progress of the reaction. Under dilute conditions spherical monodisperse nanoparticles of silica are obtained. Under alkaline conditions the synthesis pathway followed for the preparation of silica nanoparticles by dissolving surfactant in aqueous medium is shown in Figure 9.1. From the synthesis of mesoporous silica nanoparticles, it was observed that the type of surfactant, temperature and concentration have a significant effect on the self-assembling process, which in turn controls the final structure of mesoporous material. From the figure it is also illustrated that after addition of dilute silica precursor the onset of

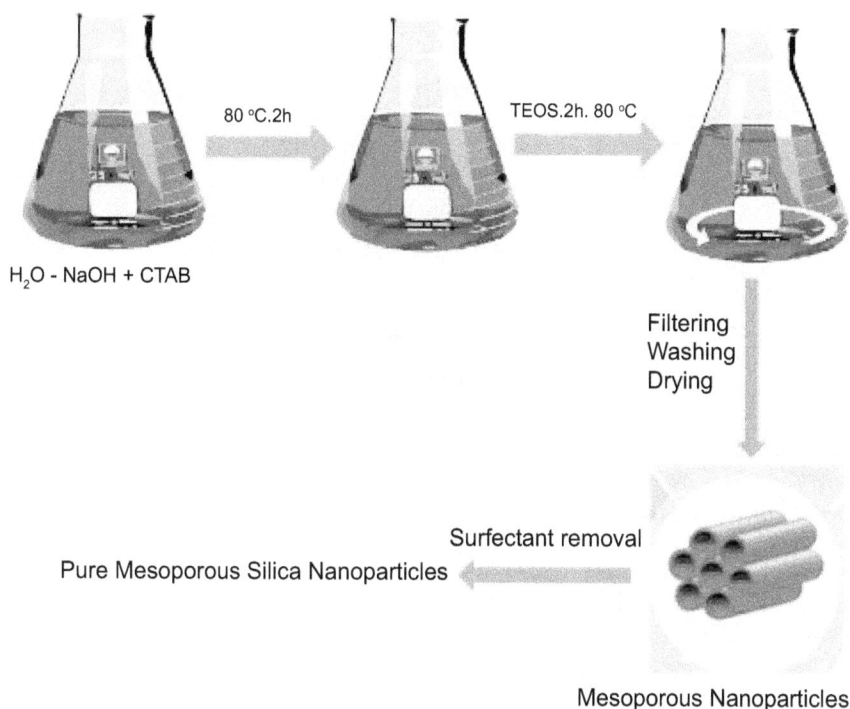

FIGURE 9.1   Synthesis of mesoporous silica nanoparticles using surfactant template in the first step followed by the addition of silica precursor.

the sol–gel process takes place, resulting in the formation of nanoparticles from the template-assisted assembled droplets. Finally, the template molecules are removed by employing a solvent extraction process, resulting in the formation of pure silica nanoparticles [51].

## 9.3.2   CHEMICAL MODIFICATIONS OF MESOPOROUS SILICA NANOPARTICLES

The presence of silanol groups on the surface of mesoporous silica nanoparticles have made MSNs open to surface modifications/functionalization. From the reported literature it was observed that the biomedical applications of mesoporous silica were considerably improved by carrying out different surface modifications. The possible modifications that can be incorporated in the hydroxyl groups present on the surface of mesoporous silica nanoparticles are shown in Figure 9.2 and include amine, organic chain, thiol and hydroxyl modifications. The main aim of such functionalization includes the development of drug systems with sustainable, controllable drug release and release of drugs based on specific stimuli/physiological changes. From various investigations it was observed that such modifications are highly useful for targeted drug delivery with high drug-loading capacity of modified MSNs and improved efficiency of therapeutic agents with low toxicity.

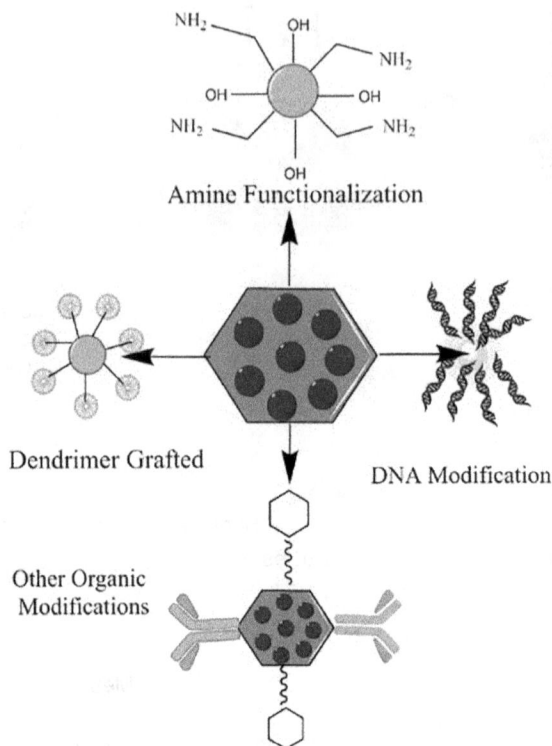

**FIGURE 9.2**   Methods used for surface modification of mesoporous silica nanoparticles.

Zeng et al. have investigated the influence of amino propyl groups on release proper-ties of modified MCM-41 mesoporous silica nanoparticles [52]. From the investiga-tions it was observed that release of aspirin by surface modified MCM-41 is directly dependent on the amount of presence of surface amino propyl groups on the surface of MSNs. Also, it was observed that no change in ordered structure and channels present in MSNs takes place during the use as a drug delivery system in a medium having similar characteristics as that of body fluid. Similarly, influence of surface modification on adsorption capacity of (MCM-41 and SBA-15) mesoporous silica nanoparticles was evaluated by Wang et al. From the obtained results it was deduced that adsorption capacity is dependent on both types of drug as well as on the type of surface modifications made to mesoporous silica nanoparticles. Wang et al. have deduced that a higher amount of ibuprofen was adsorbed over the surface of MCM-41 and SBA-15 modified with 2° amino groups and 3-amino propyl groups [53]. The improved adsorption capacity of surface-functionalized mesoporous nanopar-ticles was attributed to hydrogen bonding taking place between amine groups present on the surface-functionalized MSNs and COOH present of ibuprofen. In addition, the drug release capacity was also influenced by the type of bonding taking place between surface-functionalized groups and the adsorbed drugs. Apart from the parameters discussed earlier, Vallet-Regı et al. found that kinetics of drug release is significantly influenced by loading of the amount of drug and the drug:host ratio. Figure 9.3 represents the role and importance of functionalization of mesoporous silica nanoparticles with amine groups [54].

Based on structure, the pore size of MSNs is very large, possessing high pore volume. Therefore, for efficient utilization of these porous structures as drug delivery systems considerable advancements and efforts are being required. It was investigated that by carrying out the calcination process, the ability to absorb drugs in the porous structure of different mesoporous silica nanoparticles is improved. However, to control the release of drugs from the surface of mesoporous silica nanostructures in addition to surface functionalization, more modifications need to be incorporated. In this respect the modification is done by alkyl chains on the surface on mesoporous nanoparticles. From surface functionalization of modified silica nanoparticles, the release rate of drug is controlled by degradation of silica nanoparticles. Thus, modifying the surface by alkyl chains increases the hydrophobicity of mesoporous silica nanoparticles thus increasing their stability to the degradation process, which helps us to have considerable control over the release of drugs. Investigations of the effect of alkyl chains were evaluated, and it was observed that using such modifications (a) reduces the pore size, (b) changes the chemical interactions between drug molecule and adsorbing surface, and (c) the wettability of pore surfaces in aqueous medium is changed. It has been observed that functionalization of MSNs with organic moieties consisting of C8 to C18 chains does not affect the structural order of nanoparticles. Incorporation of organic moieties improves the biocompatibility which motivates the

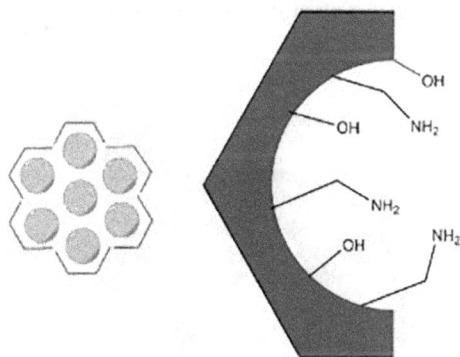

*Advantages*
Improves drug loading
Sustainable Drug release
Targeted drug release
Stimuli responsive drug release

**FIGURE 9.3**   Amine functionalized mesoporous nanoparticles and their advantages in drug delivery.

researchers to develop hybrid MSNs. The development of hybrid MSNs includes the use of polyethylene glycol (PEG), polyethyleneimine; these molecules are adsorbed on the inner or outer walls of MSNs. The advantages of these modifications using organic molecules include improved drug stability, enhanced delivery of drugs to the target sites, and enabling the MSNs to react to the environment stimulus due to the formation of smart nanoparticles. In addition to the aforementioned advantages the modification of MSNs by PEG increases the circulation time of nanoparticles with improved resistance to the attack from the immune system and proteolysis [55]. Figure 9.4 depicts the importance of different modifications of MSNs using organic molecules.

In addition to the preceding strategies, the modification of MSNs using DNA molecules is considered an important approach to improve drug delivery ability and stimuli-responsive release of MSNs. The DNA molecule acts as a switch gate in the modified hybrid MSNs. For example, cytosine-rich DNA was used to modify the MSNs where it acts as a smart molecule switch gate. In the presence of metal ions like Ag, the C-rich DNA forms a hybridized structure like C-Ag+-C. The type of hybridized structure depends on two DNA molecules (nucleobases) present on the metal; such structure also results in the pore blocking and packing of molecules. In presence of molecules like thiols the drug release is controlled by displacement reactions. During the release the DNA molecule having metal ion is converted to single stranded DNA molecules in the presence of thiols like dithriothretol (DTT), which results in the release of molecules entrapped in the host mesoporous structure. Based on the observations it was elucidated that reversible opening and closing of the structure for release of entrapped molecules can be done easily by varying the concentration of metal ions like Ag+ in the earlier case and thiol molecules. Using this approach, it was observed that even at high concentration of drugs cytotoxicity was

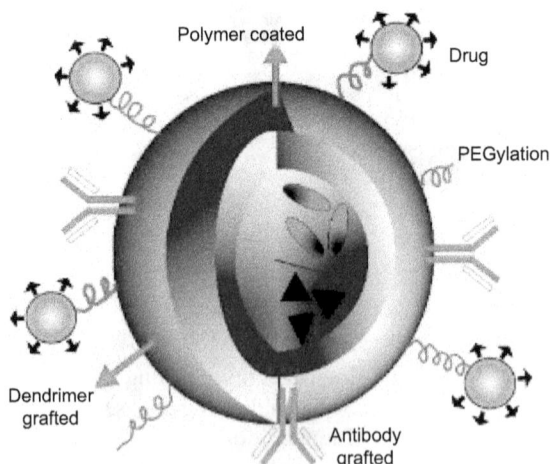

**FIGURE 9.4**  Influence of different organic chain modifications of role of MSNs as drug delivery systems.

considerably low. However, there are certain limitations, like poor water solubility associated with existing capping systems. Therefore, such molecular gate switches could be replaced by using biomolecules such as nucleotides, carbohydrates and proteins. Due to their conformational polymorphism and programmable sequence recognition properties, DNA molecules can be used as an effective molecular gate in stimuli-responsive therapeutic systems. Keeping this in view, these modifications to MSNs can be easily manipulated as efficient drug delivery and stimuli-responsive systems [56, 57].

### 9.3.3 BIODEGRADATION AND BIOSAFETY OF MESOPOROUS SILICA NANOPARTICLES

For medical practical application of any drug/material the approval from the international/national Food and Drug Administration (FDA) is necessary. Despite being explored extensively by researchers, the FDA has not approved MSNs for medical applications. Therefore, to have an approval from the administration for the use of MSNs in medical applications it is imperative that we investigate the possible biodistribution, routes and final fate of used MSNs inside the body. Generally, the performance of a drug delivery system depends on the adsorption rate and adsorption extent, distribution in the target, metabolism and final elimination of the MSNs. The distribution metabolism and destination of any drug also depends on how the drug has been incorporated inside the body [58]. The most common methods employed for injection of a drug include subcutaneous, intravenous and intratumoural methods. All these methods have their own advantages and disadvantages. The advantage of intravenous injection of a nanoparticle includes fast delivery and biodistribution into the patient's body. However, the main concern with the use of these nanoparticles is their stability under physiological operation conditions. Therefore, it is an important condition for any nanocarrier that it should protect the loaded drug during the time of travel to the target destination. In addition, it is highly desirable to have a host material which can be easily degraded after accomplishment of the target. In aqueous medium MSNs, due to the presence of $SiO_2$, are highly prone to nucleophilic attack by OH from $H_2O$. Under these hydrolytic breakdown conditions, the MSNs' network is cleaved and orthosilicic acid is produced as a by-product. The obtained by-product is biocompatible in nature and is excreted by urination. Thus, the degradability of mesoporous silica nanoparticles is controlled by a mechanism of how silicic acid is dissolved in the biological system. The acid produced during the degradation of MSNs is highly soluble in water and is considered as an important component of healthy bone; hence, for over 50 years silica has been recognized as safe by the FDA. The degradation and dissolution of MSNs is highly dependent on particle, shape, size, morphology, pore size, functionalization and characteristics of medium like temperature, pH and concentration. Based on these characteristics and final application, the dissolution and degradation properties of MSNs can be tuned from a few hours to several weeks by carrying out different modifications, as discussed in the previous section.

### 9.3.4    Application of MSNs as Smart Drug Delivery Systems

The use and success of mesoporous silica nanostructures as an important drug delivery agent/carrier depends on the already discussed physical and chemical characteristics. The use of MSNs addresses many pitfalls such as lack of specificity, high cytotoxicity, low efficacy, stability, side effects and solubility of conventional therapeutic agents. For the use of any nanostructure as a drug-delivering agent, some basic requirements needed to be fulfilled. These requirements include maximum loading of drugs, on-demand release of drugs to avoid premature release, and penetration of and release to the target diseased site.

As discussed earlier, for efficient delivery of a therapeutic agent the carrier molecule should be able to carry the drug/guest molecule to the target site without any loss. At the final/targeted destination the host molecules should be able to release the desired guest in a controlled manner. Any release before reaching the destination (premature release) is associated with numerous problems which pose a great challenge for efficient drug delivery. For example, the use and delivery of different antitumour drugs need "zero premature release" as these drugs can cause serious damage to normal cells [58]. Currently the use of many polymer-based agents for drug delivery systems shows release of encapsulated matrix upon hydrolytic erosion process of carrier molecules. These biodegradable polymer carriers immediately release the encapsulated drug molecules upon dissolution in water. Also, for loading of drugs on such host structures requires organic solvents which can induce significant undesirable changes like aggregation or denaturation in the loaded drug molecules/proteins. Therefore, to overcome these limitations surface-functionalized mesoporous silica nanoparticles are considered as important alternative target-specific carriers due to their beneficial characteristics, as discussed earlier in this chapter.

Herein, we will discuss different physical and chemical properties controlling the release properties of MSNs.

### 9.3.5    Physical and Morphological Properties
###          Controlled Release of Drugs

The primary investigation of using MSNs as carrier agents in drug delivery systems employed high pore volume and surface area of silica nanostructures for better delivering processes. In these primordial systems there is a simple adsorption of a guest molecule over the surface of mesoporous silica nanoparticles. In such systems there was no molecule which could act as a gate for controlled release of a loaded drug molecule. The release of loaded molecules in such systems was controlled by either morphology or size on the mesoporous structure. For example, two different mesoporous silica nanoparticles (MCM-41) with different pore sizes were investigated for delivery and release of ibuprofen in a body-simulated fluid [59]. From the investigations it was observed that a sample of MCM-41 with a hexagonal arrangement of channel-like pore showed high adsorption capacity of a loaded drug with a release over a longer period. However, from the release profile investigations no significant difference was observed in the samples with different pore sizes. Since from this investigation it was concluded that pore size does not have any significant role in controlled release of adsorbed molecules, the investigation regarding the effect of particle and pore morphology was carried out. To analyse the effect of these factors on controlled release, an MSN system was developed by a template-assisted approach using room temperature ionic liquid (RTIL) as a template. The as-developed RTIL-MSN was investigated for the release of antimicrobial drug, and during the investigation four morphologically different (including rods, tubes, ellipsoids and spheres) RTIL-MSNs were synthesized. The different morphologies of the pores were controlled from hexagonal mesoporous to Moiré-type rotational helical channels and to porous wormhole structure by changing the RTIL template. The controlled release of the carrier material was investigated by analysing antibacterial effects on *E. coli* K12. The comparative investigation of release of antibacterial drugs between two RTILs and two RTIL-MSNs (one having hexagonal mesoporous morphology with spherical shape and the other having wormhole-like mesopores with tubular shape) was carried out. It was observed that both RTILs have the same antibacterial activity with no significant difference, whereas the case of RTIL-MSNs having hexagonal mesopores with spherical shape shows 1000-times enhanced antibacterial activity compared to the RTIL-MSNs having wormhole mesopores (tubular shape). From this comparative investigation it was extensively elaborated that morphology of mesoporous silica nanoparticles plays a significant role in controlling the release behaviour.

Similarly, Poostforooshan et al. have synthesized hollow mesoporous silica microspheres (HMSM) using an aerosol-assisted approach using Eudragit/Triton X100 composite (EUT) nanospheres and colloidal silica nanoparticles as templates [60]. The external morphology and internal hollow structure of mesoporous silica microspheres were manipulated by varying different reaction parameters like mass ratio of silica/EUT, using silica nanoparticles having different particle size as shown in Figure 9.5.

Also, the effect of Triton X-100 on the mesoporous structure of HMSM was investigated as shown in Figure 9.6. The synthesized HMSM was investigated as a drug delivery system for amoxicillin, which is an important antibiotic compound. In this report the biocompatibility of as-developed silica microspheres was enhanced by carrying out layer-by-layer functionalization using biocompatible polymers like poly(allylamine) and alginate. The as-developed amoxicillin loaded HMSM were investigated towards E-coli bacteria and 90% inhibition of bacterial growth was observed in just 2 hours. Also, the as-developed system was investigated for treatment of pancreatic tumour cells.

For antitumour activities, chitosan molecule was used as a pH-sensitive stimuli-responsive gatekeeper. As the pH of tumorous cells is low compared to healthy cells, as-developed pH-sensitive chitosan-grafted HMSM (Chi-HMSM) surface showed pH-responsive release of proapoptotic NCL antagonist agent (N6L) (an anticancer drug) with an excellent activity towards the prevention of tumour cells. The pH-responsive release properties of Chi-HMSM are shown in Figure 9.7. From the obtained results it was observed that when the pH of the media is equal to pKa value of chitosan, no release of loaded drug was found. Under these conditions the loaded probe does not come out of the pores of HMSM particles; also, chitosan remains in deprotonated amine form due to which the collapse of chitosan polymer chain takes place, thus forming a protective layer over the surface of HMSM which prevents the release of loaded probe/drug molecules. However, at low pH chitosan becomes positively charged and undergoes swelling, which leads to the opening of mesoporous structures and results in the release of drug molecules at cancerous cells having low pH.

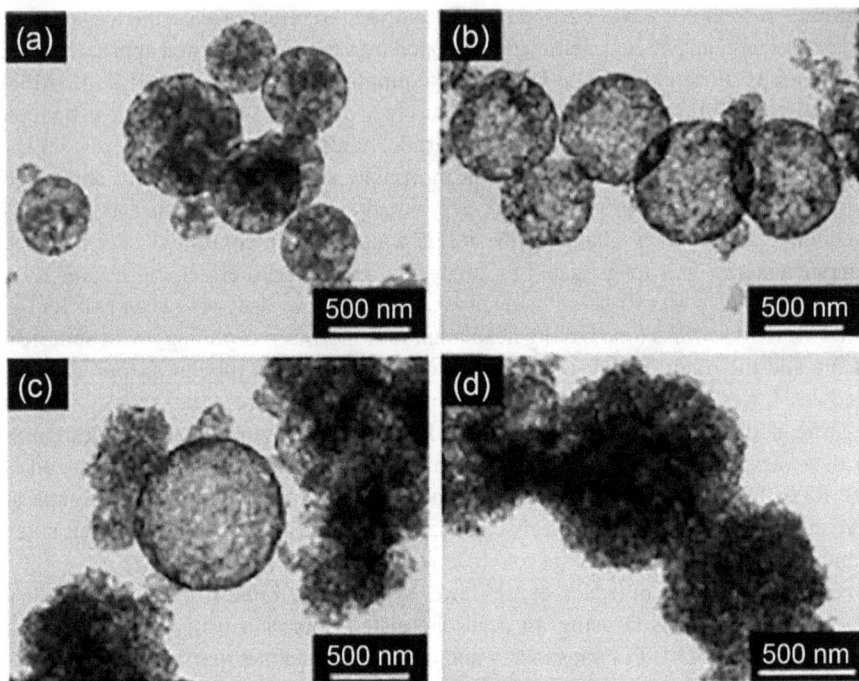

**FIGURE 9.5** TEM images of (a) HMSM-20-1, (b) HMSM-20-0.5, (c) HMSM-20-0.33 and (d) HMSM-20-0.25. Reprinted by permission and copyright of the American Chemical Society. Ref (60).

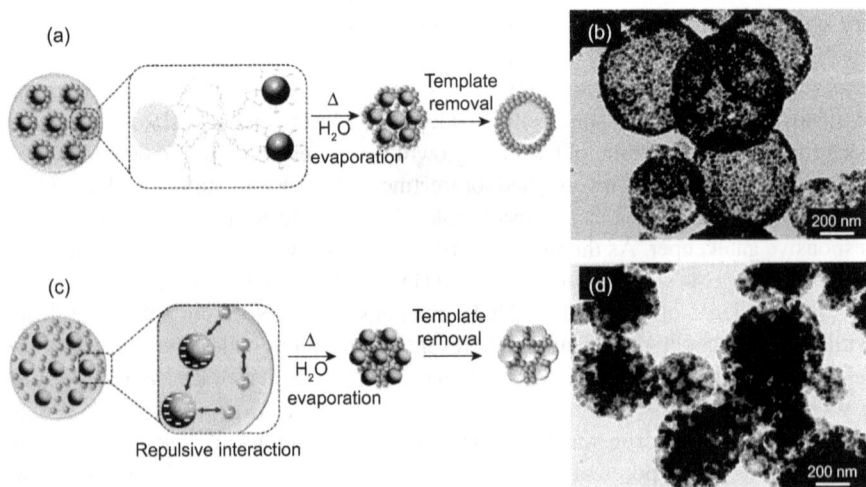

**FIGURE 9.6** Effect of presence and absence of Triton X100 on (a) hollowness and (d) porosity of HMSM, (b) TEM images of HMSM-20-0.5 and (d) HMSM 20-0.5 particles. Reprinted by permission and copyright of the American Chemical Society. Ref (60).

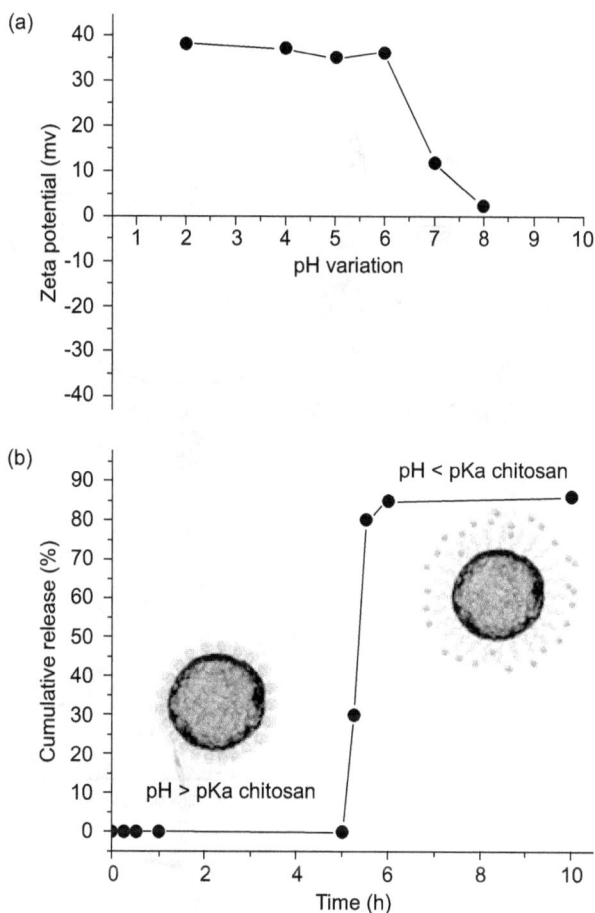

**FIGURE 9.7**    (a) Effect of pH on the zeta potential of chitosan-grafted HMSM; (b) release properties of Chi-HMSM particles over different pH condition. Reprinted by permission and copyright of the American Chemical Society. Ref (60).

### 9.3.6  CHEMICAL PROPERTIES CONTROLLED RELEASE BEHAVIOUR OF MATERIALS

As discussed earlier, efficient delivery of drugs at the target site requires zero premature release. Therefore, it is imperative to have systems that show the activation behaviour only once the system receives an external stimulus. To achieve such stimuli-responsive release of drugs, significant efforts have been made by researchers. In this respect Lin et al. have developed a stimuli-responsive MSN system showing controlled release of loaded probe molecules [61]. In this report a gated system of MSNs consisting of CdS-capped guest molecule-loaded mesoporous structure was developed as shown in Figure 9.8.

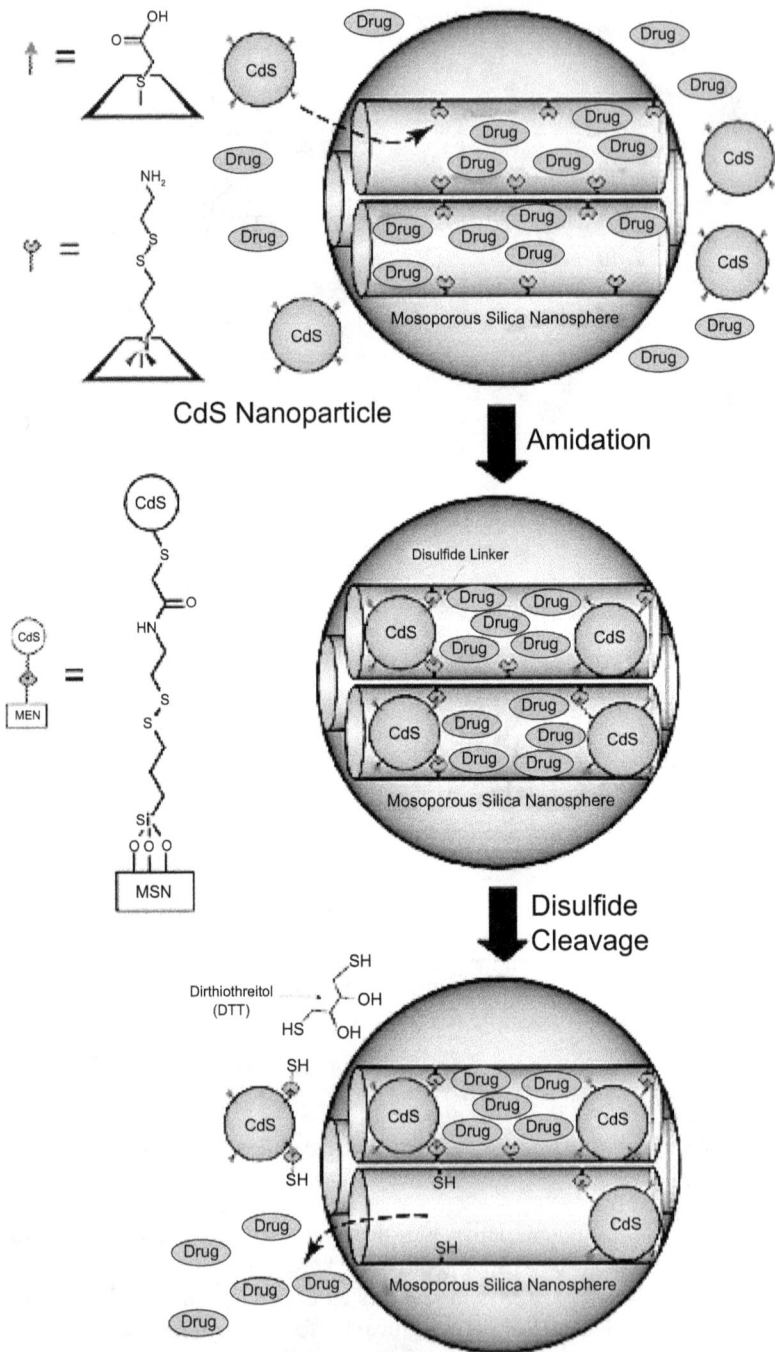

**FIGURE 9.8** Illustration of disulphide link-based CdS MSN controlled release system. Reprinted with a permission from [61]. Copyright 2003 American Chemical Society.

Due to physical blocking by CdS molecules the loaded guest substance was unable to move out of the mesoporous structure, thus preventing the premature release. In this system the release of drug molecule was triggered by the change in chemical environment which stimulated the breakdown of disulphide linkage (due to which CdS was bonded to MSN surface), which resulted in removal of capping on MSNs and thus providing a path for the release of entrapped guest molecules. For the investigation of chemical liability of as-developed MSNs capped with CdS, different chemical stimuli such as mercaptoethanol (ME) and dithiothreitol (DTT) were used, and it was observed that no premature release of loaded vancomycin drug takes place in CdS-capped MSNs. After 12 hours of continuous investigation, it was observed that release was only triggered by the addition of disulphide reducing agents like DTT and ME, and 85% of loaded drug molecules were released in 24 hours after receiving chemical stimuli.

In addition to the chemical stimulated release of carrier/drug molecules in aqueous medium as discussed earlier, researchers have tried to explore the response of host systems to other stimuli which include photochemical-based stimulus and redox-based switch gates/stimulus. Tanaka et al. have explored the photo-controlled release properties of mesoporous silica material modified with coumarin in an organic medium [62]. It was observed that adsorption, uptake, storage and release properties of modified mesoporous silica nanoparticles can be controlled by intermolecular and photoinduced dimerization. It was reported that at the pore walls of mesoporous silica MCM-41 coumarin was covalently bonded and loading of pores was done by cholestane. In this process the dimerization of coumarin takes place, which results in the formation of cyclobutane products. The formation of cyclobutane leads to the closure of MCM-41 pores, thus preventing the release of loaded organic molecules. In this developed system the release was induced by exposing it to UV radiation of wavelength equal to 250 nm, which results in the cleavage of the cyclobutane ring present in the coumarin dimer.

Similarly, an electrochemical stimulus was used to induce the controlled drug release of MSN. Such a system was developed by Zink and co-workers in which a pseudorotaxane cap was developed, and this molecular cap undergoes conformational changes when exposed to a change in redox properties of a solution. In this investigation MSNs were loaded with a fluorescent $Ir(ppy)_3$ (ppy: 2,2'-phenylpyridyl) molecule [63]. By the addition of an external reducing agent the assembled pseudorotaxane is disintegrated, causing the controlled release of organic molecules. After this investigation reversibility was introduced in such a system by switching on and off nano valves using redox chemistry.

In addition to these physical and chemical controlled properties the rich chemistry of mesoporous silica material makes this system open to different manipulations. These manipulations include magnetic manipulations. In such manipulation the directional capabilities towards the target sites were introduced mesoporous silica. In one such investigation the MSN system was manipulated/capped with iron oxide nanoparticles showing superparamagnetic behaviour [64]. By investigating such a system, it was observed that in addition to the stimuli-responsive drug release properties, modified MSN structures were directed to the targeted sites by applying an external magnetic field. In order to investigate whether the $Fe_2O_3$-modified MSN could be easily directed to the target site by applying an external magnetic field, two cuvettes were filled with magnetic MSNs loaded with fluorescein molecules. Both systems were kept in the external magnetic field. From the investigation it was observed that both of the magnetic MSNs moved toward the cuvette walls due to attraction from the tips of external magnets, as shown in Figure 9.9a. One cuvette on the right was used as a control with no DTT loaded while another on the left loaded with DTT was employed for analysis of release experiments. From the release analysis after 72 hours, fluorescence was observed in the sample loaded with DTT while in the control cuvette no such observations were made, as in Figure 9.9b. Hence, it was concluded that the addition of a triggering disulphide reducing agent induces the release of a fluorescein from $Fe_2O_3$-capped MSN system. The release kinetics of fluorescein by disulphide reducing agent and antioxidant produced in the cells show similar effect on CdS-capped MSNs. In addition to the aforementioned modifications for controlled release of loaded drugs, certain other possibilities such as the development of double delivery systems and a mesoporous silica core shell are present to modify the properties of MSNs. Different core materials like magnetite and gelatine with silica as a shell material have been developed by researchers. Other more complex structures have been developed in which metals were incorporated into a layer made up of biodegradable material. For example, gelatine core doped with gold nanoparticles was developed to track the movement of particles using luminescence properties. This type of core shell strategy using silica as a shell was also employed to provide stability to the micellar structures by preventing them from breaking during the dilution process. Another important modification for efficient drug adsorption, release properties combination of surface functionalization along with core shell, has been carried out. For example, for improved biocompatibility surface functionalization with polyethylene glycol was carried out on the surface of a core shell structure of a cobalt ferrite/silica nanostructure [64, 65].

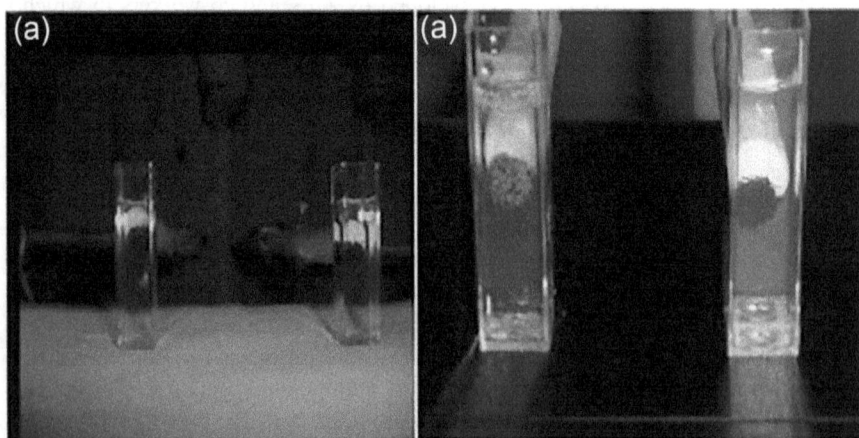

**FIGURE 9.9** Represents fluorescein loaded magnetic MSN with DTT added in left cuvette and pictures taken (a) before the addition of DTT and (b) 72 h after the addition of DTT. Reproduced with the permission from American Chemical Society [64].

## 9.4  CHALLENGES AND FUTURE PERSPECTIVES

All drug delivery systems should have the capability to continuously monitor the physiological changes taking place in patients and communicate telemetrically to the physician for timely intervention if required. The more important challenges faced by physicians for administration of a drug is patient compliance. It becomes difficult for doctors to administer drugs at regular intervals in patients who are mentally ill, aged persons and children, as they are reluctant to comply with the drug delivery routine. In another study on the human circadian cycle, it was observed that administration of a drug at a particular time of the day shows more effectiveness. A significant correlation is observed between effectiveness of the drug and the time at which it is administered. Circadian rhythm is the daily cyclic process that affects digestion, hormone secretion, heartbeat and other functions of the human body. It has been observed that chronotherapeutic designed for administration at circadian rhythm are considered more effective in disease control. For example, in the case of aspirin, peak level of the drug in the blood is observed 2–4 hours after administration. Therefore, to prevent heart attack, which usually takes place in the morning, it is more effective to take aspirin at bedtime that could be released after a few hours of administration. Therefore, to have a controlled release of drugs according to circadian rhythm, the use of stimulus-responsive drug delivery systems is considered to play an important role. In addition, the use of telemetry devices in combination with drug delivery systems could provide an important breakthrough for developing the personalized health care system in which the sensed data could be easily recorded and transferred to the physicians at any time and proper response could be taken by remotely monitoring the patients.

## 9.5  CONCLUSION

In this chapter we have highlighted the properties of mesoporous silica nanoparticles, which make them suitable candidates for targeted stimuli-responsive drug delivery systems. Discussion regarding the parameters controlling the drug adsorption capability and release properties was also included. In addition, the synthesis methods employed to synthesize morphologically different MSNs were discussed, as were possible strategies like functionalization to improve the drug adsorption capabilities. More importantly, discussed in detail were different modifications employed for efficient, stable and sustainable drug delivery followed by stimuli-responsive release. These systems are intelligent enough to respond to an external stimulus and possess an important property of detection of biomolecules for highly selective drug release. Although MSNs possess important characteristics for a suitable drug release system, still there are numerous challenges which need to be addressed like development of individualized health care systems using MSNs and the study of toxicity due to these MSNs under *in vivo* conditions.

## REFERENCES

[1] Slowing, I. I., Trewyn, B. G., Giri, S., & Lin, V. Y. (2007). Mesoporous silica nanoparticles for drug delivery and biosensing applications. *Advanced Functional Materials*, 17(8), 1225–1236.

[2] Skeel, R. T., & Khleif, S. N. (Eds.). (2011). *Handbook of cancer chemotherapy*. Lippincott Williams & Wilkins, Philadelphia, USA.

[3] Moghimi, S. M., Peer, D., & Langer, R. (2011). Reshaping the future of nanopharmaceuticals: Ad iudicium. *ACS Nano*, 5(11), 8454–8458.

[4] Meng, H., Xue, M., Xia, T., Ji, Z., Tarn, D. Y., Zink, J. I., & Nel, A. E. (2011). Use of size and a copolymer design feature to improve the biodistribution and the enhanced permeability and retention effect of doxorubicin-loaded mesoporous silica nanoparticles in a murine xenograft tumor model. *ACS Nano*, 5(5), 4131–4144.

[5] Fernando, I. R., Ferris, D. P., Frasconi, M., Malin, D., Strekalova, E., Yilmaz, M. D., & Stoddart, J. F. (2015). Esterase-and pH-responsive poly (β-amino ester)-capped mesoporous silica nanoparticles for drug delivery. *Nanoscale*, 7(16), 7178–7183.

[6] Ahmad, S., Tripathy, D. B., & Mishra, A. (2016). Sustainable nanomaterials. *Sustainable inorganic chemistry*, 205.

[7] Kroon, J., Metselaar, J. M., Storm, G., & van der Pluijm, G. (2014). Liposomal nanomedicines in the treatment of prostate cancer. *Cancer Treatment Reviews*, 40(4), 578–584.

[8] Ahmad, T., Farooq, U., & Phul, R. (2018). Fabrication and photocatalytic applications of perovskite materials with special emphasis on alkali-metal-based niobates and tantalates. *Industrial & Engineering Chemistry Research*, 57(1), 18–41.

[9] Farooq, U., Phul, R., Alshehri, S. M., Ahmed, J., & Ahmad, T. (2019). Electrocatalytic and enhanced photocatalytic applications of sodium niobate nanoparticles developed by citrate precursor route. *Scientific Reports*, 9(1), 1–17.

[10] Farooq, U., Ahmed, J., Alshehri, S. M., & Ahmad, T. (2019). High-surface-area sodium tantalate nanoparticles with enhanced photocatalytic and electrical properties prepared through polymeric citrate precursor route. *ACS Omega*, 4(21), 19408–19419.

[11] Puri, N., Gupta, A., & Mishra, A. (2021). Recent advances on nano-adsorbents and nano-membranes for the remediation of water. *Journal of Cleaner Production*, 322, 129051.

[12] Ahmad, T., Nazim, A., Farooq, U., Khan, H., Jain, S. K., Ubaidullah, M., & Ahmed, J. (2020). Biosynthesis, characterization and photo-catalytic degradation of methylene blue using silver nanoparticles. *Materials Today: Proceedings*, 29, 1039–1043.

[13] Farooq, U., Ahmed, J., Alshehri, S. M., Mao, Y., & Ahmad, T. (2022). Self-assembled interwoven nano hierarchitectures of $NaNbO_3$ and $NaNb_{1-x}Ta_xO_3$ (0.05≤ x≤ 0.20): Synthesis, structural characterization, photocatalytic applications, and dielectric properties. *ACS Omega*, 7(20), 16952–16967.

[14] Farooq, U., Chaudhary, P., Ingole, P. P., Kalam, A., & Ahmad, T. (2020). Development of cuboidal $KNbO_3@\alpha$-$Fe_2O_3$ hybrid nanostructures for improved photocatalytic and photoelectrocatalytic applications. *ACS omega*, 5(32), 20491–20505.

[15] Phul, R., Shrivastava, V., Farooq, U., Sardar, M., Kalam, A., Al-Sehemi, A. G., & Ahmad, T. (2019). One pot synthesis and surface modification of mesoporous iron oxide nanoparticles. *Nano-Structures & Nano-Objects*, 19, 100343.

[16] Stuart, M. A. C., Huck, W. T., Genzer, J., Müller, M., Ober, C., Stamm, M., & Minko, S. (2010). Emerging applications of stimuli-responsive polymer materials. *Nature Materials*, 9(2), 101–113.

[17] Muthu, M. S., Rajesh, C. V., Mishra, A., & Singh, S. (2009). Stimulus-responsive targeted nanomicelles for effective cancer therapy. Nanomedicine, 4(6), 657–667.

[18] Li, Y. Y., Dong, H. Q., Wang, K., Shi, D. L., Zhang, X. Z., & Zhuo, R. X. (2010). Stimulus-responsive polymeric nanoparticles for biomedical applications. *Sci. China Chem*, 53, 447–457.

[19] Descalzo, A. B., Martínez-Máñez, R., Sancenon, F., Hoffmann, K., & Rurack, K. (2006). The supramolecular chemistry of organic: Inorganic hybrid materials. *Angewandte Chemie International Edition*, 45(36), 5924–5948.

[20] Slowing, I. I., Trewyn, B. G., Giri, S., & Lin, V. Y. (2007). Mesoporous silica nanoparticles for drug delivery and biosensing applications. *Advanced Functional Materials*, 17(8), 1225–1236.

[21] Cotí, K. K., Belowich, M. E., Liong, M., Ambrogio, M. W., Lau, Y. A., Khatib, H. A., & Stoddart, J. F. (2009). Mechanised nanoparticles for drug delivery. *Nanoscale*, 1(1), 16–39.

[22] Yang, Q., Wang, S., Fan, P., Wang, L., Di, Y., Lin, K., & Xiao, F. S. (2005). pH-responsive carrier system based on carboxylic acid modified mesoporous silica and polyelectrolyte for drug delivery. *Chemistry of Materials*, 17(24), 5999–6003.

[23] Kim, H. J., Matsuda, H., Zhou, H., & Honma, I. (2006). Ultrasound-triggered smart drug release from a poly (dimethylsiloxane): Mesoporous silica composite. *Advanced Materials*, 18(23), 3083–3088.

[24] Natarajan, S. K., & Selvaraj, S. (2014). Mesoporous silica nanoparticles: Importance of surface modifications and its role in drug delivery. *RSC Advances*, 4(28), 14328–14334.

[25] Li, X., Shi, J., Zhu, Y., Shen, W., Li, H., Liang, J., & Gao, J. (2007). A template route to the preparation of mesoporous amorphous calcium silicate with high in vitro bone-forming bioactivity. *Journal of Biomedical Materials Research Part B: Applied Biomaterials*, 83(2), 431–439.

[26] Lin, Y. S., Tsai, C. P., Huang, H. Y., Kuo, C. T., Hung, Y., Huang, D. M., & Mou, C. Y. (2005). Well-ordered mesoporous silica nanoparticles as cell markers. *Chemistry of Materials*, 17(18), 4570–4573.

[27] Wu, S. H., Lin, Y. S., Hung, Y., Chou, Y. H., Hsu, Y. H., Chang, C., & Mou, C. Y. (2008). Multifunctional mesoporous silica nanoparticles for intracellular labeling and animal magnetic resonance imaging studies. *ChemBioChem*, 9(1), 53–57.

[28] Zhai, Q. Z., Wu, Y. Y., & Wang, X. H. (2013). Synthesis, characterization and sustaining controlled release effect of mesoporous SBA-15/ramipril composite drug. *Journal of Inclusion Phenomena and Macrocyclic Chemistry*, 77(1), 113–120.

[29] Lu, Y., Slomberg, D. L., Sun, B., & Schoenfisch, M. H. (2013). Shape-and nitric oxide flux-dependent bactericidal activity of nitric oxide-releasing silica nanorods. *Small*, 9(12), 2189–2198.

[30] Park, K. (Ed.). (1997). Controlled drug delivery: Challenges and strategies. *Acs Professional Reference Book*, USA.

[31] Reiniger, G., & Lehmann, G. (1998). Increasing nitroglycerin release from patches enables circumvention of early nitrate tolerance. *Cardiovascular Drugs and Therapy*, 12(2), 217–224.

[32] Rinaldi, C. A., & Gill, J. S. (2002). The implantable cardioverter defibrillator. *Hospital Medicine*, 63(11), 672–675.

[33] Wilson, G. S., & Hu, Y. (2000). Enzyme-based biosensors for in vivo measurements. *Chemical Reviews*, 100(7), 2693–2704.

[34] Zhang, S., Wright, G., & Yang, Y. (2000). Materials and techniques for electrochemical biosensor design and construction. *Biosensors and Bioelectronics*, 15(5–6), 273–282.

[35] Wisniewski, N., & Reichert, M. (2000). Methods for reducing biosensor membrane biofouling. *Colloids and Surfaces B: Biointerfaces*, 18(3–4), 197–219.

[36] Harwood, G. J., & Pouton, C. W. (1996). Amperometric enzyme biosensors for the analysis of drugs and metabolites. *Advanced Drug Delivery Reviews*, 18(2), 163–191.

[37] Robert, J. J. (2002). Continuous monitoring of blood glucose. *Hormone Research in Paediatrics*, 57(Suppl. 1), 81–84.

[38] Metzger, M., Leibowitz, G., Wainstein, J., Glaser, B., & Raz, I. (2002). Reproducibility of glucose measurements using the glucose sensor. *Diabetes Care*, 25(7), 1185–1191.

[39] Wright, J. C., Leonard, S. T., Stevenson, C. L., Beck, J. C., Chen, G., Jao, R. M., . . . & Skowronski, R. J. (2001). An in vivo/in vitro comparison with a leuprolide osmotic implant for the treatment of prostate cancer. *Journal of Controlled Release*, 75(1–2), 1–10.

[40] Gupta, S. K., Yih, B. M., Atkinson, L., & Longstreth, J. (1995). The effect of food, time of dosing, and body position on the pharmacokinetics and pharmacodynamics of verapamil and norverapamil. *The Journal of Clinical Pharmacology*, 35(11), 1083–1093.

[41] Miyata, T., Asami, N., & Uragami, T. (1999). A reversibly antigen-responsive hydrogel. *Nature*, 399(6738), 766–769.

[42] Deo, S. K., Moschou, E. A., Peteu, S. F., Bachas, L. G., Daunert, S., Eisenhardt, P. E., & Madou, M. J. (2003). Peer reviewed: Responsive drug delivery systems. *Analytical Chemistry*, 75(9), 206-A.

[43] Henning, A. K., Patel, S., Selser, M., & Cozad, B. A. (2003). Factors affecting silicon membrane burst strength. In *Reliability, Testing, and Characterization of MEMS/ MOEMS III*, 5343, 145–153.

[44] Yanagisawa, T., Shimizu, T., Kuroda, K., & Kato, C. (1990). The preparation of alkyltrimethylammonium: Kanemite complexes and their conversion to microporous materials. *Bulletin of the Chemical Society of Japan*, 63(4), 988–992.

[45] Kresge, A. C., Leonowicz, M. E., Roth, W. J., Vartuli, J. C., & Beck, J. S. (1992). Ordered mesoporous molecular sieves synthesized by a liquid-crystal template mechanism. *Nature*, 359(6397), 710–712.

[46] Vallet-Regi, M., Rámila, A., Del Real, R. P., & Pérez-Pariente, J. (2001). A new property of MCM-41: Drug delivery system. *Chemistry of Materials*, 13(2), 308–311.

[47] Fowler, C. E., Khushalani, D., Lebeau, B., & Mann, S. (2001). Nanoscale materials with mesostructured interiors. *Advanced Materials*, 13(9), 649–652.

[48] Nooney, R. I., Thirunavukkarasu, D., Chen, Y., Josephs, R., & Ostafin, A. E. (2002). Synthesis of nanoscale mesoporous silica spheres with controlled particle size. *Chemistry of Materials*, 14(11), 4721–4728.

[49] Brinker, C. J., & Scherer, G. W. (2013). *Sol-gel science: The physics and chemistry of sol-gel processing*. Academic press, New York, USA.

[50] Stöber, W., Fink, A., & Bohn, E. (1968). Controlled growth of monodisperse silica spheres in the micron size range. *Journal of Colloid and Interface Science*, 26(1), 62–69.

[51] Manzano, M., & Vallet-Regí, M. (2020). Mesoporous silica nanoparticles for drug delivery. *Advanced Functional Materials*, 30(2), 1902634.

[52] Zeng, W., Qian, X. F., Zhang, Y. B., Yin, J., & Zhu, Z. K. (2005). Organic modified mesoporous MCM-41 through solvothermal process as drug delivery system. *Materials Research Bulletin*, 40(5), 766–772.

[53] Wang, G., Otuonye, A. N., Blair, E. A., Denton, K., Tao, Z., & Asefa, T. (2009). Functionalized mesoporous materials for adsorption and release of different drug molecules: A comparative study. *Journal of Solid State Chemistry*, 182(7), 1649–1660.

[54] Vallet-Regi, M., Doadrio, J. C., Doadrio, A. L., Izquierdo-Barba, I., & Pérez-Pariente, J. (2004). Hexagonal ordered mesoporous material as a matrix for the controlled release of amoxicillin. *Solid State Ionics*, 172(1–4), 435–439.

[55] Jokerst, J. V., Lobovkina, T., Zare, R. N., & Gambhir, S. S. (2011). Nanoparticle PEGylation for imaging and therapy. *Nanomedicine*, 6(4), 715–728.

[56] Blumen, S. R., Cheng, K., Ramos-Nino, M. E., Taatjes, D. J., Weiss, D. J., Landry, C. C., & Mossman, B. T. (2007). Unique uptake of acid-prepared mesoporous spheres by lung epithelial and mesothelioma cells. *American Journal of Respiratory Cell and Molecular Biology*, 36(3), 333–342.

[57] Natarajan, S. K., & Selvaraj, S. (2014). Mesoporous silica nanoparticles: Importance of surface modifications and its role in drug delivery. *RSC Advances*, 4(28), 14328–14334.

[58] Lai, C. Y., Trewyn, B. G., Jeftinija, D. M., Jeftinija, K., Xu, S., Jeftinija, S., & Lin, V. S. Y. (2003). A mesoporous silica nanosphere-based carrier system with chemically removable CdS nanoparticle caps for stimuli-responsive controlled release of neurotransmitters and drug molecules. *Journal of the American Chemical Society*, 125(15), 4451–4459.

[59] Vallet-Regi, M., Rámila, A., Del Real, R. P., & Pérez-Pariente, J. (2001). A new property of MCM-41: Drug delivery system. *Chemistry of Materials*, 13(2), 308–311.

[60] Poostforooshan, J., Belbekhouche, S., Shaban, M., Alphonse, V., Habert, D., Bousserrhine, N., & Weber, A. P. (2020). Aerosol-assisted synthesis of tailor-made hollow mesoporous silica microspheres for controlled release of antibacterial and anticancer agents. *ACS Applied Materials & Interfaces*, 12(6), 6885–6898.

[61] Lai, C. Y., Trewyn, B. G., Jeftinija, D. M., Jeftinija, K., Xu, S., Jeftinija, S., & Lin, V. S. Y. (2003). A mesoporous silica nanosphere-based carrier system with chemically removable CdS nanoparticle caps for stimuli-responsive controlled release of neurotransmitters and drug molecules. *Journal of the American Chemical Society*, 125(15), 4451–4459.

[62] Mal, N. K., Fujiwara, M., & Tanaka, Y. (2003). Photocontrolled reversible release of guest molecules from coumarin-modified mesoporous silica. *Nature*, 421(6921), 350–353.

[63] Hernandez, R., Tseng, H. R., Wong, J. W., Stoddart, J. F., & Zink, J. I. (2004). An operational supramolecular nanovalve. *Journal of the American Chemical Society*, 126(11), 3370–3371.

[64] Zhao, W., Gu, J., Zhang, L., Chen, H., & Shi, J. (2005). Fabrication of uniform magnetic nanocomposite spheres with a magnetic core/mesoporous silica shell structure. *Journal of the American Chemical Society*, 127(25), 8916–8917.

[65] Giri, S., Trewyn, B. G., Stellmaker, M. P., & Lin, V. S. Y. (2005). Stimuli-responsive controlled-release delivery system based on mesoporous silica nanorods capped with magnetic nanoparticles. *Angewandte Chemie*, 117(32), 5166–5172.

# 10 Fullerenes for Anticancer Drug Delivery

*Manu Sharma and Rajat Goyal*

## 10.1 INTRODUCTION

Fullerenes were discovered in 1985 and, out of all the carbon allotropic forms, are one of the most studied for their biomedical and therapeutic applications [1–2]. The discovery of fullerenes opened new doors in the field of nanotechnology and drug delivery [3–5]. They are recognized as buckyballs, as they are crystalline and arranged as a soccer ball-like structure [6]. Fullerenes are molecular in nature, whereas other nanocarbons like nanotubes, graphene, and carbon nanohorns are not [7]. The cage-like structure of fullerenes is formed by hexagons and pentagons with alternative single (at the hexagon-pentagon junction) and double (at the hexagon-hexagon junction) bonds [8–10].The isolated pentagon rule (IPR) is followed by the fullerenes and thus avoids the cage strain and direct adjacent pentagons [11]. The increase in the cage size causes an upsurge in the number of possible structural isomers. Fullerenes are generally electronegative, and this affects their chemical reactive nature [12]. They are highly symmetric with the unique property of $C_{60}$ molecule and are composed of a hollow sphere, tube, or ellipsoid [13]. The $C_{60}$ fullerene comprises 12 pentagons and 20 hexagons with all the rings fused and the double bonds are conjugated, as described in Figure 10.1 [14].

Despite having unique conjugation, fullerenes still act as electron-deficient alkenes as that in electron-rich systems [15].They exhibit interesting and unique properties in the solution. $C_{60}$ fullerenes are less soluble in water; nevertheless, they may form aggregates in water solutions and stabilize the colloidal solutions, which consist of individual fullerenes and fullerene clusters [16–19]. The hydrophobicity of fullerenes is considered one of the major challenges in the therapeutic application of fullerenes. In the last decade, several different strategies have been applied to improve its solubility in an aqueous medium. The improved solubility of fullerene and its derivatives in biological media makes them extremely interesting candidates for biotherapeutic applications, including drug delivery carriers, gene transfer mediums, diagnostic agents, and biosensors [20, 21].

## 10.2 PHYSICAL AND CHEMICAL PROPERTIES

Fullerenes are exceedingly reactive in comparison to other forms of carbon, forming corresponding anions as a strong electron acceptor, capable of accepting up to six electrons. The presence of multiple bonds in the fullerene cage

DOI: 10.1201/9781003316398-10

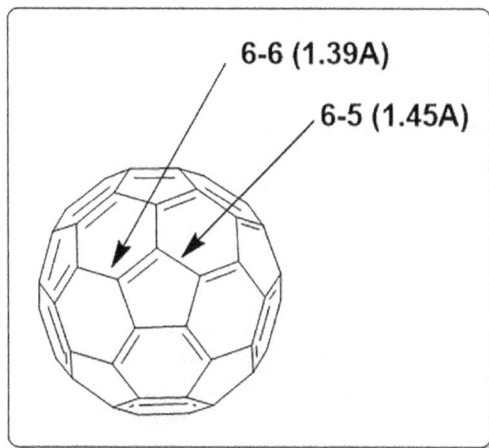

**FIGURE 10.1** $C_{60}$ molecule structure of fullerene.

makes it a strong electron-deficient molecule [22, 23]. Fullerenes react readily with nucleophiles, but like alkenes, they do not undergo electrophilic substitution reactions, but at the same time, fullerenes can also undergo cycloaddition reactions [24, 25]. Fullerenes have reported resisting extreme pressures and they can return back to their original shape, even after exposure to 3000 atmospheres of pressure, which makes them stiffer than that of diamond and steel. The non-aromatic character in the six-membered fragments of the fullerene cage is the result of a significant aberration of double bond geometry from the normal planar, as a result of pyramidalization. The pyramidization-induced strain forms the highly reactive fullerenes as compared to other carbon forms [7]. These special and interesting physicochemical characteristics of fullerenes make them a useful candidate for various applications in the field of medical, biomaterials, and engineering sciences [26–30]. The hollow cage-like structure makes them attractive molecules for the delivery of many drugs, especially anti-tumor agents, and their surface can be functionalized with several bioactive molecules for targeting drug therapy [31–33].

## 10.3 FULLERENES FOR SELECTIVE CANCER TARGETING

Cancer chemotherapy involves cytotoxic drugs, and the biggest problem associated with these drugs is that they have very little or no specificity, causing unwanted severe adverse effects and leading to systemic toxicity. Rapidly dividing tumor cells require several nutrients, vitamins, and fatty acids for their growth and survival. Therefore, a number of tumor-specific receptors are overexpressed by cancerous cells that can be utilized as targeted delivery of cytotoxic mediators into tumors. Fullerenes have been conjugated with several tumor-targeting moieties and used for the targeted delivery of cytotoxic agents.

## 10.4  FULLERENE IMMUNO/PEPTIDE-CONJUGATES

The first fullereneimmunoconjugate (1) derivative was synthesized in 2006 and developed by using a water-soluble multi-functionalized $C_{60}$ derivative. The Bingel–Hirsch reaction was used to conjugate antibody (Ab) ZME-018 via disulfide linkage to the $C_{60}$. The human melanoma cells were targeted by a single-dose drug delivery system using gp240 antigen, which is present in human melanoma cells (more than 80%) [34]. Berger et al. (2010) synthesized the metalo-fullerene-paclitaxel-antibody conjugates to be used in cancer imaging and treatment. Similarly, endohedral gado-fullerene derivatives (Gd@$C_{60}$) and paclitaxel were conjugated with antibody ZME-018 and were evaluated for both *in vivo* imaging and *in vitro* cellular labeling against melanoma cancer cells [35].The cell internalization of these immunoconjugates into A375m melanoma cells was also evaluated. The two different conjugates were prepared by using water-soluble $Gd^{3+}$ ion–filled metallofullerenes (Gd@$C_{60}$[OH]x), out of which one was conjugated with ZME-018 antibody and another was conjugated with murine IgG antibody (MuIgG). Internalization studies were conducted against antigen-positive (A375m) and antigen-negative (T24) cells, and it was observed that immuno-conjugates were either internalized inside cells or showed tight binding with the cell membrane [36].

Fullerene-peptide conjugates represented another class of interesting compounds that possess immunological properties. The conjugation of the peptide with fullerene provided diversity in structure and flexibility in charge due to long chains of amino acids and peptides that can help in structural recognition. These conjugates can be prepared using two synthetic routes, either by functionalizing $C_{60}$ using a peptide or by introducing fullerene into the amino acid peptide chain. The naturally occurring tetrapeptide is known as tufting (Thr-Lys-Pro-Arg), as this peptide possesses various pharmacological activities like stimulation and chemotaxis of phagocytes (neutrophils, macrophages, and monocytes) by binding on the specific receptors of phagocytes and having anticancer and antibiotic therapeutic efficacy. Xu et al. prepared two newer fullerene-conjugated immunomodulators from tuftsin. The conjugation of fullerene to tuftsin leads to an increase in serum stability, and $C_{60}$- tuftsin complex also had prolonged blood circulation time. Immunomodulating peptide tuftsin conjugates (2 and 3) were assayed *in vitro* on murine peritoneal macrophages against leucine-aminopeptidase to evaluate their stability and to confirm immune-stimulating properties. On comparing both conjugates with natural tuftsin, it was observed that conjugates helped to enhance chemotaxis and the phagocytosis process, and major histocompatibility complex class II (MHC II) also was formed in both cases. Conjugates were proved less toxic and resistant to enzymatic hydrolysis and, therefore, suggested that they can be used as vaccine adjuvants and immunomodulators [37]. The effects of these fullerene-peptide conjugates on the immune system were investigated by Bunz et al., who concluded that neither conjugate—N-ethyl polyamine derivatized fullerene or polyhydroxy derivative of fullerene—was able to enhance the reactivity of T-lymphocytes; however, the release of IL-6 (interleukin-6) and CD69 was significantly increased. Based on this study, it was concluded that it might be possible that T-cell reactivity was not affected by fullerene-peptide conjugates but innate immunity could be enhanced by them, as they were able to activate some killer cells [38]. In other studies, Tkatch et al. tried to compare the effects of graphene oxide and fullerenes on the immune system. The study revealed that graphene oxide caused stimulatory potential impairment in murine dendritic cells and also immuno-proteasome activity was decreased by them, which was required in antigen processing; but on the other side, fullerenes on internalization inside dendritic cells helped in the stimulation of major histocompatibility complex formation (MHC) which further restricted T-cell response. This immunological effect of fullerene was believed to be produced because of fullerene hydrophobicity, the presence of π-π bonds, and the induced fit mechanism [39]. Sofou et al. also explored these antigenic properties of fullerenes by synthesizing a proline-rich fullerene-peptide (4) by using a heptapeptide H-(Pro)2-Gly-Met-Arg-(Pro)2-OH. The reason for using this heptapeptide was that in patients with autoimmune diseases like mixed connective tissue disease (MCTD) and systemic lupus erythematosus (SLE), these peptides are the major targeting moiety for anti-Sm and anti-U1RNP auto-antibodies when checked in sera of patients, and it can also be easily recognized by anti-Ro/La positive sera. The synthesized fullerene-peptide having oxidized methionine, comprising all these properties, was able to produce specificities of anti-Sm/U1RNP and was also recognized by anti-Ro/La positive sera, which are negative to anti-Sm/U1RNP sera. The presence of fullerenes causes decreased disease specificity, which was confirmed by ELISA studies [40]. Research studies by Bianco et al. suggested that when they prepared a fullerene peptide (5) obtained by conjugation of nucleosomal histone (H3) protein with L-fulleropyrrolidino-glutamic acid residue, it also possessed antigenic properties. Findings of molecular modeling showed that a major histocompatibility complex was formed by the conjugate with retaining the same binding capacity possessed by the parent peptide molecule, but its T-cell responses needed to be investigated further [41]. All these studies by different researchers revealed that fullerene-peptide conjugates can play an important role in immunity induction and thus they should be explored to enhance their biological properties. The chemical structures of fullerene immunoconjugates are described in Figure 10.2.

## 10.5   FULLERENE-BIOTIN CONJUGATES

For the first time, a biotinylated lipofullerene was synthesized by Braun et al. in 2000. It can be employed as a transmembrane anchor, in which biotin worked as a biofunctional group of the lipofullerene and had a higher affinity for streptavidin protein. This conjugate (6) was mainly designed to cross biological membranes, especially lecithin membranes, and it could cross a monolayer and distribute lipofullerenes homogeneously in a lipid bilayer [42]. This study was further explored by Capaccio et al., who developed novel biotinylated fullerenes using streptavidin as a molecular adapter (7). Fullerenes were attached to streptavidin very easily because of the strong biotin-streptavidin interaction and also the presence of quatern biotin-binding spots on the streptavidin molecule. The activity of the conjugate was confirmed against biotinylated alkaline phosphatase, and it was also suggested that suspensions of $C_{60}$ alone were settled in minutes, but this complex increased the stability of

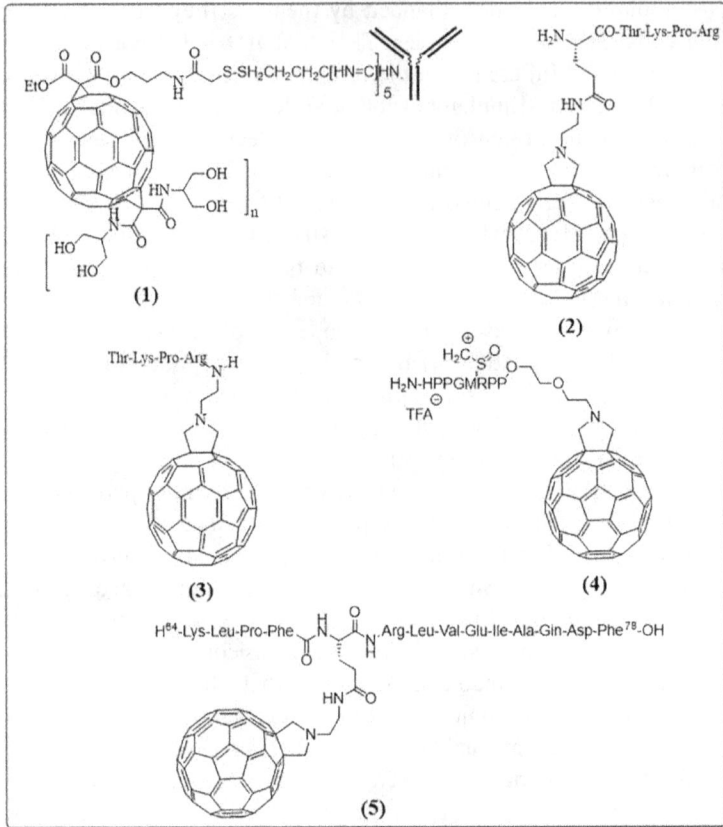

**FIGURE 10.2**    Structures of fullerene immunoconjugates.

suspension [43]. These studies helped to explore biotin-conjugated fullerenes for their drug delivery applications. In 2013, Yamada et al. developed a few photo-labeling agents by using methanol fullerenes. Among them, biotinylated methano-fullerene derivative (8) was formed via amidation. The binding ability of the biotin molecule in conjugate with avidin was checked by performing HABA (4'-hydroxyazobenzene-2-carboxylic acid) assays in DMSO [44]. The chemical structures of fullerene-biotin conjugates are described in Figure 10.3.

## 10.6    FULLERENE-HYALURONIC ACID CONJUGATES

Hyaluronic acid, (HA) is a naturally occurring linear polysaccharide, biocompatible and biodegradable in nature, and it was used as a hydrophilic backbone polymer for specifically targeting CD44 receptors which are overexpressed on tumor surfaces [45, 46]. HA plays a vital role in many biological activities and also performs some biomedical and therapeutic functions, such as tissue engineering and

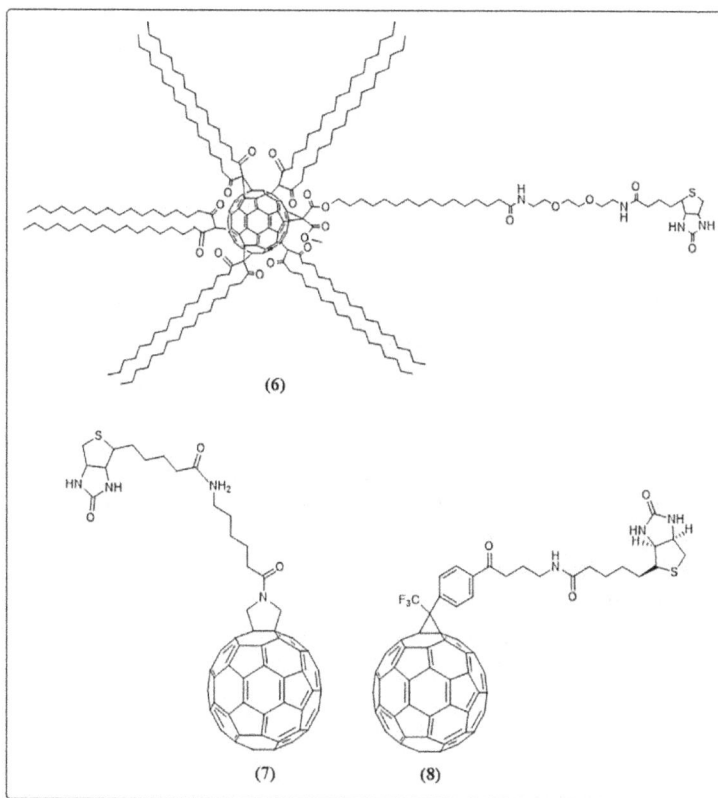

**FIGURE 10.3** Structures of fullerene-biotin conjugates.

drug delivery [47]. Utilizing these properties of HA, in 2013 Kwang et al. reported some hyaluronate fullerene conjugates, HA-g-C$_{60}$ (**9**), that can be used as a novel photosensitizer and also possess antitumor properties. The photoluminescent property of conjugate was used in tumor diagnosis and *in vivo* studies showed that the conjugate was able to suppress HCT-116 tumors more than free C$_{60}$ did because of the higher affinity of HA towards the overexpressed CD44 receptors. Also, the conjugation of fullerene with HA increased the stability of C$_{60}$ in PBS, and the conjugate can be hydrolyzed in the presence of hyaluronidase (HYAL) [48]. To enhance the dispersive property of C60 fullerenes in water and biological fluids, Zhang et al. developed another hyaluronate fullerene conjugate in which artesunate (AS) was used as a cytotoxic agent. A multifunctional drug delivery system — HA-C$_{60}$-Tf—was synthesized and had both photodynamic and selective tumor targeting capacity. The anticancer property of AS-loaded HA-C$_{60}$-Tf (**10**) conjugate was compared with free AS both (MCF-7 cells) *in vitro* and (murine model) *in vivo*, and results confirmed that the conjugate had enhanced antitumor effects because of augmented intracellular accretion of AS inside the tumor cells [49]. In 2015, Kim et al. focused on improving the anticancer activity by utilizing photodynamic tumor therapy. They employed Hoechst 33258 as a targeting moiety and hyaluronic acid to bind with the CD44 receptor, Ho-Hf (11).

This conjugate consisted of both hydrophilic (Hoechst 33258 and hyaluronic acid) and hydrophobic moieties (fullerene). The conjugate showed enhanced anticancer activity in HCT-116 cells and the tumor suppression model (in vivo) [50]. To accomplish the controlled drug delivery, and more powerful intelligent targeting, Wang et al. synthesized a dual-responsive nano-vehicle that can utilize magnetic and reductive glutathione (GSH) properties. For the first time fullerenes were modified with magnetic surfactant Cetyltrimethylammonium Fluoride (CTAF) using hydrophobic interaction and paramagnetic fullerene ($C_{60}$@CTAF) nano-vehicles. The DNA molecule was assembled with this complex, serving as an electrostatic framework that can load the anticancer medication Dox, and formed the conjugate $C_{60}$@CTAF/DNA [51]. The most commonly used carbon nanomaterials—GO, SWNT, and $C_{60}$—have been deliberated in the medicinal field by several researchers and their teams. To compare the phototherapy anticancer conjugation effect of HA with different carbon nanomaterials, Hou et al. carried out research, and it was observed that HA-$C_{60}$ had the highest reactive oxygen yield among them [52].The chemical structures of fullerene-hyaluronic acid conjugates are described in Figure 10.4.

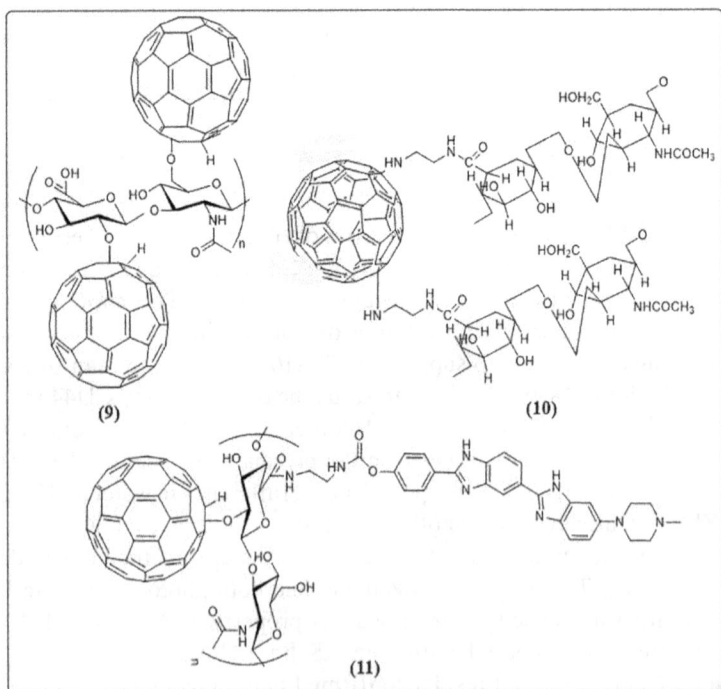

**FIGURE 10.4**  Structures of fullerene-hyaluronic acid conjugates.

## 10.7 FULLERENE-FOLIC ACID CONJUGATES

Folic acid (FA) is highly biocompatible, nontoxic, stable, biodegradable, and used as an essential artificial food additive to supplement folate loss during food processing. FA is essential because it is required in the body to synthesize purine and pyrimidine needed to build and repair nucleotides. FA is water-soluble and has higher bioavailability than naturally occurring folate. FA is reported to target folate receptors, and human cancers such as breast cancer, lung cancer, and prostate cancer are known to have the alpha folate receptor (FR-α) which is overexpressed on the cell surface [53]. The FR-α has a high affinity for folate, and it reduces the form of folic acid. Considering all of these facts, Hu et al. (2010) synthesized two amphiphilic derivatives using fullerenes, one of which was conjugated with folic acid (folic acid-$C_{60}$, FDD) and another was oxidized glutathione conjugate of fullerene (oxidized glutathione-$C_{60}$, OGED), in the form of spherical aggregates. They had the ability of cell membrane penetration and possessed cytotoxic activity against rat pheochrome cells (PC12) because of their antioxidant properties. Also, the size of FDD-aggregates **(12)** was much smaller than previously synthesized amino acid–fullerene conjugates, due to which their intracellular accumulation was increased and better apoptotic action was observed [54].

Using the strategy of prodrug therapy, Fan et al. developed a target-specific drug delivery system **(13)** using water-dispersible fullerenes in which doxorubicin was used as an anticancer agent and folic acid as a targeting ligand. The major advantage of this system was that it can be used as pH-responsive tumor-targeting therapy and as well as in photodynamic therapy (PDT). The synthesized nanoparticles had the size of 135 nm and were easily available to penetrate tumor cells via FR-positive receptor cells [55]. Guan et al. also worked on the same strategy and synthesized multifunctional upconversion–nanoparticles–polyoxyethylene bis (amine)–tri-methyl pyridyl porphyrin–fullerene nanocomposite (UCNP–PEG–FA/PC70), and these nanoparticles could be used in the treatment of malignant and non-malignant tumors. The presence of folic acid in the conjugate enhanced its accumulation properties, and *in vitro* as well as *in vivo* studies suggested that this nanocomposite could be used as a theranostic imaging agent which overcomes the limitations of photodynamic therapy like less penetration in the oxygen-deficient microenvironment [56]. Another targeting system was developed by Knežević et al. in which they used fullerene molecules and synthesized folic acid-functionalized mesoporous silica nanoparticles loaded with antitumor agent vinblastine. Their activity was determined against breast cancer MCF-7, healthy MRC-5, and cervical cancer HeLa cells to evaluate the efficacy for all these tumor cells because each one of them has overexpressed folate receptors on their surfaces [57]. Xu et al. worked to improve the selectivity and decreased the toxicity of doxorubicin (DOX). They also used fullerenols and designed a drug delivery system as folic acid -grafted FU-DOX conjugates **(14)**. Anti-tumor studies were conducted against human hepatocellular carcinoma (HCC) cell lines (BEL-7402 and HepG2) and the immortalized normal human hepatocytes (L02). Results showed that the presence of folic acid in the conjugate enhanced the cell internalization of the drug and also enhanced the toxicity of HCC cells [58].The chemical structures of fullerene-folic acid conjugates are described in Figure 10.5.

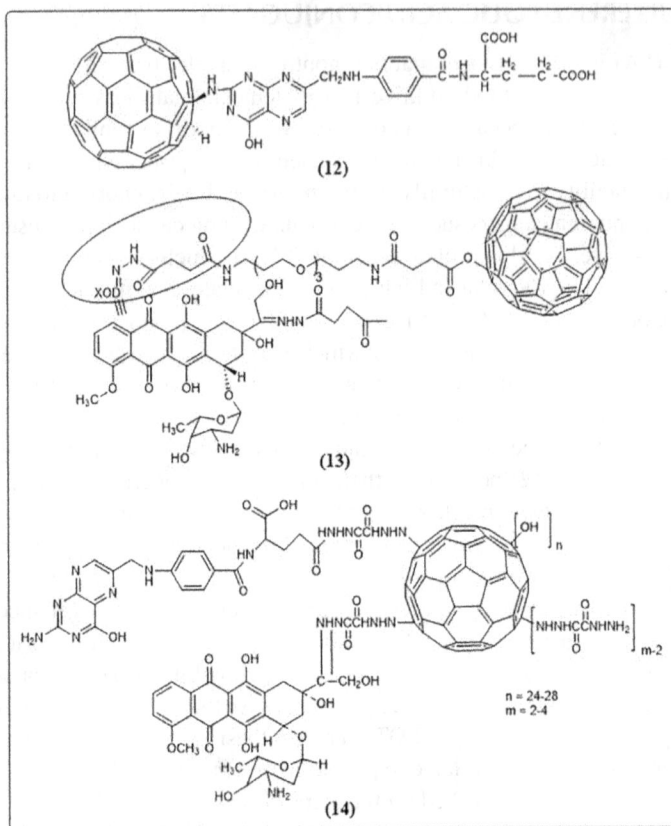

FIGURE 10.5    Structures of fullerene-folic acid conjugates.

## 10.8    FULLERENES AS ANTICANCER DRUG DELIVERY CARRIER

Fullerene has been conjugated with a number of anticancer drugs on their surface by using different types of linkers such as disulfide, esters, and amides, and these conjugates have been extensively evaluated for their anticancer properties. These conjugates have shown better efficacy and potency than the parent drug molecule. Several anticancer drugs like paclitaxel, camptothecin, doxorubicin, and cisplatin have been directly attached to fullerenes and are extensively studied for their antitumor efficacy.

Zakharian et al. (2005) developed the first fullerene-conjugated antitumor agent (15) by covalently attaching fullerene to paclitaxel via an ester bond. The conjugate was found to have retention of the therapeutic efficacy of paclitaxel with the release of the drug inside tumor cells due to hydrolysis of the ester bond. In the conjugate alone and as liposome aerosol formulation, it was observed that $IC_{50}$ values obtained were 1.6 times more than the preparation consisting of only paclitaxel [59]. In another study, Partha et al. mentioned the synthesis of self-assembled spherical shaped

nanostructures and their preliminary (*in vitro*) studies, which consisted of amphiphilic fullerenes, (AF-1) known as "buckysomes." In this study, it was observed that
at elevated temperature self-assembly occurred of the conjugate, which resulted in
the formation of buckysomes with diameter of 100–200 nm. In these spherical nano-
constructs, paclitaxel was embedded and these buckysomes were found to be more
potent against MCF-7 cells than other marketed products [60]. Mackeyev et al. synthesized paclitaxel–[$C_{60}$]-immunoconjugates (**16, 17**), and these compounds showed
similar tumor volume reduction as that of FDA-approved medication, Abraxane, but
without any loss of weight in animals, which indicated the better tolerance of this
new formulation [61].The chemical structures of fullerene-paclitaxel conjugates are
described in Figure 10.6.

Doxorubicin was conjugated with polyhydroxylated fullerenes (fullerenols) by
using a carbamate linker. The doxorubicin-fullerenol conjugate (**18**) was observed
to be comparatively stable in the phosphate buffer solutions, but on incubation with
tumor cell lysate, it released the active drug. The conjugate inhibited the proliferation
of cancer cell lines *in vitro* and induced apoptosis. At the same time, the conjugate

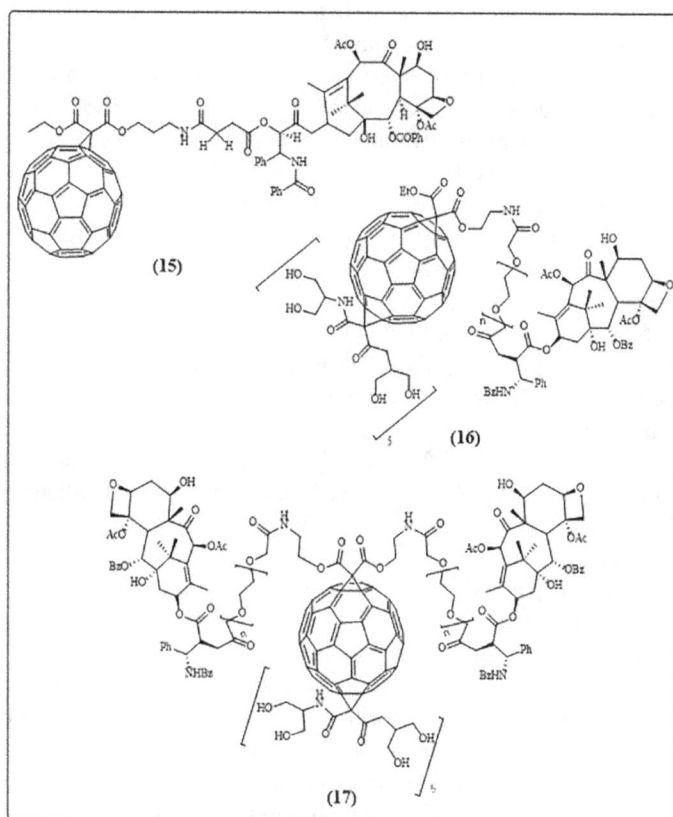

**FIGURE 10.6**  Structures of fullerene-paclitaxel conjugates.

was found to be less toxic in the murine tumor model (*in vivo*) than the free drug was [62]. The conjugated fullerene with doxorubicin was studied by Liu et al. for its uptake, and distribution in breast cancer cell lines. In fullerene-DOX conjugate (**19**), two DOX moieties were linked covalently via hydrophilic spacers and structurally built upon a methano $C_{60}$ derivative for aqueous compatibility [63].

The comparative study of the anticancer effect of DOX and pristine $C_{60}$ fullerene was carried out by Prylutska et al. In this study, malignant tumor growth was accessed in the presence of pristine $C_{60}$ fullerenes, doxorubicin, and also the cumulative effect was examined by using $C_{60}$ and doxorubicin together in mice. The results revealed that when $C_{60}$ combined with DOX was injected, it efficiently inhibited malignant tumor growth and also Lewis lung carcinoma metastases. This study further showed that drugs such as $C_{60}$ fullerenes, DOX, and $C_{60}$+DOX (physical mixture) at low therapeutic doses can produce antitumor activity. The physical mixture of $C_{60}$+DOX enhanced the antitumor activity by the inhibition of tumor growth as well as metastases in male mice having Lewis lung carcinoma induced by using the C57Bl/6J cell line. The results demonstrated tumor growth inhibition of 34%, and the metastasis inhibition index was found to be 79% while the animal life span increase was 24.4%. These findings may be an outcome of higher antioxidant properties possessed by $C_{60}$ fullerenes [64], neutralization of excessive reactive oxygen species (ROS), and possible antagonizing receptors like endothelial growth factor [65–68].

Zhang et al. investigated a drug delivery system using polyethyleneimine-derivatized fullerenes onto which doxorubicin was loaded ($C_{60}$-PEI-DOX). The conjugation was carried out by using pH-sensitive hydrazine linkage between DOX, and $C_{60}$-PEI, and the pH values determined the DOX release from $C_{60}$-PEI-DOX. The $C_{60}$-PEI-DOX (**20**) in comparison to free DOX produced better antitumor effects when examined in a murine model and did not show any side effects on normal organs with 2.4 times more DOX released in the cancer cells when compared to the other tissues. The conjugate also could provide photodynamic therapy along with enhanced tumor targeting and specific treatment effect [69]. In another study, Blazkova et al. conjugated fullerene with doxorubicin with pH-regulated release to lower the severe effects of the DOX and enhance the selectivity and drug cellular uptake by cancer cells. *Staphylococcus aureus* was used to observe the cellular toxic effects of the synthesized conjugate by administering different concentrations in chicken embryos and to measure the applicability for (*in vivo*) imaging. Results showed that the presence of fullerenes slightly diminishes the antimicrobial effect of DOX. The study determined that the fullerenes exhibit substantial protective effects against DOX toxicity [70].

Afanasieva et al. explained the genotoxic properties of DOX in human lymphocytes by performing *in vitro* studies using atomic force microscopy. The study was carried out on DOX alone, $C_{60}$ fullerene, and precipitates obtained from an aqueous solution of their mixture. *In vitro* human lymphocytes evaluation was done for genotoxic effects of DOX conjugated with $C_{60}$. The toxicity of DOX in normal cells was prevented by $C_{60}$ fullerene, and it was suggested that for biomedical application $C_{60}$+DOX complex might be useful [71]. The possible mechanism that can be followed by fullerene-conjugated doxorubicin ($C_{60}$+DOX) was explained by Prylutska et al. by primary screening, as this conjugate can decrease Lewis lung carcinoma metastasis and growth in mice. The possible mechanism of antitumor effect of

$C_{60}$+DOX was either immunomodulation or tumor cell death by enhancing the stress sensitivity [72]. The cellular distribution of drug conjugate was studied by Butowska et al. using the confocal microscopy of cells that were exposed to the conjugate (**21**) or drug alone. The confocal microscopy suggested that free DOX was able to penetrate the nuclei but conjugated instead of penetrating, accumulating on the membranes of the nucleus [73]. The chemical structures of fullerene-doxorubicin conjugates are described in Figure 10.7.

$C_{60}$-PEI-FA/Docetaxel (DTX) conjugate was synthesized and its antitumor efficacy was evaluated against PC3 cells by Shi et al. PEI-derivatized fullerene ($C_{60}$-PEI) (**22**) with folic acid was carried out by using an amide linker and further DTX was coupled to this complex using ultrasonication. In the *in vitro* studies on PC3 cells, results showed that conjugate in comparison to free DTX can pass cell membranes more efficiently and induce apoptosis, and thus they were able to produce a

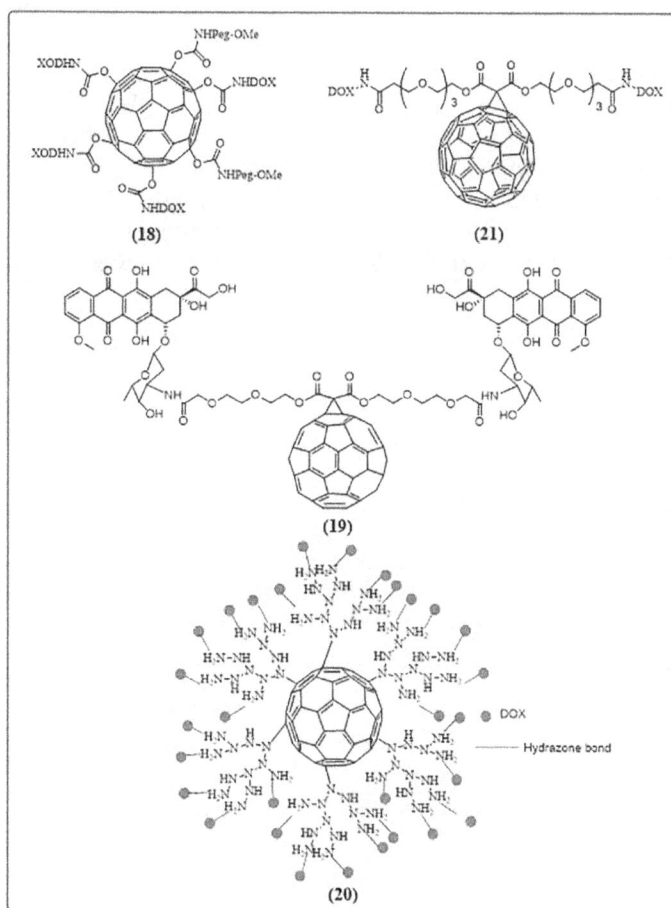

**FIGURE 10.7**   Structures of fullerene-doxorubicin conjugates.

better antitumor effect [74]. The effective therapeutic potential of di-adduct malonic acid-fullerene (DMA-$C_{60}$) as a chemo-phototherapeutic agent was synthesized and studied by Guo et al. There was an improvement in the stability of DTX resulting in prevention from hemolysis on the addition of DMA-$C_{60}$. It was observed that micelle formation helped to improve the antitumor activity by 2.25 times than DTX-MC alone, while it was 4.6 times higher than the marketed drug Duopafei® when given intravenously. Also, it was observed that drug distribution was improved towards tumor tissues and less to other organs without any weight loss [75]. In another study, Raza et al. developed docetaxel conjugate with carboxylated fullerenes without using the spacer. The conjugate showed better bioavailability of the drug (4.2 times than the parent drug with reduced body clearance). The conjugate showed controlled release and was observed to be compatible with erythrocytes. The cytotoxicity of conjugate against MDAMB231 and MCF-7 cell lines was increased several fold, thus indicating the reduction in dose along with high efficacy [76].The chemical structures of fullerene-docetaxel conjugates are described in Figure 10.8.

Chaudhuri et al. synthesized cisplatin-fullerenol (23) complex, and its clusters were found to be less than 100 nm. Fullerenol-cisplatin exhibited a greater anti-proliferative effect on the Lewis lung cancer cell line as compared to free cisplatin, and thus showed that the nanoparticles might be considered beneficial to the cisplatin chemotherapy [62]. In another study, Kuznietsova et al. carried out a comparison between toxicities produced on honeybee *Apis mellifera L.* by a fullerene-cisplatin conjugate, cisplatin, and $C_{60}$ fullerene alone. The water-soluble pristine $C_{60}$ fullerene did not show any toxic effect on a honeybee when given at different doses. Cisplatin alone produces severe toxicity when two doses (60 mg/kg) were given to honeybees, resulting in 56%

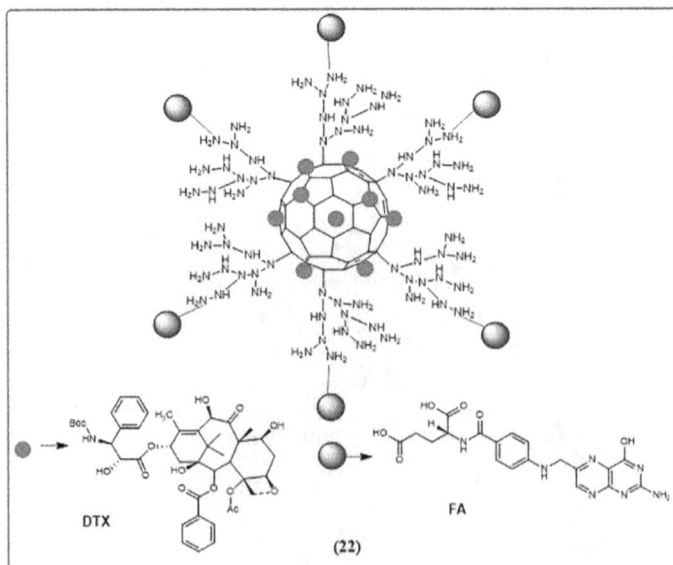

**FIGURE 10.8** Structures of fullerene-docetaxel conjugates.

honeybee death. On the other hand, the conjugate of fullerene and cisplatin was also proven more toxic than cisplatin alone, as it caused 50% of honeybee death even at a dose of 40 mg/kg [77]. Franskevych et al. also carried out *in vivo* studies by employing normal Wistar rats with thymocytes and transformed L1210 cells. The concentration of 1 μg/ml cisplatin resulted in the formation of ROS and lessened the potential of the mitochondrial membrane in both of them, but when the same cells received combined treatment with fullerene $C_{60}$ (7.2 μg/ml) with cisplatin (1 μg/ml), L1210 cells showed intensified ROS formation; on the other hand, it was lower in the case of thymocytes. It was concluded that these effects could be due to the combination of fullerene with cisplatin [78]. Prylutska et al. estimated the possible toxic effects that can be produced by $C_{60}$-cisplatin conjugate when examined towards Lewis lung cancerous cells (LLC) and compared with free drug. The cytotoxic effect was higher in the case of a free drug, as it can be confirmed through $IC_{50}$ values of free drug and conjugate. From their studies, Prylutska et al. assumed that enhanced antitumor effect and less cytotoxicity of the conjugate could be due to the presence of $C_{60}$, which promoted the entry of cisplatin into the cells and thus produced their cytotoxic effects [79]. The chemical structure of fullerene-cisplatin conjugate is described in Figure 10.9.

## 10.9  ANTIOXIDANT-FULLERENE CONJUGATES AS ANTICANCER AGENTS

Nanotechnology has sparked a prodigious interest in recent decades and been utilized in a variety of fields. The carbon nanomaterials, such as carbon nanotubes, graphene, and fullerenes, have piqued the curiosity of researchers because of their broad range of applications in several fields due to their good optical, thermal, and mechanical properties [80]. The functionalized fullerenes are one of the diversified classes of compounds being studied in the hastily developing field of nanomedicines. The interesting electrochemical, photochemical, and physical assets of fullerene compounds, combined with the enormous likelihood of chemical derivatization, biological acceptance, and symmetrical nature, are the primary advantages of this compound for usage in a variety of medicinal fields [81].

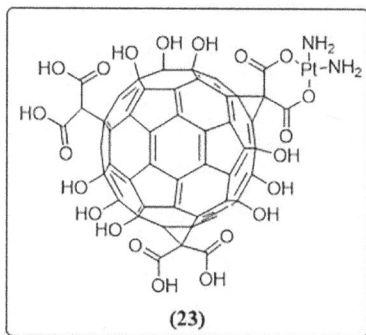

(23)

**FIGURE 10.9**  Fullerene-cisplatin conjugate.

Recently, fullerene has been known to be an auspicious compound for usage in anticancer therapy. The $C_{60}$ fullerene and their conjugates are biocompatible, display no toxic or adverse effects on normal tissues at lower concentration, and possess robust free radical-scavenging and antioxidant activities [82]. Fullerenes are amongst the sturdiest antioxidants and are categorized as "radical sponges" [83]. They are shown to generate singlet oxygen in higher quantum yields on exposure to light. The generation of singlet oxygen, combined with the direct transfer of e⁻ from the excited state of DNA bases and fullerenes, is used for the DNA targeting of rapidly growing cancerous cells; thus, these are intended for use as potent antitumor medications [84]. It is recognized that fullerenes and their conjugates have an exclusive capacity for ROS scavenging [85].

Prylutska et al. investigated the anticancer, and antioxidant enzyme activity of combined $C_{60}$ fullerene molecule and DOX (i.e., $C_{60}$+DOX) in the heart and liver of mice by using Lewis lung carcinoma cells as compared to the treatment with DOX alone. It was found that the combination of DOX, (a total dose of 2.5 mg/kg), and $C_{60}$ fullerene compound (a total dose of 25 mg/kg) led to the inhibition of cancer growth and metastasis, perpetuation of life, and increased apoptotic cancer cell numbering, and it was found to be more efficacious than the treatment with DOX alone. Also, $C_{60}$ fullerene compounds exhibited the protective effects the inhibition of glutathione peroxidase, and superoxide dismutase persuaded via DOX-dependent, oxidative insult [82].

Yang et al. investigated the antioxidant activity of several fullerene-based derivatives. The two compounds—fullerene α-aspartate acid (24) and fullerene α-glutamic acid (25)—showed the exclusive ability to confiscate the oxygen-free radicals. The $C_{60}$-substituted phenylalanine (26), along with its N-Boc-protected analog and lysine derivatives, exhibited similar antioxidant effects in a dose-dependent manner. A sulfur-containing dihydro pyrrolo $C_{60}$ derivative (27) is of particular interest because of its structure and its tendency to act as an antioxidant compound or probably as an amino acid mimic [83].

Yin et al. examined the protective properties of fullerene nanomaterials as the inhibitors of ROS (i.e., free radical scavengers and antioxidants). The ESR spin-trap method provided in vitro indicated that three fullerene conjugates such as fullerenol ($C_{60}(OH)_{22}$), gadolinium endohedral metallofullerene ($Gd@C_{82}(OH)_{22}$), and carboxy fullerene ($C_{60}(C(COOH)_2)_2$), can proficiently scavenge the free radicals and cause the inhibition of lipid peroxidation. Both ROS and nitrogen-centered free radicals were interrupted by these fullerene conjugates. These key findings deliver a molecular basis for several diseases, in which oxidative stress (OS) plays a vital role [85]. The chemical structures of antioxidant-fullerene conjugates are described in Figure 10.10.

## 10.10 PHARMACOKINETIC STUDIES OF FULLERENES AND CONJUGATES

In drug delivery, $C_{60}$ fullerenes (CF) are being investigated for the delivery of several antitumor medications to the target sites. These conjugated systems, with or without linkers, are superior in terms of controlled drug release, efficacy augmentation, and toxicity of cancer cells. Several scientists and investigators described the

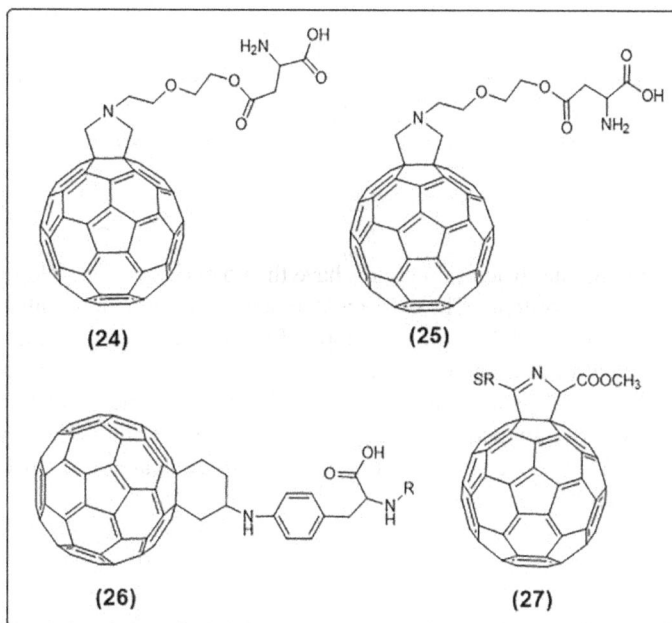

**FIGURE 10.10**  Structures of antioxidant-fullerene conjugates.

pharmacokinetic studies and safety profiles of anticancer medications loaded onto these novel compounds [76–85].

The fullerenes and their conjugates are usually functionalized to overwhelm the major challenge of aqueous insolubility. Due to the attachment of hydrophilic moieties, fullerenes and their conjugates become water-soluble and carry the drugs and genes for cellular delivery. Functionalized fullerene can cross the cellular membrane, and binds to the mitochondria [86]. The incorporation of a fullerene-based amino acid delivers a pathway for the intracellular delivery of peptides, resulting in the formation of a newer class of cell-penetrating peptides [87].

## 10.11  SAFETY PROFILE OF FULLERENE AND CONJUGATES

As fullerenes are one of the most common nanomaterials, there is a growing-concern about their safety profiles. Thus, several toxicity studies on fullerenes and their conjugates have been conducted around the world in recent years via the usage of aquatic organisms, laboratory animals, and *in vitro* and *in vivo* microorganismal systems [88].

It has been reported that the water-soluble fullerenes protect the human skin keratinocytes from ROS produced by peroxy lipids or ultrasound (UV) irradiation [89]. Also, fullerenes inhibit UVA-induced melanogenesis by decreasing the expression of tyrosinase in human melanocytes, and skin tissues [90]. Moreover, a water-soluble fullerene with oligo ethylene glycol bridges has been shown potential in the treatment of secondary progressive multiple sclerosis [91]. Several carboxy fullerenes have been shown to have cytoprotective and neuroprotective properties [92]. Fullerenol $(C_{60}OH_{24})$ prevents oxidative damage, and mitochondrial dysfunction in a cellular model of Parkinson's disease [93]. A water-soluble fullerene prevents the articular cartilage degeneration in osteoarthritis by decreasing the chondrocyte catabolic activity, and inhibiting the degeneration of cartilage during disease progression [94].

## 10.12  CONCLUSION

Fullerenes with unique structural features have the potential to be developed as anti-cancer drug delivery systems. The efficacy of anticancer drugs can be enhanced with the pharmacokinetic modulations. The scope of the surface modifications in fuller-enes make them attractive nanomaterial for targeted drug delivery into cancer cells by using various bioactive molecules. However, the safety profile of fullerenes needs to be more rigorously studied to make them drug-like molecules. The applications of fullerenes in drug delivery, diagnostics, and biomaterial cannot be ruled out, and extensive research is required to bring them from lab to clinics.

## REFERENCES

[1] Nakanishi, W., Minami, K., Shrestha, L. K., Ji, Q., Hill, J. P., & Ariga, K. (2014). Bioactive nanocarbon assemblies: Nanoarchitectonics and applications. *Nano Today*, *9*(3), 378–394.
[2] Semenov, K. N., Charykov, N. A., Postnov, V. N., Sharoyko, V. V., Vorotyntsev, I. V., Galagudza, M. M., & Murin, I. V. (2016). Fullerenols: Physicochemical properties and applications. *Progress in Solid State Chemistry*, *44*(2), 59–74.
[3] Kroto, H. W., Heath, J. R., O'Brien, S. C., Curl, R. F., & Smalley, R. E. (1985). C60: Buckminsterfullerene. *Nature*, *318*(6042), 162–163.
[4] Hirsch, A., Chen, Z., & Jiao, H. (2000). Spherical aromaticity in Ih symmetrical fullerenes: The 2 (N+ 1) 2 rule. *Angewandte Chemie International Edition*, *39*(21), 3915–3917.
[5] Taylor, R., & Walton, D. R. (1993). The chemistry of fullerenes. *Nature*, *363*(6431), 685–693.
[6] Troshin, P. A. E., & Lyubovskaya, R. N. (2008). Organic chemistry of fullerenes: The major reactions, types of fullerene derivatives and prospects for practical use. *Russian Chemical Reviews*, *77*(4), 323.
[7] David, W. I., Ibberson, R. M., Matthewman, J. C., Prassides, K., Dennis, T. J. S., Hare, J. P., . . . & Walton, D. R. (1991). Crystal structure and bonding of ordered C60. *Nature*, *353*(6340), 147–149.
[8] Liu, S., Lu, Y. J., Kappes, M. M., & Ibers, J. A. (1991). The structure of the C60 molecule: X-ray crystal structure determination of a twin at 110 K. *Science*, *254*(5030), 408–410.
[9] Bickelhaupt, F., & de Wolf, W. H. (1998). Unusual reactivity of highly strained cyclo-phanes. *Journal of Physical Organic Chemistry*, *11*(5), 362–376.
[10] Montellano, A., Da Ros, T., Bianco, A., & Prato, M. (2011). Fullerene C 60 as a multi-functional system for drug and gene delivery. *Nanoscale*, *3*(10), 4035–4041.
[11] Georgakilas, V., Pellarini, F., Prato, M., Guldi, D. M., Melle-Franco, M., & Zerbetto, F. (2002). Supramolecular self-assembled fullerene nanostructures. *Proceedings of the National Academy of Sciences*, *99*(8), 5075–5080.
[12] Murry, R. L., & Scuseria, G. E. (1994). Theoretical evidence for a C60 "window" mecha-nism. *Science*, *263*(5148), 791–793.

[13] Beckhaus, H. D., Rüchardt, C., Kao, M., Diederich, F., & Foote, C. S. (1992). The stability of buckminsterfullerene (C60): Experimental determination of the heat of formation. *Angewandte Chemie International Edition in English, 31*(1), 63–64.

[14] Ruoff, R. S., Tse, D. S., Malhotra, R., & Lorents, D. C. (1993). Solubility of fullerene (C60) in a variety of solvents. *The Journal of Physical Chemistry, 97*(13), 3379–3383.

[15] Johnson, R. D., Meijer, G., Salem, J. R., & Bethune, D. S. (1991). 2D nuclear magnetic resonance study of the structure of the fullerene C70. *Journal of the American Chemical Society, 113*(9), 3619–3621.

[16] Ruoff, R. S., & Ruoff, A. L. (1991). The bulk modulus of C60 molecules and crystals: A molecular mechanics approach. *Applied Physics Letters, 59*(13), 1553–1555.

[17] Xie, Q., Perez-Cordero, E., & Echegoyen, L. (1992). Electrochemical detection of C606-and C706-: Enhanced stability of fullerides in solution. *Journal of the American Chemical Society, 114*(10), 3978–3980.

[18] Fowler, P. W., & Ceulemans, A. (1995). Electron deficiency of the fullerenes. *The Journal of Physical Chemistry, 99*(2), 508–510.

[19] Csencsitsi, R., Gruen, D. M., Krauss, A. R., & Zuiker, C. (1995). TEM study of diamond films grown from fullerene precursors. *MRS Online Proceedings Library (OPL), 403*.

[20] Rather, J. A., & De Wael, K. (2013). Fullerene-C60 sensor for ultra-high sensitive detection of bisphenol-A and its treatment by green technology. *Sensors and Actuators B: Chemical, 176*, 110–117.

[21] Torres, V. M., Posa, M., Srdjenovic, B., & Simplício, A. L. (2011). Solubilization of fullerene C60 in micellar solutions of different solubilizers. *Colloids and Surfaces B: Biointerfaces, 82*(1), 46–53.

[22] Scrivens, W. A., Tour, J. M., Creek, K. E., & Pirisi, L. (1994). Synthesis of 14C-labeled C60, its suspension in water, and its uptake by human keratinocytes. *Journal of the American Chemical Society, 116*(10), 4517–4518.

[23] Raza, K., Kumar, D., Kiran, C., Kumar, M., Guru, S. K., Kumar, P., . . . & Katare, O. P. (2016). Conjugation of docetaxel with multiwalled carbon nanotubes and codelivery with piperine: Implications on pharmacokinetic profile and anticancer activity. *Molecular Pharmaceutics, 13*(7), 2423–2432.

[24] Ying, L., Fang, J., Yang, Q., Wei, L., Ying, Q., Chixia, T., . . . & Zhifang, C. (2009). Immunostimulatory properties and enhanced TNF-{alpha} mediated cellular immunity for tumor therapy by C {sub 60}(OH){sub 20} nanoparticles. *Nanotechnology (Print), 20*.

[25] Jiao, F., Liu, Y., Qu, Y., Li, W., Zhou, G., Ge, C., . . . & Chen, C. (2010). Studies on antitumor and antimetastatic activities of fullerenol in a mouse breast cancer model. *Carbon, 48*(8), 2231–2243.

[26] Tang, J., Chen, Z., Sun, B., Dong, J., Liu, J., Zhou, H., . . . & Liu, Y. (2016). Polyhydroxylated fullerenols regulate macrophage for cancer adoptive immunotherapy and greatly inhibit the tumor metastasis. *Nanomedicine: Nanotechnology, Biology and Medicine, 12*(4), 945–954.

[27] Yang, D., Zhao, Y., Guo, H., Li, Y., Tewary, P., Xing, G., . . . & Zhang, N. (2010). [Gd@ C82 (OH) 22] n nanoparticles induce dendritic cell maturation and activate Th1 immune responses. *ACS Nano, 4*(2), 1178–1186.

[28] Kang, S. G., Zhou, G., Yang, P., Liu, Y., Sun, B., Huynh, T., . . . & Zhou, R. (2012). Molecular mechanism of pancreatic tumor metastasis inhibition by Gd@ C82 (OH) 22 and its implication for de novo design of nanomedicine. *Proceedings of the National Academy of Sciences, 109*(38), 15431–15436.

[29] Kuroki, M., Voest, E. E., Amano, S., Beerepoot, L. V., Takashima, S., Tolentino, M., . . . & Adamis, A. P. (1996). Reactive oxygen intermediates increase vascular endothelial growth factor expression in vitro and in vivo. *The Journal of Clinical Investigation, 98*(7), 1667–1675.

[30] Rofstad, E. K., & Halsør, E. F. (2000). Vascular endothelial growth factor, interleukin 8, platelet-derived endothelial cell growth factor, and basic fibroblast growth factor promote angiogenesis and metastasis in human melanoma xenografts. *Cancer Research*, *60*(17), 4932–4938.

[31] Wang, J., Chen, C., Li, B., Yu, H., Zhao, Y., Sun, J., . . . & Wan, L. (2006). Antioxidative function and biodistribution of [Gd@ C82 (OH) 22] n nanoparticles in tumor-bearing mice. *Biochemical Pharmacology*, *71*(6), 872–881.

[32] Meng, J., Xing, J., Ma, X., Cao, W., Lu, J., Wang, Y., . . . & Zhao, Y. (2013). Metallofullerol nanoparticles with low toxicity inhibit tumor growth by induction of G0/G1 arrest. *Nanomedicine*, *8*(2), 203–213.

[33] Ye, S., Chen, M., Jiang, Y., Chen, M., Zhou, T., Wang, Y., . . . & Ren, L. (2014). Polyhydroxylated fullerene attenuates oxidative stress-induced apoptosis via a fortifying Nrf2-regulated cellular antioxidant defence system. *International Journal of Nanomedicine*, *9*, 2073.

[34] Ashcroft, J. M., Tsyboulski, D. A., Hartman, K. B., Zakharian, T. Y., Marks, J. W., Weisman, R. B., . . . & Wilson, L. J. (2006). Fullerene (C 60) immunoconjugates: Interaction of water-soluble C 60 derivatives with the murine anti-gp240 melanoma antibody. *Chemical Communications*, (28), 3004–3006.

[35] Berger, S., Wilson, L., Bolskar, R. D., Collier, J., Rosenblum, M., & Marks, W. (2010, February). Fullerene immunoconjugates for cancer imaging and treatment. *ECS Meeting Abstracts*, (33), 1539. IOP Publishing.

[36] Berger, C. S., Marks, J. W., Bolskar, R. D., Rosenblum, M. G., & Wilson, L. J. (2011). Cell internalization studies of gadofullerene-(ZME-018) immunoconjugates into A375m melanoma cells. *Translational Oncology*, *4*(6), 350-IN2.

[37] Xu, Y., Zhu, J., Xiang, K., Li, Y., Sun, R., Ma, J., . . . & Liu, Y. (2011). Synthesis and immuno-modulatory activity of [60] fullerene: Tuftsin conjugates. *Biomaterials*, *32*(36), 9940–9949.

[38] Bunz, H., Plankenhorn, S., & Klein, R. (2012). Effect of buckminsterfullerenes on cells of the innate and adaptive immune system: An in vitro study with human peripheral blood mononuclear cells. *International Journal of Nanomedicine*, *7*, 4571.

[39] Tkach, A. V., Yanamala, N., Stanley, S., Shurin, M. R., Shurin, G. V., Kisin, E. R., . . . & Shvedova, A. A. (2013). Graphene oxide, but not fullerenes, targets immunoproteasomes and suppresses antigen presentation by dendritic cells. *Small*, *9*(9–10), 1686–1690.

[40] Sofou, P., Elemes, Y., Panou-Pomonis, E., Stavrakoudis, A., Tsikaris, V., Sakarellos, C., . . . & Toniolo, C. (2004). Synthesis of a proline-rich [60] fullerene peptide with potential biological activity. *Tetrahedron*, *60*(12), 2823–2828.

[41] Ahmad, S., Tripathy, D. B., & Mishra, A. (2016). Sustainable nanomaterials. In *Sustainable Inorganic Chemistry* (p. 205).

[42] Braun, M., Camps, X., Vostrowsky, O., Hirsch, A., Endreß, E., Bayerl, T. M., . . . & Gauglitz, G. (2000). Synthesis of a biotinated lipofullerene as a new type of transmembrane anchor. *European Journal of Organic Chemistry*, *2000*(7), 1173–1181.

[43] Capaccio, M., Gavalas, V. G., Meier, M. S., Anthony, J. E., & Bachas, L. G. (2005). Coupling biomolecules to fullerenes through a molecular adapter. *Bioconjugate Chemistry*, *16*(2), 241–244.

[44] Yamada, M., Harada, K., Maeda, Y., & Hasegawa, T. (2013). A versatile approach to functionalisation of [60] fullerene using 3-trifluoromethyl-3-phenyldiazirine derivatives as photolabelling reagents. *New Journal of Chemistry*, *37*(11), 3762–3769.

[45] Kogan, G., Šoltés, L., Stern, R., & Gemeiner, P. (2007). Hyaluronic acid: A natural biopolymer with a broad range of biomedical and industrial applications. *Biotechnology Letters*, *29*(1), 17–25.

[46] Zhong, Y., Zhang, J., Cheng, R., Deng, C., Meng, F., Xie, F., & Zhong, Z. (2015). Reversibly crosslinked hyaluronic acid nanoparticles for active targeting and intelligent delivery of doxorubicin to drug resistant CD44+ human breast tumor xenografts. *Journal of Controlled Release*, *205*, 144–154.

[47] Ito, T., Iida-Tanaka, N., Niidome, T., Kawano, T., Kubo, K., Yoshikawa, K., . . . & Koyama, Y. (2006). Hyaluronic acid and its derivative as a multi-functional gene expression enhancer: Protection from non-specific interactions, adhesion to targeted cells, and transcriptional activation. *Journal of Controlled Release*, *112*(3), 382–388.

[48] Kwag, D. S., Park, K., Oh, K. T., & Lee, E. S. (2013). Hyaluronated fullerenes with photoluminescent and antitumoral activity. *Chemical Communications*, *49*(3), 282–284.

[49] Zhang, H., Hou, L., Jiao, X., Ji, Y., Zhu, X., & Zhang, Z. (2015). Transferrin-mediated fullerenes nanoparticles as $Fe^{2+}$-dependent drug vehicles for synergistic anti-tumor efficacy. *Biomaterials*, *37*, 353–366.

[50] Kim, S., Park, J., Youn, Y. S., Oh, K. T., Bae, J. H., & Lee, E. S. (2015). Hoechst 33258: Conjugated hyaluronated fullerene for efficient photodynamic tumor therapy and necrotic tumor targeting. *Journal of Bioactive and Compatible Polymers*, *30*(3), 275–288.

[51] Wang, L., Wang, Y., Hao, J., & Dong, S. (2017). Magnetic fullerene-DNA/hyaluronic acid nanovehicles with magnetism/reduction dual-responsive triggered release. *Biomacromolecules*, *18*(3), 1029–1038.

[52] Hou, L., Yuan, Y., Ren, J., Zhang, Y., Wang, Y., Shan, X., . . . & Zhang, Z. (2017). In vitro and in vivo comparative study of the phototherapy anticancer activity of hyaluronic acid-modified single-walled carbon nanotubes, graphene oxide, and fullerene. *Journal of Nanoparticle Research*, *19*(8), 1–18.

[53] Boss, S. D., & Ametamey, S. M. (2020). Development of folate receptor-targeted PET radiopharmaceuticals for tumor imaging—a bench-to-bedside journey. *Cancers*, *12*(6), 1508.

[54] Hu, Z., Liu, S., Wei, Y., Tong, E., Cao, F., & Guan, W. (2007). Synthesis of glutathione C60 derivative and its protective effect on hydrogen peroxide-induced apoptosis in rat pheochromocytoma cells. *Neuroscience Letters*, *429*(2–3), 81–86.

[55] Fan, J., Fang, G., Zeng, F., Wang, X., & Wu, S. (2013). Water-dispersible fullerene aggregates as a targeted anticancer prodrug with both chemo-and photodynamic therapeutic actions. *Small*, *9*(4), 613–621.

[56] Guan, M., Dong, H., Ge, J., Chen, D., Sun, L., Li, S., . . . & Shu, C. (2015). Multifunctional upconversion-nanoparticles-trismethylpyridylporphyrin-fullerene nanocomposite: A near-infrared light-triggered theranostic platform for imaging-guided photodynamic therapy. *NPG Asia Materials*, *7*(7), e205–e205.

[57] Knežević, N. Ž., Mrđanović, J., Borišev, I., Milenković, S., Janaćković, Đ., Cunin, F., & Djordjevic, A. (2016). Hydroxylated fullerene-capped, vinblastine-loaded folic acid-functionalized mesoporous silica nanoparticles for targeted anticancer therapy. *RSC Advances*, *6*(9), 7061–7065.

[58] Xu, B., Yuan, L., Hu, Y., Xu, Z., Qin, J. J., & Cheng, X. D. (2021). Synthesis, characterization, cellular uptake, and in vitro anticancer activity of fullerenol-doxorubicin conjugates. *Frontiers in Pharmacology*, *11*, 1685.

[59] Zakharian, T. Y., Seryshev, A., Sitharaman, B., Gilbert, B. E., Knight, V., & Wilson, L. J. (2005). A fullerene-paclitaxel chemotherapeutic: Synthesis, characterization, and study of biological activity in tissue culture. *Journal of the American Chemical Society*, *127*(36), 12508–12509.

[60] Partha, R., Mitchell, L. R., Lyon, J. L., Joshi, P. P., & Conyers, J. L. (2008). Buckysomes: Fullerene-based nanocarriers for hydrophobic molecule delivery. *Acs Nano*, *2*(9), 1950–1958.

[61] Mackeyev, Y., Raoof, M., Cisneros, B., Koshkina, N. S. B. C., Berger, C. S., Wilson, L. J., & Curley, S. A. (2014). Toward paclitaxel-[60] Fullerene immunoconjugates as a targeted prodrug against cancer. *Наносистемы: физика, химия, математика*, *5*(1), 67–75.

[62] Chaudhuri, P., Paraskar, A., & Soni, S. (2009). Fullerenol-cytotoxic conjugates for cancer chemotherapy. *ACS Nano*, *3*, 2505–2514.

[63] Liu, J. H., Cao, L., Luo, P. G., Yang, S. T., Lu, F., Wang, H., . . . & Sun, Y. P. (2010). Fullerene-conjugated doxorubicin in cells. *ACS Applied Materials & Interfaces*, 2(5), 1384–1389.

[64] Prylutska, S. V., Grynyuk, I. I., Matyshevska, O. P., Prylutskyy, Y. I., Ritter, U., & Scharff, P. (2008). Anti-oxidant properties of C60 fullerenes in vitro. *Fullerenes, Nanotubes and Carbon Nanostructures*, 16(5–6), 698–705.

[65] Murugesan, S., Mousa, S. A., O'Connor, L. J., Lincoln II, D. W., & Linhardt, R. J. (2007). Carbon inhibits vascular endothelial growth factor-and fibroblast growth factor-promoted angiogenesis. *FEBS Letters*, 581(6), 1157–1160.

[66] Meng, H., Xing, G., Sun, B., Zhao, F., Lei, H., Li, W., . . . & Zhao, Y. (2010). Potent angiogenesis inhibition by the particulate form of fullerene derivatives. *ACS Nano*, 4(5), 2773–2783.

[67] Injac, R., Radic, N., Govedarica, B., Perse, M., Cerar, A., Djordjevic, A., & Strukelj, B. (2009). Acute doxorubicin pulmotoxicity in rats with malignant neoplasm is effectively treated with fullerenol C60 (OH) 24 through inhibition of oxidative stress. *Pharmacological Reports*, 61(2), 335–342.

[68] Prylutska, S. V., Prylutskyy, Y. I., Ritter, U., & Scharff, P. (2011). Comparative study of antitumor effect of pristine C60 fullerenes and doxorubicin. *Biotechnology*, 4(6), 82–87.

[69] Shi, J., Liu, Y., Wang, L., Gao, J., Zhang, J., Yu, X., . . . & Zhang, Z. (2014). A tumoral acidic pH-responsive drug delivery system based on a novel photosensitizer (fullerene) for in vitro and in vivo chemo-photodynamic therapy. *Acta Biomaterialia*, 10(3), 1280–1291.

[70] Blazkova, I., Viet Nguyen, H., Kominkova, M., Konecna, R., Chudobova, D., Krejcova, L., . . . & Kizek, R. (2014). Fullerene as a transporter for doxorubicin investigated by analytical methods and in vivo imaging. *Electrophoresis*, 35(7), 1040–1049.

[71] Afanasieva, K. S., Prylutska, S. V., Lozovik, A. V., Bogutska, K. I., Sivolob, A. V., Prylutskyy, Y. I., . . . & Scharff, P. (2015). C60 fullerene prevents genotoxic effects of doxorubicin in human lymphocytes in vitro. *The Ukrainian Biochemical Journal*, 87(1), 91–98.

[72] Prylutska, S. V., Skivka, L. M., Didenko, G. V., Prylutskyy, Y. I., Evstigneev, M. P., Potebnya, G. P., . . . & Scharff, P. (2015). Complex of C60 fullerene with doxorubicin as a promising agent in antitumor therapy. *Nanoscale Research Letters*, 10(1), 1–7.

[73] Butowska, K., Kozak, W., Zdrowowicz, M., Makurat, S., Rychłowski, M., Hać, A., . . . & Rak, J. (2019). Cytotoxicity of doxorubicin conjugated with C60 fullerene: Structural and in vitro studies. *Structural Chemistry*, 30(6), 2327–2338.

[74] Shi, J., Zhang, H., Wang, L., Li, L., Wang, H., Wang, Z., . . . & Zhang, Z. (2013). PEI-derivatized fullerene drug delivery using folate as a homing device targeting to tumor. *Biomaterials*, 34(1), 251–261.

[75] Guo, X., Ding, R., Zhang, Y., Ye, L., Liu, X., Chen, C., . . . & Zhang, Y. (2014). Dual role of photosensitizer and carrier material of fullerene in micelles for chemo: Photodynamic therapy of cancer. *Journal of Pharmaceutical Sciences*, 103(10), 3225–3234.

[76] Raza, K., Thotakura, N., Kumar, P., Joshi, M., Bhushan, S., Bhatia, A., . . . & Katare, O. P. (2015). C60-fullerenes for delivery of docetaxel to breast cancer cells: A promising approach for enhanced efficacy and better pharmacokinetic profile. *International Journal of Pharmaceutics*, 495(1), 551–559.

[77] Kuznietsova, H. M., Ogloblya, O. V., Cherepanov, V. V., Prylutskyy, Y. I., & Rybalchenko, V. K. (2015). Effects of C60 Fullerene: Cisplatin complex on honeybee Apis mellifera L. *Biotechnologia Acta*, 8(4), 108–112.

[78] Franskevych, D. V., Grynyuk, I. I., Prylutska, S. V., & Matyshevska, O. P. (2016). Modulation of cisplatin-induced reactive oxygen species production by fullerene C60 in normal and transformed lymphoid cells. *The Ukrainian Biochemical Journal*, 88(1), 44–50.

[79] Prylutska, S., Grynyuk, I., Skaterna, T., Horak, I., Grebinyk, A., Drobot, L., . . . & Frohme, M. (2019). Toxicity of C60 fullerene: Cisplatin nanocomplex against Lewis lung carcinoma cells. *Archives of Toxicology, 93*(5), 1213–1226.

[80] Cui, X., Xu, S., Wang, X., & Chen, C. (2018). The nano-bio interaction and biomedical applications of carbon nanomaterials. *Carbon, 138,* 436–450.

[81] Partha, R., & Conyers, J. L. (2009). Biomedical applications of functionalized fullerene-based nanomaterials. *International Journal of Nanomedicine, 4,* 261.

[82] Prylutska, S., Grynyuk, I., Matyshevska, O., Prylutskyy, Y., Evstigneev, M., Scharff, P., & Ritter, U. (2014). C60 fullerene as synergistic agent in tumor-inhibitory doxorubicin treatment. *Drugs in R&D, 14*(4), 333–340.

[83] Yang, X., Ebrahimi, A., Li, J., & Cui, Q. (2014). Fullerene: Biomolecule conjugates and their biomedicinal applications. *International Journal of Nanomedicine, 9,* 77.

[84] Mroz, P., Pawlak, A., Satti, M., Lee, H., Wharton, T., Gali, H., . . . & Hamblin, M. R. (2007). Functionalized fullerenes mediate photodynamic killing of cancer cells: Type I versus type II photochemical mechanism. *Free Radical Biology and Medicine, 43*(5), 711–719.

[85] Yin, J. J., Lao, F., Fu, P. P., Wamer, W. G., Zhao, Y., Wang, P. C., . . . & Chen, C. (2009). The scavenging of reactive oxygen species and the potential for cell protection by functionalized fullerene materials. *Biomaterials, 30*(4), 611–621.

[86] Foley, S., Crowley, C., Smaihi, M., Bonfils, C., Erlanger, B. F., Seta, P., & Larroque, C. (2002). Cellular localisation of a water-soluble fullerene derivative. *Biochemical and Biophysical Research Communications, 294*(1), 116–119.

[87] Yang, J., Wang, K., Driver, J., Yang, J., & Barron, A. R. (2007). The use of fullerene substituted phenylalanine amino acid as a passport for peptides through cell membranes. *Organic & Biomolecular Chemistry, 5*(2), 260–266.

[88] Aoshima, H., Saitoh, Y., Ito, S., Yamana, S., & Miwa, N. (2009). Safety evaluation of highly purified fullerenes (HPFs): Based on screening of eye and skin damage. *The Journal of Toxicological Sciences, 34*(5), 555–562.

[89] Xiao, L., Takada, H., Maeda, K., Haramoto, M., & Miwa, N. (2005). Antioxidant effects of water-soluble fullerene derivatives against ultraviolet ray or peroxylipid through their action of scavenging the reactive oxygen species in human skin keratinocytes. *Biomedicine & Pharmacotherapy, 59*(7), 351–358.

[90] Xiao, L., Matsubayashi, K., & Miwa, N. (2007). Inhibitory effect of the water-soluble polymer-wrapped derivative of fullerene on UVA-induced melanogenesis via down-regulation of tyrosinase expression in human melanocytes and skin tissues. *Archives of Dermatological Research, 299*(5), 245–257.

[91] Basso, A. S., Frenkel, D., Quintana, F. J., Costa-Pinto, F. A., Petrovic-Stojkovic, S., Puckett, L., . . . & Weiner, H. L. (2008). Reversal of axonal loss and disability in a mouse model of progressive multiple sclerosis. *The Journal of Clinical Investigation, 118*(4), 1532–1543.

[92] Beuerle, F., Lebovitz, R., & Hirsch, A. (2008). Antioxidant properties of water-soluble fullerene derivatives. In *Medicinal Chemistry and Pharmacological Potential of Fullerenes and Carbon Nanotubes*, Springer Dordrecht, (pp. 51–78).

[93] Cai, X., Jia, H., Liu, Z., Hou, B., Luo, C., Feng, Z., . . . & Liu, J. (2008). Polyhydroxylated fullerene derivative C60 (OH) 24 prevents mitochondrial dysfunction and oxidative damage in an MPP+-induced cellular model of Parkinson's disease. *Journal of Neuroscience Research, 86*(16), 3622–3634.

[94] Yudoh, K., Shishido, K., Murayama, H., Yano, M., Matsubayashi, K., Takada, H., . . . & Nishioka, K. (2007). Water-soluble C60 fullerene prevents degeneration of articular cartilage in osteoarthritis via down-regulation of chondrocyte catabolic activity and inhibition of cartilage degeneration during disease development. *Arthritis & Rheumatism: Official Journal of the American College of Rheumatology, 56*(10), 3307–3318.

# Index

For Product Safety Concerns and Information please contact our EU
representative  GPSR@taylorandfrancis.com
Taylor & Francis Verlag GmbH, Kaufingerstraße 24, 80331 München, Germany

* 9 7 8 1 0 3 2 3 2 7 2 3 5 *